D0759456

Practical Numerical Algorithms for Chaotic Systems

Thomas S. Parker   Leon O. Chua

Practical Numerical Algorithms for Chaotic Systems

With 152 Illustrations

Springer-Verlag
New York Berlin Heidelberg
London Paris Tokyo Hong Kong

Thomas S. Parker
Hewlett Packard
Santa Rosa, CA 95401, USA

Leon O. Chua
Department of Electrical Engineering
and Computer Sciences
University of California,
Berkeley, CA 94720, USA

Front cover: The photograph on the cover shows a strange attractor internationally known as the "double scroll from Chua's circuit."

Library of Congress Cataloging-in-Publication Data
Parker, Thomas S.
Practical numerical algorithms for chaotic systems / Thomas S.
Parker, Leon O. Chua.
p. cm.
ISBN 0-387-96688-9 (U.S. : alk. paper)
1. Chaotic behavior in systems. 2. Algorithms. 3. Nonlinear
theories. I. Chua, Leon O., 1936– . II. Title.
Q172.5.C45P37 1989
003—dc20 89-11307

Printed on acid-free paper.

Camera-ready copy prepared by the authors using LATEX
Printed and bound by Edwards Brothers Incorporated, Ann Arbor, Michigan.
Printed in the United States of America.
9 8 7 6 5 4 3 2 1

ISBN 0-387-96689-7 Springer-Verlag New York Berlin Heidelberg
ISBN 3-540-96689-7 Springer-Verlag Berlin Heidelberg New York

To Don and Betty for their
love and support over the years.

T.S. Parker

To Diana.

L.O. Chua

Preface

One of the basic tenets of science is that deterministic systems are completely predictable—given the initial condition and the equations describing a system, the behavior of the system can be predicted for all time.[1] The discovery of chaotic systems has eliminated this viewpoint. Simply put, a chaotic system is a deterministic system that exhibits random behavior.

Though identified as a robust phenomenon only twenty years ago, chaos has almost certainly been encountered by scientists and engineers many times during the last century only to be dismissed as physical noise. Chaos is such a wide-spread phenomenon that it has now been reported in virtually every scientific discipline: astronomy, biology, biophysics, chemistry, engineering, geology, mathematics, medicine, meteorology, plasmas, physics, and even the social sciences.

It is no coincidence that during the same two decades in which chaos has grown into an independent field of research, computers have permeated society. It is, in fact, the wide availability of inexpensive computing power that has spurred much of the research in chaotic dynamics. The reason is simple: the computer can calculate a solution of a nonlinear system. This is no small feat. Unlike linear systems, where closed-form solutions can be written in terms of the system's eigenvalues and eigenvectors, few nonlinear systems and virtually no chaotic systems possess closed-form solutions.

The computer allows nonlinear systems to be studied in a way that was undreamt of by the pioneers of nonlinear dynamics. One can perform numerical "experiments" quickly and easily. Parameters can be changed, system equations modified, and solutions displayed—all at the touch of a key. Before the advent of low-cost computing, the

[1]There are random events in nature, but science has traditionally used probabilistic models to describe them.

ability to perform such simulations was restricted to researchers with access to a large computing facility. Today, anyone with a personal computer can simulate a nonlinear system.

Simulations are a powerful tool for gaining intuition about nonlinear systems and for exploring the exciting terrain of chaotic dynamics, but they do have their limitations. Computers have finite precision and inevitably generate errors when evaluating floating-point expressions. Furthermore, computers are discrete-time in nature and there are unavoidable errors when they are used to simulate continuous-time systems. Finally, a simulation is of little or no help in proving theoretical results; even if the result of a simulation were completely accurate, it is just one solution of one system from one initial condition using one set of parameter values.

The moral is: though simulations are a useful tool, simulation data must be interpreted carefully, checked against intuition and theory, and used only for purposes for which it is suited.

The book The purpose of this book is to present robust, reliable algorithms for simulating nonlinear dynamics, with an emphasis on chaotic behavior. In addition, we present the theoretical underpinnings of the algorithms. This allows the reader to understand in which situations an algorithm is applicable and how to interpret its results.

Mathematics is used throughout the book to describe the solutions and properties of dynamical systems, but we rarely state theorems and even less often prove them. The few proofs that are included in the book are relegated to the appendices. No prior knowledge of nonlinear systems is assumed though we do assume that the reader is familiar with the basic concepts from linear systems—equilibrium points, eigenvalues, eigenvectors, sinusoidal steady state, stability. Much of the language of dynamical systems theory is couched in the terms of differential topology, so we have included an appendix that reviews the concepts of diffeomorphisms, manifolds, and transversality.

We explain the basic theory behind chaotic systems and present algorithms for simulating and characterizing chaotic systems. Thus, most of the chapters contain two sections, the first section introduces and explains some aspect of dynamical systems theory and the second section presents algorithms that implement the ideas from the first section.

Pseudo-code It has been our experience that a verbal description of an algorithm often makes complete sense until one actually sits down to implement it. To alleviate this problem, we give detailed pseudo-code for the algorithms.

There are a few points to remember while reading the pseudo-code. Local variables are not explicitly declared. Thus, any variable that is not global and not an argument to the function in which it occurs is a local variable defined only in the function in which it occurs. The type of a variable should be clear from its usage and the context. Vectors are indicated by brackets. For example, $x[\,]$ is an array and $a[\,][\,]$ is a matrix. The ith row of $a[\,][\,]$ is $a[i][\,]$ and the jth column is $a[\,][j]$.

We have also included a chapter devoted to programming techniques and style in which we share the experience we have gained from writing simulation programs.

INSITE Most of the algorithms presented in this book are available in the software package INSITE.[2] INSITE is a collection of interactive, graphically-oriented programs for the simulation and investigation of nonlinear systems, with an emphasis on chaotic systems. INSITE is written in C and runs under the UNIX and PC-DOS operating systems. For information on ordering INSITE, please write

INSITE
P.O. Box 9662
Berkeley, CA 94709-9662.

Acknowledgments We would like to thank Peter Kennedy for his thorough and thoughtful reading of the manuscript. The clear, concise portions of this book owe much to him. We also deeply appreciate the comments and suggestions on the manuscript provided by Dr. Alistair Mees. Thanks are gratefully extended to Dr. Gregory Bernstein for his comments on Chapter 4 and to Dr. Kenneth Kundert for his comments on Chapter 5. Special thanks are due Dr. Stephen Boyd for suggesting that the book be written.

We would also like to thank Dr. Klaus Kählert, Dr. Rene Lozi, Professor Gui-nian Lin, and Orla Feely for their careful reading of the final manuscript. We gratefully acknowledge the Office of Naval Re-

[2] Interactive Nonlinear Systems Investigative Toolkit for Everyone

search and the National Science Foundation, without whose research support this book would not have been possible.

T. Parker would like to thank Bill's Bed and Breakfast for providing accommodations during the final phases of writing. He would also like to thank Bruce Donecker and David Sharrit of Hewlett-Packard for their understanding and patience while he was working on this project. Finally, he would like to thank Nicholas, Ginger, and Luther for their constant support and understanding.

The book was typeset by T. Parker using the LaTeX document preparation system written by Leslie Lamport and Donald Knuth.

Contents

Chapter 1

Steady-State Solutions and Limit Sets

1.1 Systems

In this section, we define dynamical systems and present some useful facts from the theory of differential equations.

1.1.1 Autonomous continuous-time dynamical systems

An *nth-order autonomous continuous-time dynamical system* is defined by the state equation

$$\dot{x} = f(x), \qquad x(t_0) = x_0 \tag{1.1}$$

where $\dot{x} := dx/dt$, $x(t) \in \mathbb{R}^n$ is the state at time t, and $f: \mathbb{R}^n \to \mathbb{R}^n$ is called the *vector field*. Since the vector field does not depend on time, the initial time may always be taken as $t_0 = 0$.

To show explicitly the dependence on the initial condition, the solution to (1.1) is often written as $\phi_t(x_0)$. The one-parameter family of mappings $\phi_t: \mathbb{R}^n \to \mathbb{R}^n$ satisfies the two relationships, $\phi_{t_1+t_2} = \phi_{t_1} \circ \phi_{t_2}$ and $\phi_0(x) = x$, and is called the *flow*.

The set of points $\{\phi_t(x_0) : -\infty < t < \infty\}$ is called the *trajectory* through x_0.

The dynamical system (1.1) is linear if the vector field f is linear.

1.1.2 Non-autonomous continuous-time dynamical systems

An *nth-order non-autonomous continuous-time dynamical system* is defined by the state equation

$$\dot{x} = f(x, t), \qquad x(t_0) = x_0. \tag{1.2}$$

For non-autonomous systems, the vector field depends on time and unlike the autonomous case, the initial time cannot, in general, be set to 0. The solution to (1.2) passing through x_0 at time t_0 is denoted by $\phi_t(x_0, t_0)$.

The dynamical system (1.2) is linear if the vector field $f(x,t)$ is linear with respect to x.

If there exists a $T > 0$ such that $f(x,t) = f(x, t+T)$ for all x and t, then the system is said to be *time periodic* with period T. The smallest such T is called the *minimum period.*

Assumption: Unless otherwise stated, all non-autonomous systems are assumed to be time periodic.

1.1.3 Relationship between autonomous and non-autonomous systems

An nth-order time-periodic non-autonomous system with period T can always be converted into an $(n+1)$th-order autonomous system by appending an extra state $\theta := 2\pi t/T$. The autonomous system is given by

$$\begin{aligned} \dot{x} &= f(x, \theta T/2\pi), \qquad & x(t_0) &= x_0 \\ \dot{\theta} &= 2\pi/T, & \theta(t_0) &= 2\pi t_0/T. \end{aligned} \tag{1.3}$$

Since f is time periodic with period T, the system (1.3) is periodic in θ with period 2π. Hence, the planes, $\theta = 0$ and $\theta = 2\pi$, may be identified and the state space transformed from the Euclidean space $\mathbb{R}^{n+1}$ to the cylindrical space $\mathbb{R}^n \times S^1$ where $S^1 := [0, 2\pi)$ denotes the circle.

The solution of (1.3) in the cylindrical state space is

$$\begin{bmatrix} x(t) \\ \theta(t) \end{bmatrix} = \begin{bmatrix} \phi_t(x_0, t_0) \\ 2\pi t/T \bmod 2\pi \end{bmatrix} \tag{1.4}$$

where the modulo function restricts $0 \leq \theta < 2\pi$. Using this transformation, the theory of autonomous systems can be applied to time-periodic non-autonomous systems.

Remark: A non-autonomous system that is not time periodic can also be converted to an autonomous system using (1.3) by choosing any $T > 0$. The solution, however, is necessarily unbounded ($\theta \to \infty$ as $t \to \infty$) and, thus, many of the theoretical results concerning the steady-state behavior of autonomous systems do not apply.

1.1.4 Useful facts regarding continuous-time dynamical systems

There are several standard and useful facts concerning the existence and uniqueness of solutions to continuous-time systems. Precise statements of these theorems can be found in any text on differential equations (e.g., Hirsch and Smale [1974]).

Assumption: For any t, ϕ_t and $\phi_t(\cdot, t_0)$ are assumed to be diffeomorphisms.[1]

This is not a restrictive assumption and has several consequences.

1. The solution of the continuous-time system exists for all t.

2. For any time t, $\phi_t(x) = \phi_t(y)$ if and only if $x = y$. From this fact and from the fact that $\phi_{t_1+t_2} = \phi_{t_1} \circ \phi_{t_2}$, it follows that a trajectory of an autonomous system is uniquely specified by its initial condition and that distinct trajectories cannot intersect.

3. For any times, t and t_0, $\phi_t(x, t_0) = \phi_t(y, t_0)$ if and only if $x = y$. This implies that given the initial time, the solution of a non-autonomous system is uniquely specified by the initial state.

   Note, however, that if $t_0 \neq t_1$, it is possible that there exists a time t such that $\phi_t(x, t_0) = \phi_t(y, t_1)$ and a time t' such that $\phi_{t'}(x, t_0) \neq \phi_{t'}(y, t_1)$, showing that unlike autonomous systems, distinct solutions of nonautonomous systems can intersect.

4. The derivative of a trajectory with respect to the initial condition exists and is nonsingular. It follows that for t and t_0 fixed, $\phi_t(x_0)$ and $\phi_t(x_0, t_0)$ are continuous with respect to the initial state x_0. More specifically, given x_0, a finite time interval

[1]See Appendix C for a definition of diffeomorphism.

$[t_0, t_1]$, and $\epsilon > 0$, there exists a $\delta > 0$ such that

$$\|\phi_{t_0+\tau}(x_0, t_0) - \phi_{t_0+\tau}(x_1, t_0)\| < \epsilon \tag{1.5}$$

for all $\tau \in [t_0, t_1]$ and all x_1 satisfying $\|x_1 - x_0\| < \delta$.

1.1.5 Discrete-time systems

Any map $P: \mathbb{R}^n \to \mathbb{R}^n$ defines a *discrete-time dynamical system* by the state equation

$$x_{k+1} = P(x_k), \qquad k = 0, 1, 2 \ldots \tag{1.6}$$

where $x_k \in \mathbb{R}^n$ is the state and P maps the state x_k to the next state x_{k+1}. Starting with an initial condition x_0, repeated applications of P generate a sequence of points $\{x_k\}_{k=0}^{\infty}$ called an *orbit*. In the literature, the term "orbit" is often used as a synonym for "trajectory." To avoid confusion, we will use "orbit" to refer only to the solution of a discrete-time system.

Although this book focuses on continuous-time systems, we will find discrete-time systems useful for two reasons. First, the Poincaré map—a technique that replaces the analysis of the flow of a continuous-time system with the analysis of a discrete-time system—is a powerful tool for studying continuous-time dynamical systems. Second, maps will be used to illustrate important concepts without getting bogged down in the details of solving differential equations.

1.2 Limit sets

In this section, continuous-time dynamical systems are classified in terms of their steady-state behavior and their limit sets. The discussion of the steady-state behavior of discrete-time systems is reserved for Chapter 2.

Steady state refers to the asymptotic behavior of a system as $t \to \infty$. To make sense, it is required that the steady state be bounded. The difference between the trajectory and its steady state is called the *transient*.

The concepts of transient and steady state are familiar from linear system theory and are quite useful in that setting. In the realm of nonlinear systems, where the principle of superposition does not hold and closed-form solutions are hard to come by, these time-based concepts lose some of their usefulness. We now present some definitions related to limit sets, the state-space equivalent of the steady state.

A point y is an *ω-limit point* of x if, for every neighborhood U of y, $\phi_t(x)$ repeatedly enters U as $t \to \infty$.

The set $L(x)$ of all ω-limit points of x is called the *ω-limit set* of x. As the notation indicates, the ω-limit set depends on x.

An ω-limit set L is *attracting* if there exists an open neighborhood U of L such that $L(x) = L$ for all $x \in U$. Attracting limit sets are of special interest since non-attracting limit sets cannot be observed in physical systems or simulations.

The *basin of attraction* B_L of an attracting set L is defined as the union of all such neighborhoods U. B_L is the set of all initial conditions that tend toward L as $t \to \infty$. It is an open set.

Remarks:

1. Although these definitions are stated in terms of autonomous, continuous-time systems, they apply to autonomous, discrete-time systems as well. Since a non-autonomous vector field is time-dependent, limit sets of non-autonomous differential equations are not meaningful unless the non-autonomous system is converted to an autonomous system via transformation (1.3) or via the Poincaré map (Chapter 2).
2. There exist reverse-time limit points and limit sets called *α-limit points* and *α-limit sets*.[2] The definitions are identical to those for ω-limit points and ω-limit sets except that $t \to \infty$ is replaced by $t \to -\infty$.

In an asymptotically stable linear system, there is only one limit set. Hence, it makes sense to speak, for example, of *the* sinusoidal steady state. Furthermore, the basin of attraction of this limit set is the entire state space. In other words, the steady state is independent of the initial condition. By way of contrast, in a typical nonlinear system there may be several limit sets. In particular, there can be several attracting limit sets, each with a different basin of attraction. The initial condition determines which limit set is eventually reached.

We now present four different types of steady-state behavior starting with the simplest and moving to the most complex. Each steady

[2]The names arise from the fact that α and ω are the first and last letters of the Greek alphabet.

state will be described from three points of view: in the time domain as a time waveform, in the frequency domain as a spectrum, and in the state-space domain as a limit set.

1.2.1 Equilibrium points

Definition An *equilibrium point* x_{eq} of an autonomous system is a constant solution of (1.1), that is, $x_{eq} = \phi_t(x_{eq})$ for all t. At an equilibrium point the vector field vanishes and, except for a few pathological cases (Koksal [1986]), $f(x) = 0$ implies that x is an equilibrium point.

Except for pathological cases, there is no corresponding steady-state behavior for non-autonomous systems.

Spectrum The spectrum of each non-zero component of $\phi_t(x_{eq})$ contains a single spike at zero frequency.

Limit set The limit set corresponding to an equilibrium point is simply the equilibrium point itself.

Examples A simple example of a system exhibiting multiple equilibria is the damped pendulum equation

$$\begin{aligned} \dot{x} &= y \\ \dot{y} &= -\epsilon y - \sin(x) \end{aligned} \tag{1.7}$$

which is a second-order autonomous system with an infinity of equilibrium points at $(x, y) = (k\pi, 0)$ for $k = 0, \pm 1, \pm 2, \ldots$. The equilibrium points with k odd are attracting. Some of the equilibrium points and their basins of attraction are shown in Fig. 1.1(a).

Another example is shown in Fig. 1.1(b). This phase portrait is for the simple predator-prey equation

$$\begin{aligned} \dot{x} &= -3x + 4x^2 - xy/2 - x^3 \\ \dot{y} &= -2.1y + xy \end{aligned} \tag{1.8}$$

where x is the population of the prey and y is the population of the predator.

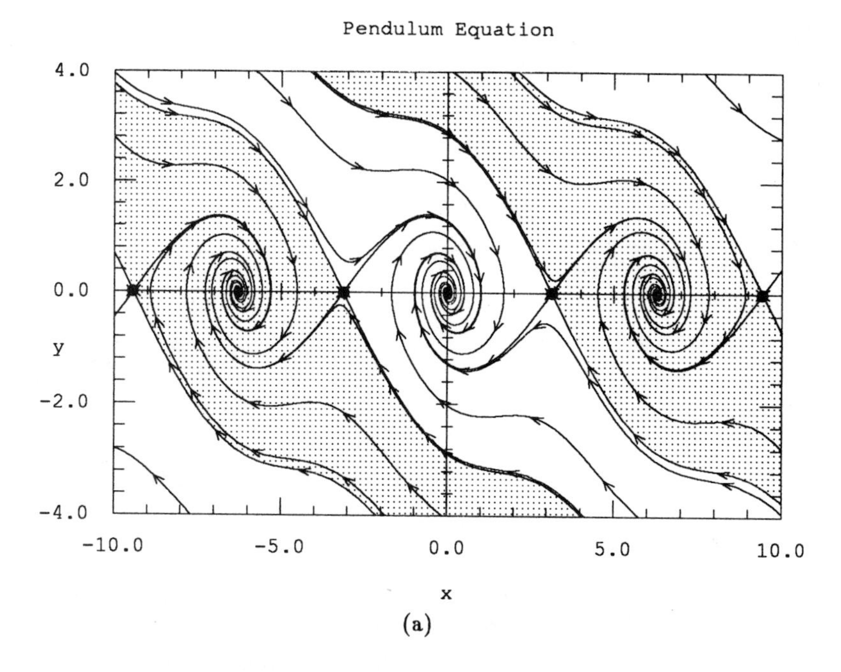

(a)

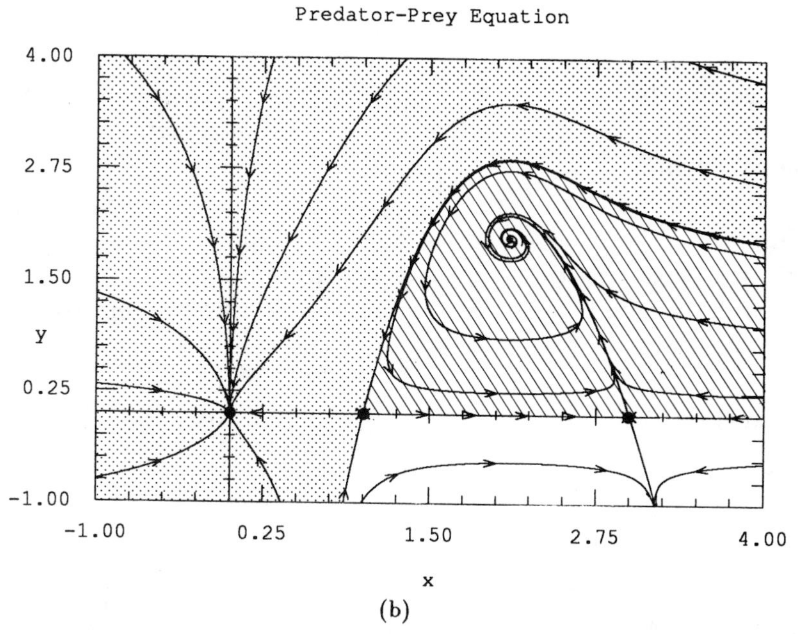

(b)

Figure 1.1: Phase portraits of (a) the damped pendulum equation with $\epsilon = 0.4$; (b) the predator-prey equation (1.8).

1.2.2 Periodic solutions

The autonomous case

Definition $\phi_t(x^*)$ is a *periodic solution* of an autonomous system if, for all t,

$$\phi_t(x^*) = \phi_{t+T}(x^*) \tag{1.9}$$

for some minimum period $T > 0$. The restriction $T > 0$ is required to prevent the classification of an equilibrium point as a periodic solution. Also note that x^* is not unique since any point lying on the periodic solution will do. Changing x^* corresponds to changing the time origin.

A periodic solution is *isolated* if it possesses a neighborhood that contains no other periodic solution. In the autonomous case, an isolated periodic solution is called a *limit cycle*.

Spectrum The spectrum of each component of a limit cycle contains a spike at frequency 0 and spikes spaced at integer multiples of the fundamental frequency $f = 1/T$. Some of these frequency components may have zero amplitude and it is even possible for the fundamental to be absent. Thus, the frequency of a periodic waveform cannot be determined solely from the frequency of the first non-zero spike; the spacing of the remaining non-zero spikes must also be examined.

Limit set The limit set corresponding to a limit cycle is the closed curve traced out by $\phi_t(x^*)$ over one period. This limit set is a diffeomorphic copy of the circle S^1 and it is often convenient to think of limit cycles as circles.

Example The classic example of a limit cycle is found in van der Pol's equation

$$\begin{aligned} \dot{x} &= y \\ \dot{y} &= (1 - x^2)y - x. \end{aligned} \tag{1.10}$$

The existence of the van der Pol limit cycle, shown in Fig. 1.2, can be explained by casting (1.10) into a scalar differential equation:

$$\ddot{x} + (x^2 - 1)\dot{x} + x = 0. \tag{1.11}$$

The damping term, $x^2 - 1$, is negative for $|x| < 1$ implying an expanding solution while it is positive for $|x| > 1$ indicating a contracting

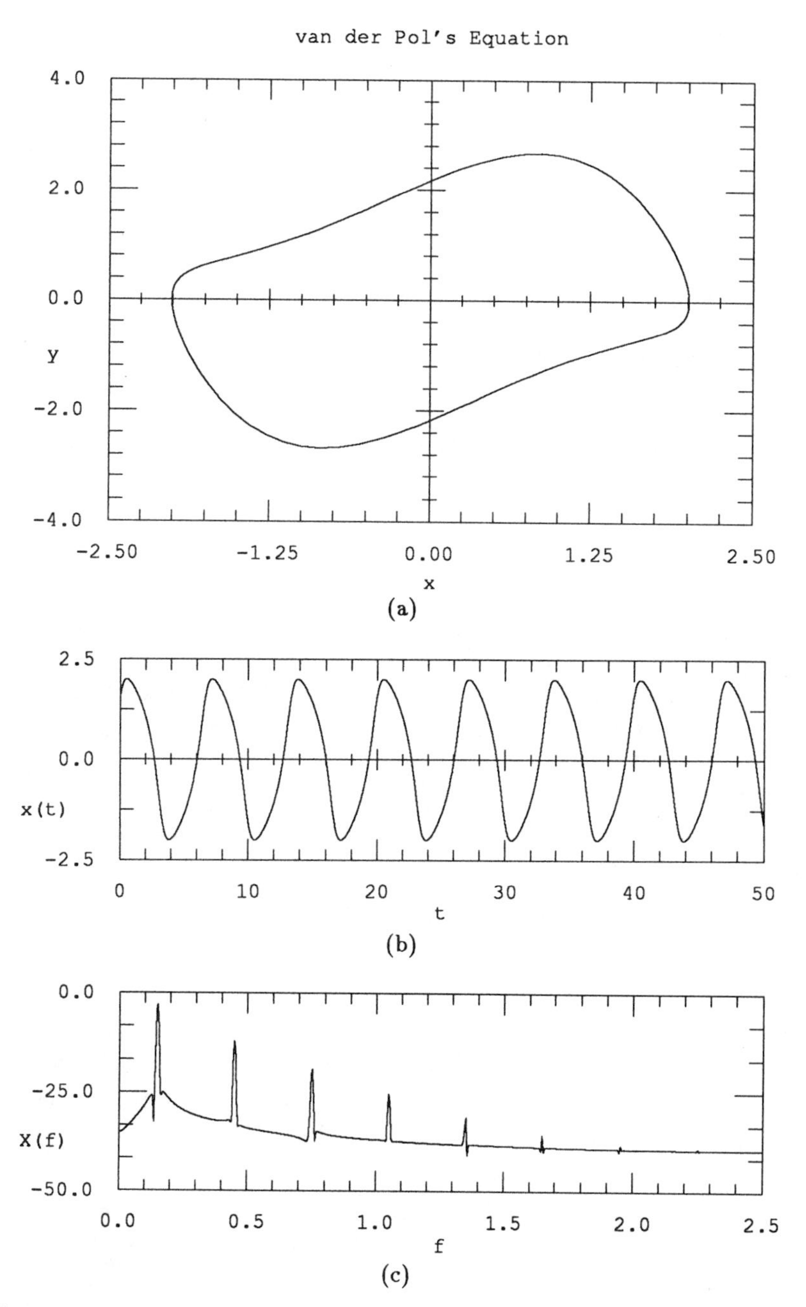

Figure 1.2: Solution of the van der Pol equation (1.10). (a) The limit cycle; (b) the time waveform of the x-component; (c) the spectrum of the waveform in (b). Due to symmetry, the even harmonics of the spectrum are missing.

solution. Since trajectories near the origin expand, trajectories from the outer regions contract, and the only equilibrium point is at $(0,0)$, there must be a limit cycle encircling the origin; otherwise, the trajectory, constrained to lie on the plane, would intersect itself.

The non-autonomous case

A non-autonomous system can be considered a special case of an autonomous system due to (1.3), but since non-autonomous systems are commonplace, it is useful to examine this case separately.

Definition $\phi_t(x^*, t_0)$ is a *periodic solution* of a non-autonomous system if, for all t,

$$\phi_t(x^*, t_0) = \phi_{t+T}(x^*, t_0) \tag{1.12}$$

for some minimum period $T > 0$.

When a time-periodic non-autonomous system with minimum period T_f is converted to an autonomous system via (1.3), the periodic solution becomes a limit cycle in the cylindrical state space. It follows that T is some integer multiple K of the forcing period T_f. The solution $\phi_t(x^*, t_0)$ is called a *period-K solution.* Using terminology borrowed from Fourier analysis, a period-one solution is often called a *fundamental solution* and if $K > 1$, a period-K solution is referred to as a *Kth-order subharmonic.*

Spectrum The spectrum of each component of a period-one solution consists of a spike at frequency 0 and spikes at integer multiples of $1/T_f$. The spectrum of a Kth-order subharmonic contains a spike at frequency 0 and spikes at integer multiples of $1/KT_f$. Thus, the higher the order of the subharmonic, the more closely spaced are the non-zero frequency components. The amplitudes of some of the frequency components may be zero.

Example Fig. 1.3 shows a fundamental solution and Fig. 1.4, a third-order subharmonic for Duffing's equation

$$\begin{aligned} \dot{x} &= y \\ \dot{y} &= x - x^3 - \epsilon y + \gamma \cos(\omega t). \end{aligned} \tag{1.13}$$

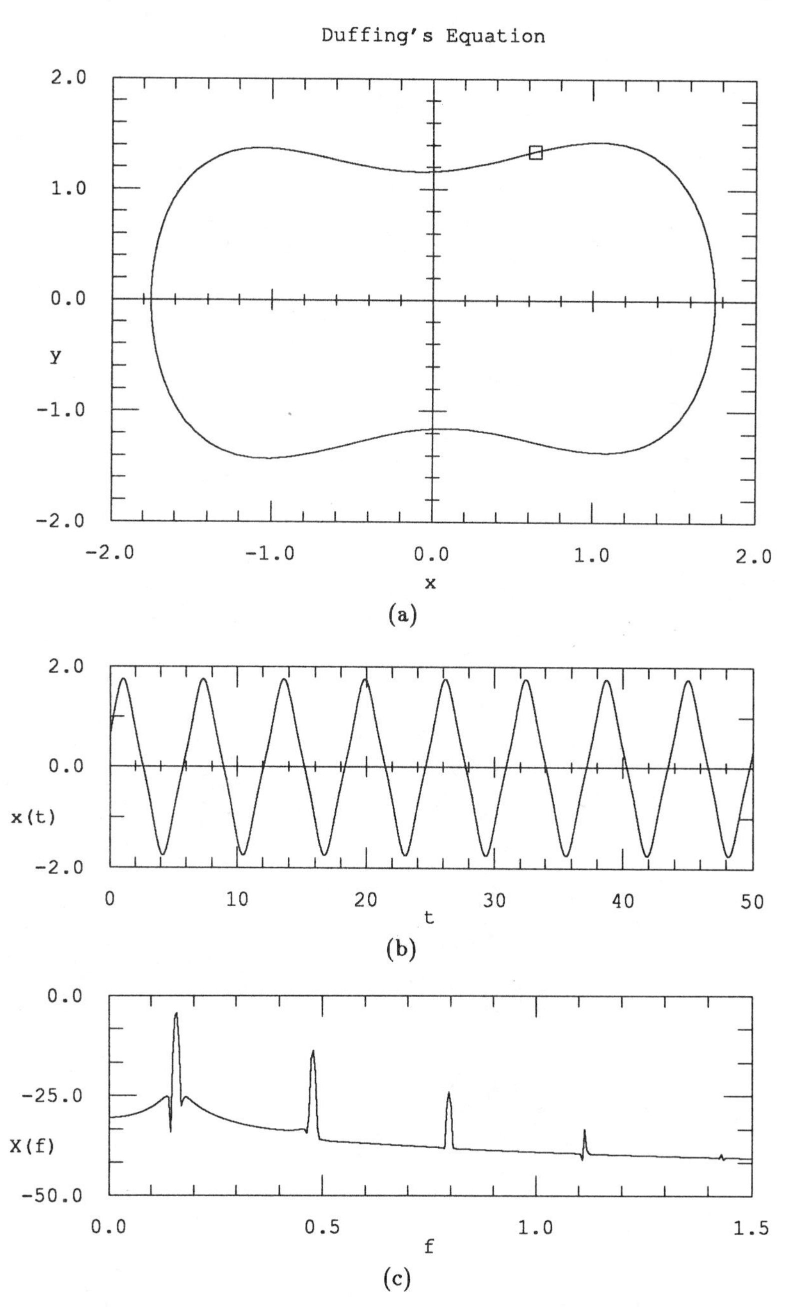

Figure 1.3: Solution of Duffing's equation for $\epsilon = 0.15$, $\gamma = 0.3$ and $\omega = 1$. (a) A period-1 trajectory; (b) the time waveform of the x-component; (c) the spectrum of the x-component. Due to symmetries in the waveform, only the odd harmonics are present.

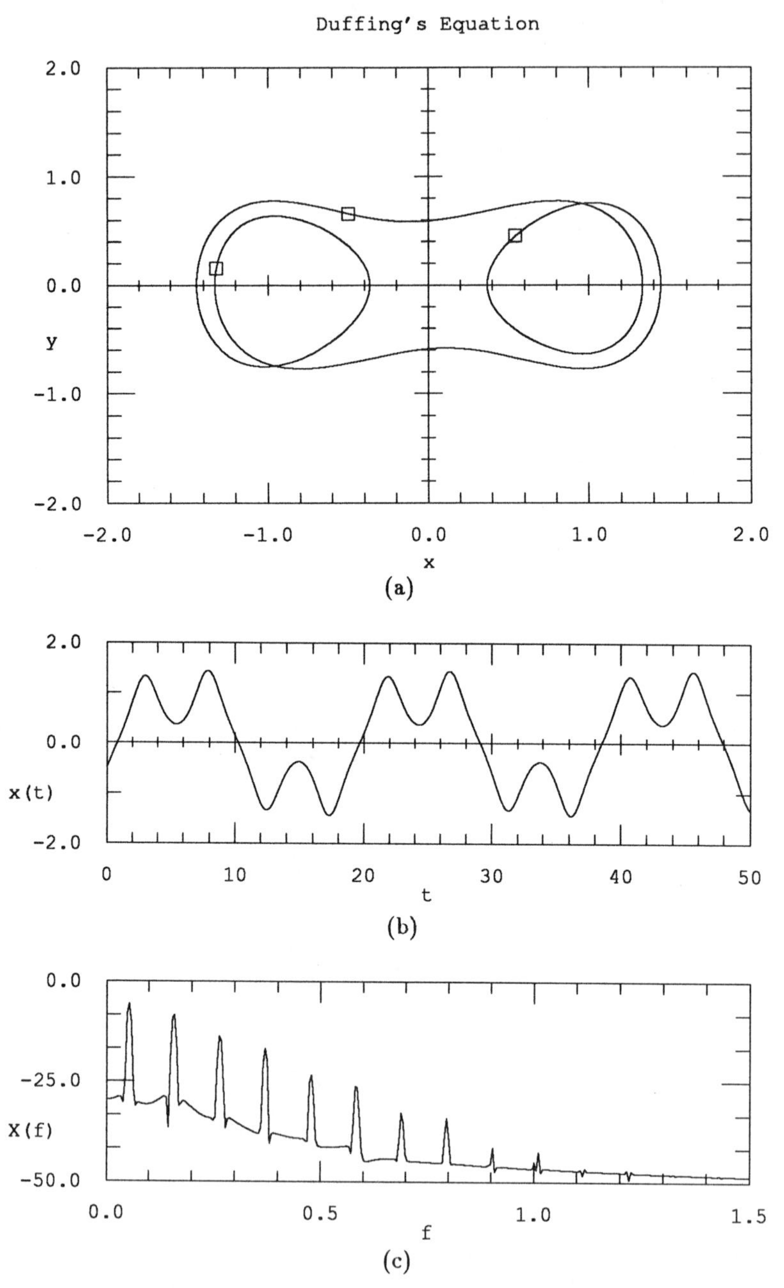

Figure 1.4: Solution of Duffing's equation for $\epsilon = 0.22$, $\gamma = 0.3$ and $\omega = 1$. (a) A period-3 trajectory; (b) the time waveform of the x-component; (c) the spectrum of the x-component. Due to symmetries in the waveform, only the odd harmonics are present.

Note that for a fundamental solution, x^* is unique. It is the point on the periodic solution that is passed at $t = t_0$. x^* is represented by the small box in Fig. 1.3(a). Given an x^* for a Kth-order subharmonic, there are $K-1$ other values

$$x_k^* = \phi_{t_0+(k-1)T_f}(x^*, t_0), \qquad k = 1, \ldots, K. \tag{1.14}$$

that can be used to identify the subharmonic. Each corresponds to a shift in time origin of T_f, that is,

$$\phi_t(x^*, t_0) = \phi_t(x_k^*, t_0 + kT_f), \qquad k = 1, \ldots, K-1 \tag{1.15}$$

for all t. The x_k^* are represented by small boxes in Fig. 1.4(a).

It is important to mention that subharmonics of *different* orders can exist in Duffing's equation for the *same* set of parameter values. The initial condition determines which steady state is reached.

Finally, we note that subharmonics make no sense for most autonomous systems since there is no reference frequency. In the case of period-doubling bifurcations (Chapter 8), however, the subharmonic concept can be salvaged, since one can say that this solution has twice the period of that solution and is, in some relative sense, a second-order subharmonic.

1.2.3 Quasi-periodic solutions

Definition A *quasi-periodic* function is one that can be expressed as a countable sum of periodic functions

$$x(t) = \sum_i h_i(t) \tag{1.16}$$

where h_i has minimum period T_i and frequency $f_i = 1/T_i$. Furthermore, there must exist a finite set of *base frequencies* $\{\hat{f}_1, \ldots, \hat{f}_p\}$ with the following two properties:

1. It is linearly independent, that is, there does not exist a nonzero set of integers $\{k_1, \ldots, k_p\}$ such that $k_1\hat{f}_1 + \cdots + k_p\hat{f}_p = 0$.

2. It forms a *finite integral base* for the f_i, that is, for each i, $f_i = |k_1\hat{f}_1 + \cdots + k_p\hat{f}_p|$ for some integers $\{k_1, \ldots, k_p\}$.

In words, a quasi-periodic waveform is the sum of periodic waveforms each of whose frequency is one of the various sums and differences of a finite set of base frequencies.

Note that the base frequencies are not defined uniquely—$\{\hat{f}_1, \hat{f}_2\}$ spans the same set of frequencies as $\{\hat{f}_1, \hat{f}_1 + \hat{f}_2\}$—but that p is. A quasi-periodic solution with p base frequencies is called *p-periodic.* Be careful not to confuse a p-periodic solution with a period-K solution. The former is quasi-periodic; the latter is periodic.

Spectrum The spectrum of each component of a quasi-periodic waveform consists of a spike at frequency 0 and spikes at the frequencies kf_i for $k = 1, 2, \ldots$. Some of the frequency components may have zero amplitude. Theoretically, the spectrum of a quasi-periodic waveform of order two or higher can be distinguished from that of a periodic waveform since the spikes of the quasi-periodic spectrum are not spaced at integer multiples of one particular frequency. In practice, due to the impossibility of determining whether a measured value is rational or irrational, a spectrum that appears to be quasi-periodic may actually be periodic with an extremely long period.

Examples A periodic waveform is a quasi-periodic waveform with $p = 1$. If the minimum period is T, then $\{1/T\}$ is a finite integral base as is $\{1/kT\}$ for any integer $k > 0$.

A very simple two-periodic function is

$$x(t) = \cos(2\pi t) + \cos(2\pi\sqrt{2}\,t). \tag{1.17}$$

A frequency base for this waveform is $\{1, \sqrt{2}\}$. The irrationality of $\sqrt{2}$ guarantees that condition 1 is satisfied. The spectrum of this waveform consists of two spikes, one at 1 and one at $\sqrt{2}$.

A slightly more complicated two-periodic function is

$$x(t) = h_1(t) + h_2(t) \tag{1.18}$$

where h_1 and h_2 are arbitrary periodic functions with periods T_1 and T_2. Here we assume T_1 and T_2 are *incommensurate*, that is, T_1/T_2 is irrational. The spectrum of this waveform consists of two sets of harmonics. The first set, positioned at frequencies k/T_1 for $k = 0$, $1, \ldots$, corresponds to $h_1(t)$ and the second set, with fundamental frequency $1/T_2$, corresponds to $h_2(t)$. Note that if T_1 and T_2 are commensurate, there exist two integers, p and q, such that $pT_1 = qT_2$, and it follows that $x(t)$ is periodic with period $T = pT_1$.

Here are two, less contrived examples that occur often in electrical engineering:

Amplitude Modulation

$$x(t) = m(t)\cos(2\pi f_c t) \tag{1.19}$$

where the message $m(t)$ is periodic with frequency f_m. It is well-known that the spectrum of $x(t)$ consists of spikes at frequencies $|f_c \pm k f_m|$ where $k = 0, 1, \ldots$. If f_m and f_c are incommensurate, then $x(t)$ is two-periodic with base frequencies $\{f_c, f_m\}$.

Phase modulation

$$x(t) = \cos(2\pi f_c t + m(t)). \tag{1.20}$$

If $m(t)$ is periodic with frequency f_m, then the spectrum has spikes at the same locations as the AM spectrum. If f_m and f_c are incommensurate, $x(t)$ is two-periodic with frequency base $\{f_c, f_m\}$.

Neither of these examples has been presented as a solution of a dynamical system but they do illustrate one important point: quasi-periodic functions can be generated when two or more periodic functions with incommensurate frequencies interact nonlinearly. In the amplitude modulation case, the nonlinearity is $g(u, v) := uv$ and the time waveform is

$$x(t) = g(m(t), \cos(2\pi f_c t)) \tag{1.21}$$

For phase modulation, the time waveform is $x(t) = g(m(t), t)$, and the nonlinearity is the time-periodic function

$$g(u, t) := \cos(2\pi f_c t + u). \tag{1.22}$$

To see how quasi-periodic solutions arise in dynamical systems, consider the van der Pol equation (1.10). It possesses a limit cycle whose natural period T_1 depends on the system parameters. Now add a sinusoidal forcing term as follows:

$$\begin{aligned} \dot{x} &= y \\ \dot{y} &= (1 - x^2)y - x + A\cos(2\pi t / T_2). \end{aligned} \tag{1.23}$$

The solution of the forced system could synchronize with some multiple of the input period T_2 resulting in a subharmonic. This is precisely what happens in a phase-locked loop. It is also possible that, in the conflict between T_1 and T_2, neither period "wins" and quasi-periodic behavior is observed.

A two-periodic trajectory of the forced van der Pol system is shown in Fig. 1.5. The trajectory is confined to an annular-like region of the state space and is uniformly distributed in this region. Not all two-periodic trajectories exhibit these properties, but their presence is a good indication of two-periodic behavior. The time waveform is clearly amplitude modulated and, since the zero crossings are not uniformly spaced, it is frequency modulated as well. The spectrum consists of the spectrum of the unforced system (Fig. 1.2(c)) plus tightly spaced sidebands due to the relatively slow modulation. To within measurement error, the spacing of the sidebands is the difference between the natural frequency and the forcing frequency.

Two-periodic behavior can occur in autonomous systems, too. Consider two oscillators with incommensurate periods T_1 and T_2. If they are weakly coupled, we are back in the position of two frequencies fighting for dominance in a single system. If they each lock onto the same period, typically one that is close to some multiple of T_1 and close to some other multiple of T_2, then the result is a periodic solution. If there is no single period satisfactory to both oscillators, quasi-periodic behavior can result.[3]

Limit set In an autonomous system (or converted non-autonomous system), a two-periodic trajectory lies on a diffeomorphic copy of the two-torus $T^2 := S^1 \times S^1$, where each S^1 represents one of the base frequencies (Fig. 1.6). Since a trajectory is a curve and the two-torus a surface, not every point of the torus lies on the trajectory; however, it can be shown that the trajectory repeatedly passes arbitrarily close to every point on the torus and, therefore, the two-torus is the limit set of the two-periodic behavior. This is the first example we have seen where a limit set is not a single trajectory.

We now present a geometrical interpretation why two-periodic behavior requires incommensurate frequencies. Consider a trajectory traveling on a torus looping in the ω_1 direction with period T_1 and in the ω_2 direction with period T_2. If T_1 and T_2 are commensurate,

[3] A third possibility is chaos.

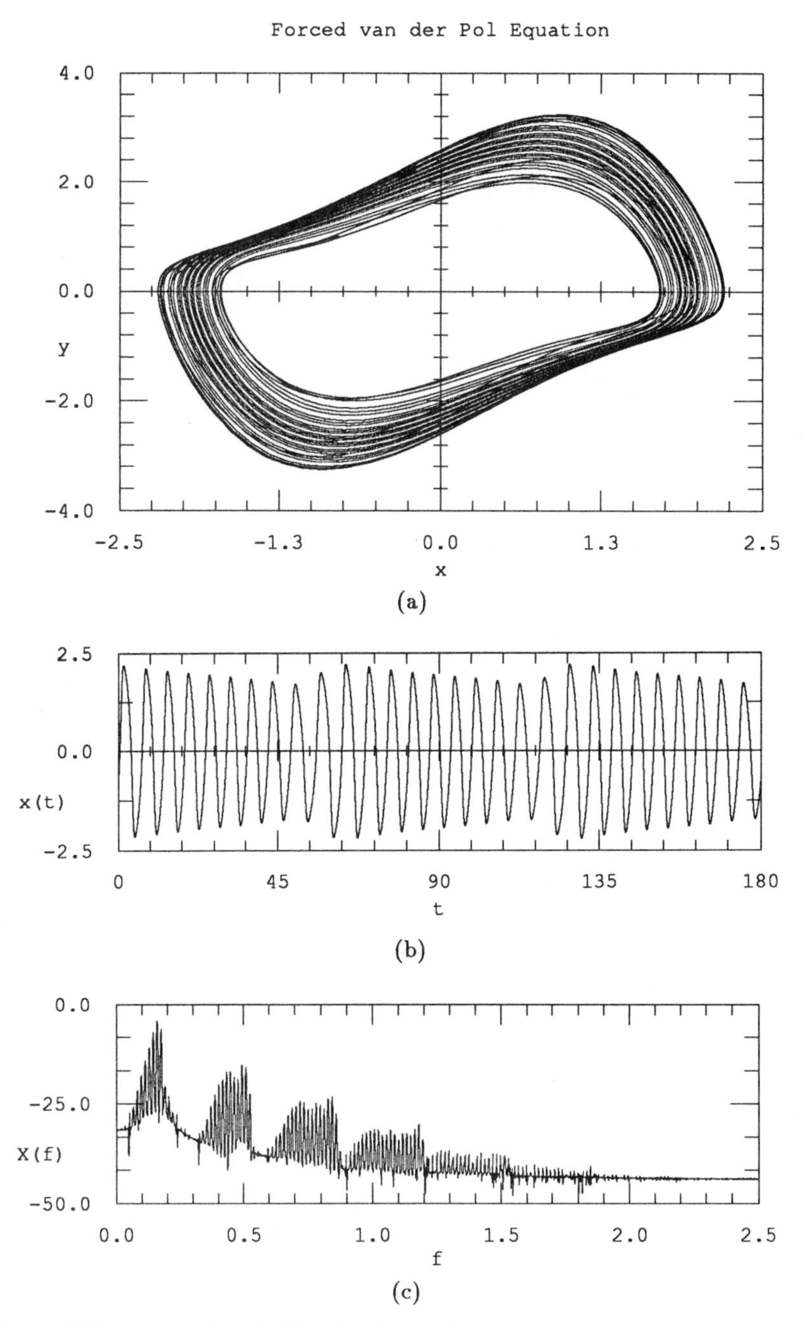

Figure 1.5: A quasi-periodic solution of the forced van der Pol equation for $A = 0.5$ and $T_2 = 2\pi/1.1$. (a) The quasi-periodic trajectory; (b) the time waveform of the x-component; (c) the spectrum of the waveform in (b).

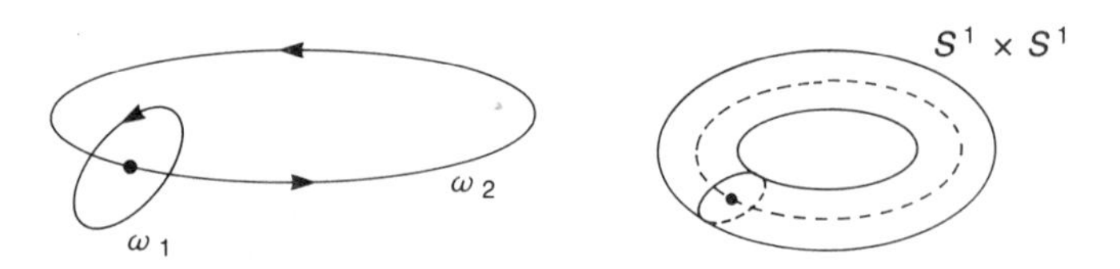

Figure 1.6: Two-periodic behavior lies on a two-torus $S^1 \times S^1$. Each S^1 represents one periodic component.

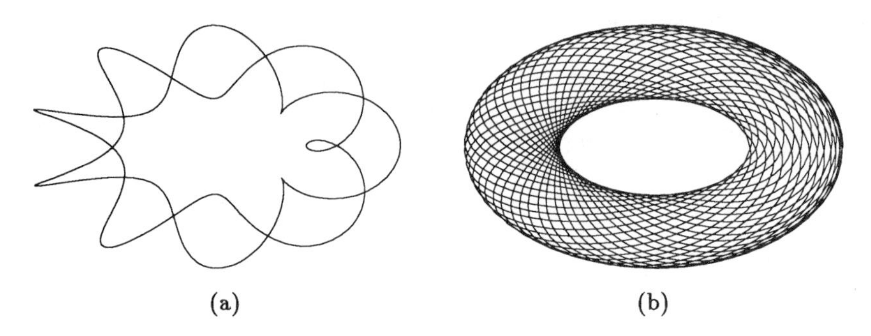

Figure 1.7: Trajectories on a two-torus. (a) A periodic trajectory with $T_1 = 2$ and $T_2 = 9$; (b) a two-periodic trajectory with $T_1 = 1$ and $T_2 = \sqrt{2}$. If the simulations were run longer, (a) would not change, but (b) would become more densely filled in.

there exist positive integers p and q such that $pT_1 = qT_2$. It follows that in pT_1 seconds, the trajectory will close on itself since it will have made exactly p loops in the first direction and exactly q loops in the second (Fig. 1.7(a)). The result is a limit cycle with period pT_1. However, if T_1 and T_2 are incommensurate, no such p and q exist and the trajectory never closes on itself (Fig. 1.7(b)).

Higher-order quasi-periodic trajectories can occur in higher-order systems. A quasi-periodic solution with base dimension p possesses a limit set that is diffeomorphic to a p-torus.

1.2.4 Chaos

Definition There is no widely accepted definition of chaos. From a practical point of view, it can be defined as "none of the above," that is, as bounded steady-state behavior that is not an equilibrium point, not periodic, and not quasi-periodic.

Examples Fig. 1.8 shows a chaotic trajectory from Duffing's equation (1.13), and Fig. 1.9 illustrates chaos in Chua's circuit (Matsu-mo-to [1984]), a simple, electronic circuit described by the double scroll equation

$$\begin{aligned} \dot{x} &= \alpha(y - h(x)) \\ \dot{y} &= x - y + z \\ \dot{z} &= -\beta y \end{aligned} \tag{1.24}$$

where h is the piecewise-linear function

$$h(x) = \begin{cases} m_1 x + (m_0 - m_1), & x \geq 1 \\ m_0 x, & |x| \leq 1 \\ m_1 x - (m_0 - m_1), & x \leq -1. \end{cases} \tag{1.25}$$

It is evident from these pictures that the trajectories are, indeed, bounded and that they are not periodic. It is difficult to tell from the figures that they are not quasi-periodic, but when the evolution of the trajectories is viewed on a graphics terminal, they exhibit a feature common to many chaotic systems: the trajectory evolves as a random mixture of two types of behavior. In Fig. 1.8(a), the trajectory loops for a while in the region $x > 0$ and then passes over to the region $x < 0$ where it loops for a while before returning to the first region. The number and size of loops between transitions follow no set pattern; they appear to be random. In the double scroll, the same type of mixing behavior occurs, with random transitions between the disks in the upper-right and lower-left of Fig. 1.9(a).

Spectrum As shown in the figures, a chaotic spectrum is quite different from at of a periodic or quasi-periodic waveform. It has a continuous, broad-band nature. This noise-like spectrum is a characteristic exhibited by all chaotic systems. In addition to the broad-band component, it is quite common for chaotic spectra to contain spikes indicating the predominant frequencies of the solution.

Limit set The geometrical object in state space to which chaotic trajectories are attracted is called a *strange attractor*. There are good theoretical reasons why the definition of an attracting limit set does not suffice for chaotic systems (see Eckmann and Ruelle [1985] and Milnor [1985]), but as of yet there is no generally agreed upon definition of a strange attractor. From a practical point of view,

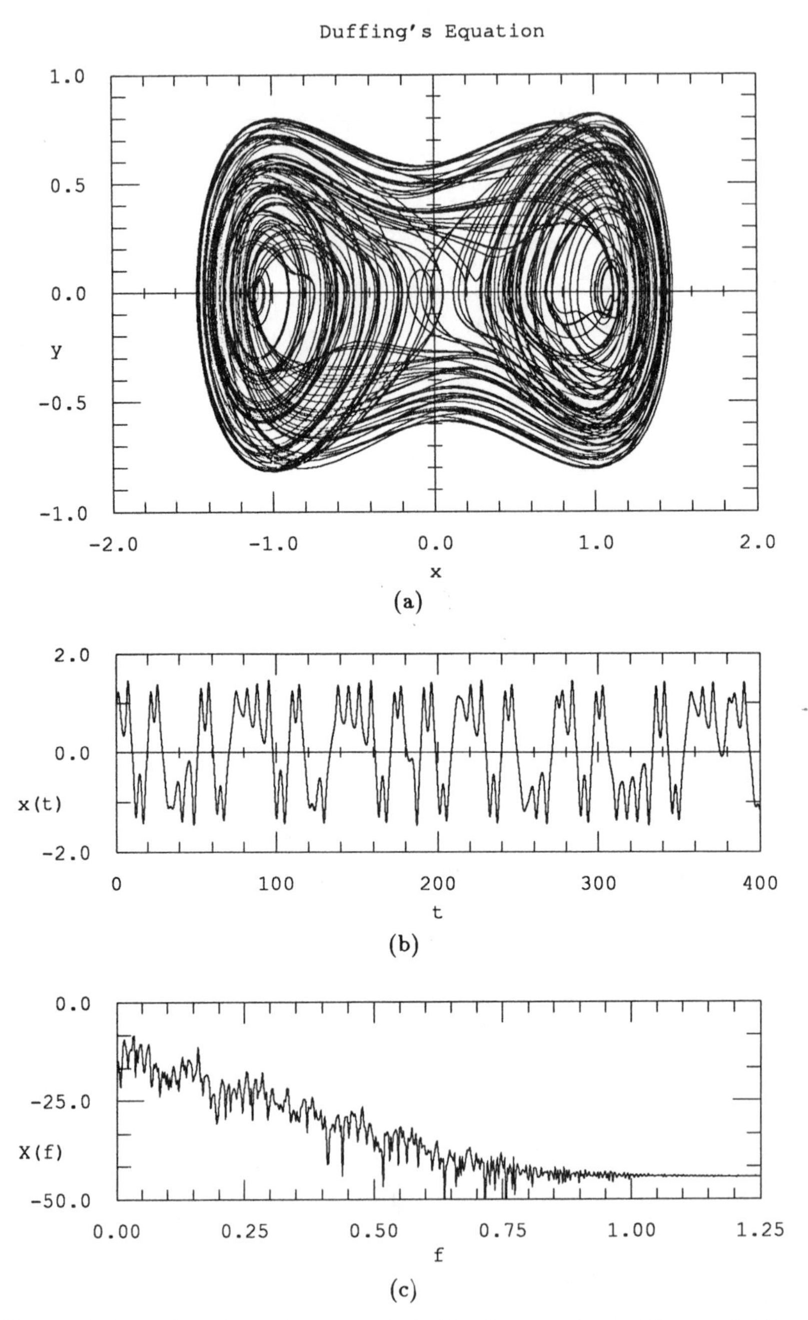

(a)

(b)

(c)

Figure 1.8: Chaotic trajectory of Duffing's equation with $\epsilon = 0.25$, $\gamma = 0.3$, and $\omega = 1.0$. (a) The trajectory; (b) the time waveform of the x-component; (c) the spectrum of the waveform in (b).

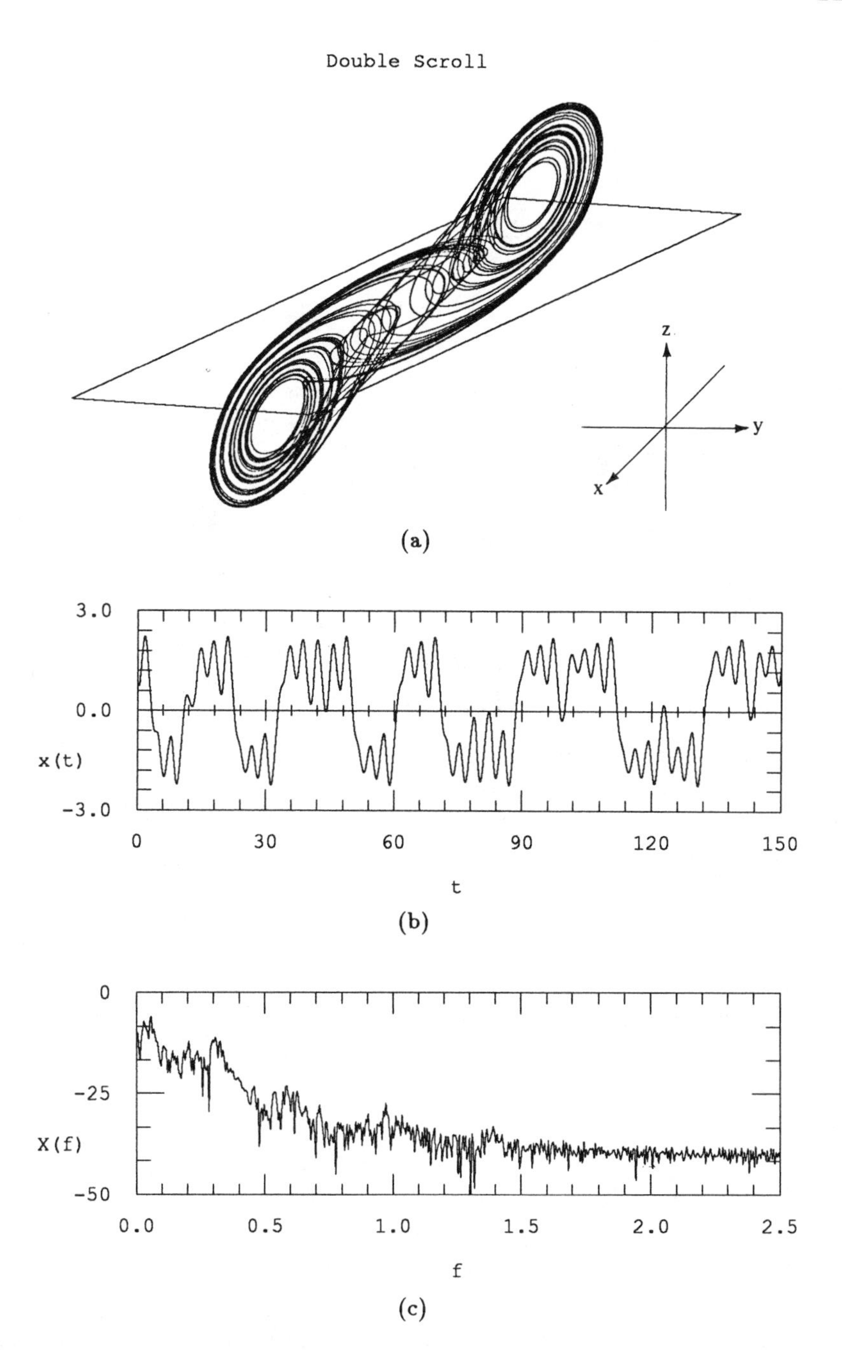

Figure 1.9: Chaotic trajectory of the double scroll equation with $\alpha = 9$, $\beta = 100/7$, $m_0 = -1/7$, and $m_1 = 2/7$. (a) The trajectory; (b) the time waveform of the x-component; (c) the spectrum of the waveform in (b).

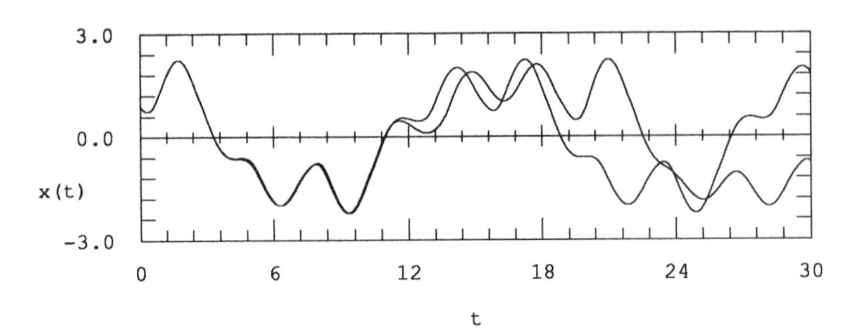

Figure 1.10: Sensitive dependence on initial conditions. Shown are the x-coordinates of two trajectories plotted versus time for the double scroll equation. The initial conditions differ by 0.01%.

there is little harm in ignoring the difference between attractors and attracting limit sets and that is precisely what we will do.

The attractor of a chaotic system is not a simple geometrical object like a circle or a torus; in fact, it is not even a manifold. Strange attractors are quite complicated geometrically—they are related to Cantor sets (see Appendix F) and they possess fractional dimension (Chapter 7). Just how complicated strange attractors are will become clear throughout the rest of the book.

Sensitive dependence on initial conditions Besides possessing a broad-band spectrum and a complex limit set, most chaotic systems exhibit *sensitive dependence on initial conditions*: given two distinct initial conditions arbitrarily close to one another, the trajectories emanating from these initial conditions diverge, at a rate characteristic of the system, until for all practical purposes, they become uncorrelated (Fig. 1.10).

Sensitive dependence on initial conditions has an important implication. In practice, there is always some error in measuring or specifying the state of a system. In addition, there is always noise present, be it thermal noise in a physical system or computational noise in a simulation. Due to sensitive dependence, these errors, however small, will eventually alter the macroscopic behavior of a chaotic system. Thus, in a very real sense, chaotic systems are unpredictable.

In Section 1.1.4, we mentioned that a trajectory is continuous with respect to its initial condition. At first glance, the continuity property may seem to forbid sensitive dependence on initial conditions because continuity requires that nearby trajectories stay nearby while sensitive dependence requires that nearby trajectories eventually separate. This paradox is resolved by noting that continuity is guaranteed only over a *finite* interval of time and, thus, the continuity property implies nothing about the long-term behavior of the trajectories. What continuity does imply, however, is that if two trajectories exhibit sensitive dependence on initial conditions, then the closer their initial conditions, the longer it takes for the trajectories to diverge.

To illustrate these concepts, picture two twigs drifting downstream on a gently flowing river. As time passes, the twigs stay near one another and the position of one twig is a good indication of the position of the other. This is non-sensitive dependence.

Now imagine the same twigs in a turbulent river. The twigs are initially near one another but quickly become separated in the swirling water. As time passes, the position of one twig yields little information on the position of the other. This is sensitive dependence on initial conditions and explains why researchers believe chaotic systems may hold the key to understanding turbulent fluid flow. Also note that, as continuity leads us to expect, the closer the twigs are at the start, the longer it takes for them to separate.

Two twigs in a river may illustrate the concept of sensitive dependence on initial conditions, but since it is not clear that a river is a deterministic system, they do not satisfactorily explain how a deterministic system can exhibit random behavior.

Though random behavior in a deterministic system may seem surprising at first, there are many simple examples. For instance, pseudo-random number generators have been used for decades. As another example, consider the following discrete-time system that maps the unit interval into itself:

$$x_{k+1} = P(x_k) := 10\,x_k \bmod 1, \qquad x_0 \in [0,1]. \tag{1.26}$$

If x_0 is written as a decimal number

$$x_0 = \sum_{i=1}^{\infty} d_i 10^{-i} = 0.d_1 d_2 d_3 \ldots \tag{1.27}$$

where $d_i \in \{0, \ldots, 9\}$, then the action of P is to drop d_1 and shift the remaining digits one to the left. The kth iterate is

$$x_k = \sum_{i=1}^{\infty} d_{i+k} 10^{-i} = 0.d_{k+1} d_{k+2} d_{k+3} \ldots. \tag{1.28}$$

Hence, under repeated applications of P, the most significant digits are lost and less significant digits become more significant until they, too, are discarded.

To show that this system is unpredictable, assume the exact initial condition is $x_0 = 1/\sqrt{2} = 0.70710678\ldots$, but that an observer can measure the state to only three significant figures: 0.707. The observer can predict that

$$\begin{aligned} 0.7065 &< x_0 < 0.7075, \\ 0.065 &< x_1 < 0.075, \\ 0.65 &< x_2 < 0.75, \end{aligned} \tag{1.29}$$

but can say nothing about x_3 and beyond. For this system, the observed initial condition *loses* predictive power as time increases.

Now suppose that an observer observes the state after every iteration of P. The first five observations are 0.707, 0.071, 0.711, 0.107, and 0.068. If the observer takes the most significant digit of each observation, he can reconstruct the initial condition to five decimal places: 0.70710. Clearly if n observations are made, the initial condition is known to n decimal places. By repeating measurements, the observer *gains* information about the state as the system evolves.

The rate of loss of predictive power is equal to the rate of information gain and, in fact, these are just two different descriptions of the same phenomenon. In this example, the rate is one digit per iteration. Note that the rate is system dependent—if the 10 in P is replaced by 100, the rate becomes two digits per iteration.

This example is a non-invertible map and since ϕ_t is invertible, it could be argued that the example has no direct bearing on continuous-time systems. We now examine how the concepts of information gain and loss of predictive power carry over to the continuous-time case.

1.2.5 Predictive power

The purpose of this section is to define what is meant by an unpredictable deterministic system and to present the underlying features that cause a continuous-time system to act unpredictably.

Definition Consider two observers observing a dynamical system. Observer A observes the system at time t_1 and observer B at time

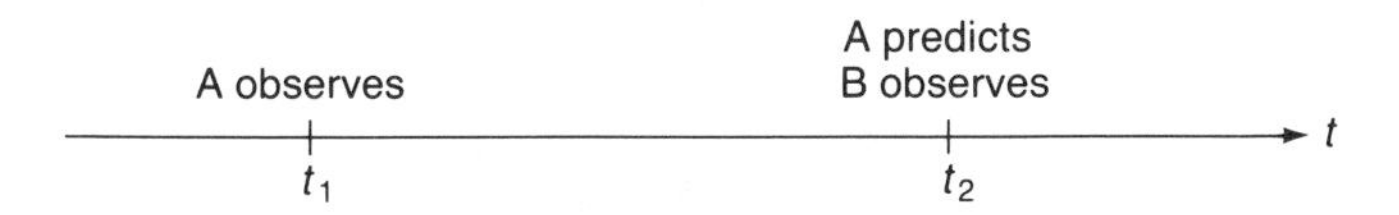

Figure 1.11: Observer A observes at time t_1 and observer B at time t_2. Observer A uses his observation to predict the state at time t_2.

$t_2 > t_1$ (Fig. 1.11). The uncertainty of observation is the same for each observer. Furthermore, given an initial condition at t_1, observer A can predict the state at t_2 with infinite precision, that is, the only error is due to the imprecise observation of the initial condition. The question is: Which observer knows more precisely the state at time t_2, observer A using observation plus prediction or observer B using only observation?

If observer A knows more precisely the state at t_2, the system is said to be *predictive.* For a predictive system, earlier observations convey more information than later ones. In other words, predictions are more accurate than observations. Information is destroyed in predictive systems and they can be thought of as *information sinks.*

If observer B knows more precisely the state at t_2, the system is called *unpredictive.* In an unpredictive system, the later the observation, the more information is gained. Equivalently, observations are more accurate than predictions. An unpredictive system creates information and is an *information source.*

When we say a deterministic system is random, we mean it is unpredictive. An unpredictive system exhibits sensitive dependence on initial conditions and vice versa.

Examples As an example of a predictive system, consider a contracting flow, that is, a system where all the trajectories approach one another:

$$\|\phi_{t_2}(y) - \phi_{t_2}(x)\| < \|\phi_{t_1}(y) - \phi_{t_1}(x)\| \tag{1.30}$$

for any x and y, $x \neq y$, and any t_1 and t_2, $t_1 < t_2$. Call observer A's observation x_1 (see Fig. 1.12(a)). Observer A knows that the actual state at time t_1 lies somewhere in the ϵ-ball $B_\epsilon(x_1)$ centered at x_1 and, therefore, that the actual state at time t_2 is contained in the set $\phi_{t_2-t_1}(B_\epsilon(x_1))$. If observer B's observation is x_2, he knows that the actual state at time t_2 lies in the ϵ-ball $B_\epsilon(x_2)$. Since the flow

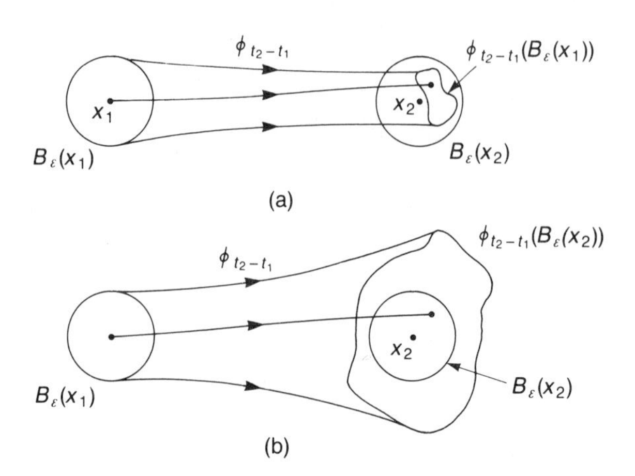

Figure 1.12: Simple examples of predictive and unpredictive systems. x_1 is observer A's observation. x_2 is observer B's observation. (a) A contracting system; (b) an expanding system;

is contracting, the diameter[4] of $\phi_{t_2-t_1}(B_\epsilon(x_{1A}))$ is smaller than that of $B_\epsilon(x_{2B})$ and observer A knows more precisely the position of tha actual state at time t_2.

Now consider an expanding flow, that is, one where all trajectories diverge:

$$\|\phi_{t_2}(y) - \phi_{t_2}(x)\| > \|\phi_{t_1}(y) - \phi_{t_1}(x)\| \tag{1.31}$$

for any x and y, $y \neq x$, and any t_1 and t_2, $t_2 > t_1$. For an expanding system (Fig. 1.12(b)), the diameter of $\phi_{t_2-t_1}(B_\epsilon(x_1))$ is larger than that of $B_\epsilon(x_2)$ and observer B knows more precisely the position of the actual state at time t_2. Thus, an expanding system is unpredictive.

This example shows that an expanding flow leads to an unpredictive system; however, since an expanding flow is unbounded, it can never achieve a steady state. The question of how the steady-state behavior of a dynamical system can be unpredictive remains unanswered.

To answer this question, we present the Smale horseshoe, an unpredictive one-to-one discrete-time system that maps the unit square into itself (Fig. 1.13(a)). As Fig. 1.13(b) shows, one iteration of the Smale horseshoe comprises three actions:

[4]The *diameter* of a set S is $\sup_{x,y \in S} \|x - y\|$.

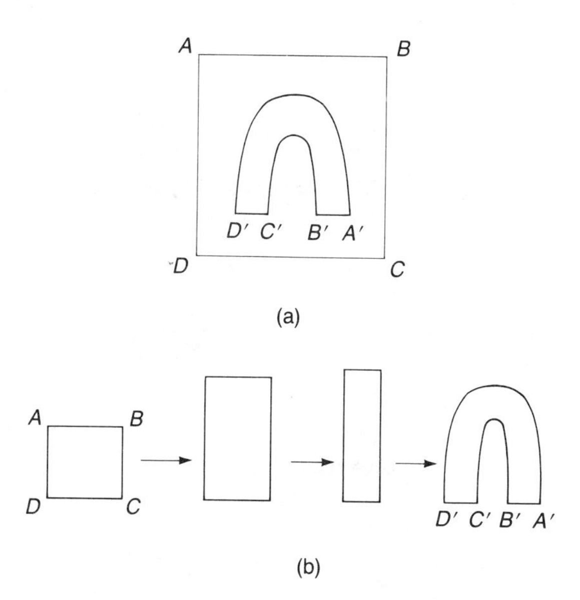

Figure 1.13: The Smale horseshoe. (a) The unit square $ABCD$ is mapped into the horseshoe $A'B'C'D'$; (b) three basic operations comprise the map: stretching, squeezing, and folding.

1. *Stretching*: As we have seen, the stretching makes the map unpredictive.

2. *Squeezing*: Squeezing is required to make the map one-to-one. To understand this, observe that the area of the unit square is 1. After stretching, the area is greater than 1. Since the Smale horseshoe maps into the unit square, if there were no squeezing, the image of the unit square would necessarily overlap itself and, in the region of overlap, the map would not be one-to-one. It follows that for the map to be one-to-one, the area of the image of the unit square must be less than or equal to 1. In a dissipative system like the Smale horseshoe, the contraction outweighs the expansion, that is, the area of the image of the unit square is less than 1.

3. *Folding*: Folding is required to map the unit square back into the unit square. Clearly, without the folding, portions of the image of the unit square would lie outside of the unit square.

In Chapter 2, we present a useful technique that transforms a continuous-time system into an invertible discrete-time map called the

Poincaré map. As we shall see, Smale horseshoes can appear in Poincaré maps of chaotic systems, so even though the Smale horseshoe may seem extremely simple, far removed from the complicated flows that can appear in continuous-time systems, it is actually quite useful in understanding the chaotic behavior of flows. We will not discuss the dynamics of the Smale horseshoe further. Interested readers are referred to Guckenheimer and Holmes [1983].

1.3 Summary

- *Autonomous continuous-time system*: A differential equation $\dot{x} = f(x)$ where the vector field f does not depend on time. The solution that passes through x_0 at time 0 is denoted $\phi_t(x_0)$.

- *Non-autonomous continuous-time system*: A differential equation $\dot{x} = f(x, t)$ where the vector field f depends on time. The solution that passes through x_0 at time t_0 is denoted $\phi_t(x_0, t_0)$.

- *Trajectory*: The locus of points traversed by $\phi_t(x)$ or $\phi_t(x, t_0)$ over all time.

- *Discrete-time system*: Defined by the iteration $x_{k+1} = P(x_k)$. The solution starting at initial condition x_0 is the sequence $\{x_k\}_{k=0}^{\infty}$.

- *Orbit*: The solution $\{x_k\}_{k=0}^{\infty}$ of a discrete-time system.

- *Steady state*: The asymptotic behavior as $t \to \infty$. The steady state must be bounded to make sense.

- *Limit set*: The set of points in state space that a trajectory repeatedly visits. A limit set is defined only for autonomous systems, be they discrete- or continuous-time.

- *Equilibrium point*: A constant solution of an autonomous differential equation. The spectrum contains a single spike at zero frequency. It is its own limit set.

- *Periodic solution*: A non-constant solution of a differential equation that repeats itself. The spectrum contains spikes at integer multiples of the fundamental frequency.

- *Subharmonic*: A periodic solution of a non-autonomous system whose period is an integer multiple (> 1) of the forcing period.

- *Limit cycle*: An isolated periodic solution of an autonomous system. The points on the limit cycle constitute the limit set.

- *Quasi-periodic solution*: A solution that can be written as the sum of a countable number of periodic functions each of whose periods is an integer combination of frequencies taken from a finite base set. The spectrum consists of spikes at the various sum and difference frequencies of the base set. In an autonomous system, the limit set is a p-torus where p is the minimum number of base frequencies.

- *Chaos*: A steady-state solution that is none of the above. The spectrum may have spikes but always has a broad-band, noise-like component. Most chaotic systems exhibit sensitive dependence on initial conditions. In the autonomous case, the attractor is not a manifold; it is a complicated geometrical structure with fractional dimension.

- *Sensitive dependence on initial conditions*: Nearby trajectories diverge and eventually become uncorrelated (to working precision).

- *Predictive system*: Earlier observations convey more information about the state of the system than later ones. Predictions are more accurate than observations.

- *Unpredictive system*: Later observations convey more information about the state of the system than earlier ones. Observations are more accurate than predictions. Equivalent to sensitive dependence on initial conditions. Three mechanisms are required for continuous-time systems to be unpredictive: stretching, squeezing, and folding.

Chapter 2

Poincaré Maps

A classical technique for analyzing dynamical systems is due to Poincaré. It replaces the flow of an nth-order continuous-time system with an $(n-1)$th-order discrete-time system called the Poincaré map. The definition of the Poincaré map ensures that its limit sets correspond to limit sets of the underlying flow. The Poincaré map's usefulness lies in the reduction of order and the fact that it bridges the gap between continuous- and discrete-time systems.

2.1 Definitions

The definition of the Poincaré map is different for autonomous and non-autonomous systems. We present the two cases separately.

2.1.1 The Poincaré map for non-autonomous systems

Recall that a time-periodic nth-order non-autonomous system with minimum period T can be transformed into an $(n+1)$th-order autonomous system in the cylindrical state space $\mathbb{R}^n \times S^1$ via (1.3). Consider the n-dimensional hyperplane $\Sigma \in \mathbb{R}^n \times S^1$ defined by

$$\Sigma := \{(x, \theta) \in \mathbb{R}^n \times S^1 : \theta = \theta_0\}. \tag{2.1}$$

Every T seconds, the trajectory (1.4) intersects Σ (Fig. 2.1). The resulting map $P_N : \Sigma \to \Sigma$ ($\mathbb{R}^n \to \mathbb{R}^n$) is defined by

$$P_N(x) := \phi_{t_0+T}(x, t_0). \tag{2.2}$$

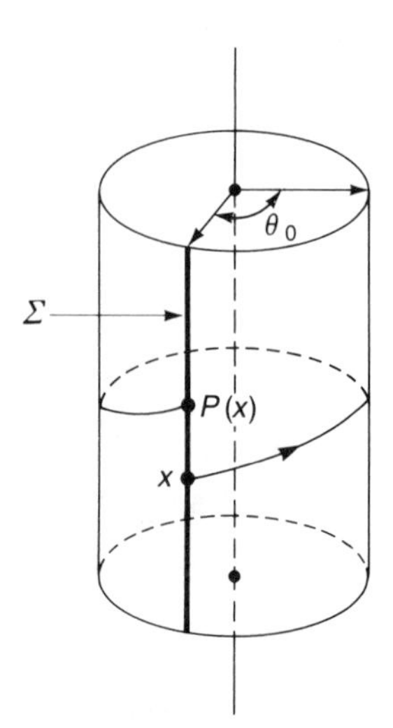

Figure 2.1: The Poincaré map of a first-order non-autonomous system.

P_N is called the *Poincaré map* for the non-autonomous system. The subscript N stands for *non-autonomous*, and is used to distinguish this map from the Poincaré maps that are used with autonomous systems. Note that for t fixed, ϕ_t is a diffeomorphism and, therefore, P_N is invertible and differentiable.

P_N can be thought of in two ways:

1. $P_N(x)$ indicates where the flow takes x after T seconds. This is called a *T-advance mapping*.

2. The orbit $\{P_N^k(x)\}_{k=1}^{\infty}$ is a sampling of a single trajectory every T seconds, that is,

$$P_N^k(x) = \phi_{t_0+kT}(x, t_0), \qquad k = 0, 1, \ldots. \tag{2.3}$$

   This is similar to the action of a stroboscope flashing with period T.

2.1.2 The Poincaré map for autonomous systems

Consider an nth-order autonomous system with the limit cycle Γ shown in Fig. 2.2. Let x^* be a point on the limit cycle and let Σ be an $(n-1)$-dimensional hyperplane transversal to Γ at x^*. The trajectory emanating from x^* will hit Σ at x^* in T seconds, where

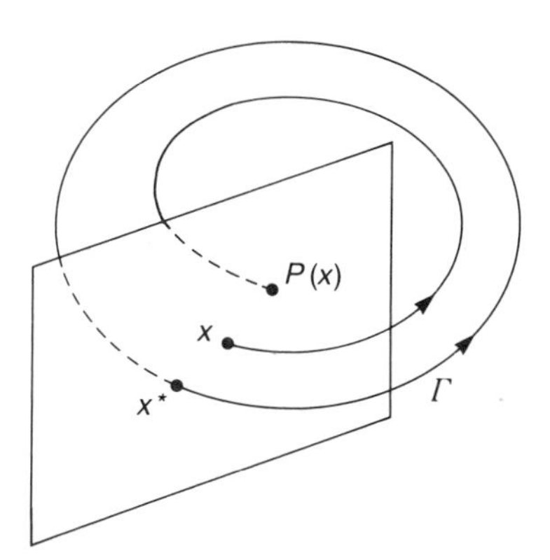

Figure 2.2: The Poincaré map for a third-order autonomous system.

T is the minimum period of the limit cycle. Due to the continuity of ϕ_t with respect to the initial condition, trajectories starting on Σ in a sufficiently small neighborhood of x^* will, in approximately T seconds, intersect Σ in the vicinity of x^*. Hence, ϕ_t and Σ define a mapping P_A of some neighborhood $U \subset \Sigma$ of x^* onto another neighborhood $V \subset \Sigma$ of x^*. P_A is a *Poincaré map* of the autonomous system.

Remarks:

1. P_A is defined locally, that is, in a neighborhood of x^*. Unlike the non-autonomous case, it is not guaranteed that the trajectory emanating from any point on Σ will intersect Σ.
2. For a Euclidean state space, the point $P_A(x)$ is not the first point where $\phi_t(x)$ intersects Σ; $\phi_t(x)$ must pass through Σ at least once before returning to V. Contrast this situation with the cylindrical state space of Fig. 2.1.
3. P_A is a diffeomorphism and is, therefore, invertible and differentiable. A proof is presented in Appendix D.

The definition of the Poincaré map just presented is the standard one from the theory of dynamical systems, but it is rarely used in simulations and experimental settings because it requires advance knowledge of the position of a limit cycle.

In practice, one chooses an $(n-1)$-dimensional hyperplane Σ, which divides $\mathbb{R}^n$ into two regions:

$$\Sigma_+ := \{x : \langle h, x - x_\Sigma \rangle > 0\} \tag{2.4}$$

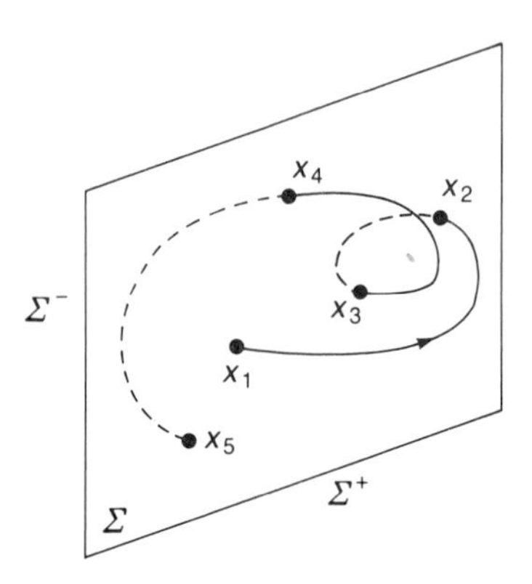

Figure 2.3: A typical trajectory intersecting a cross-section Σ. The sequence $\{x_1, x_3, x_5, \ldots\}$ is an orbit of the one-sided Poincaré map P_+, and $\{x_2, x_4, \ldots\}$ is an orbit of P_-. The complete sequence $\{x_1, x_2, \ldots\}$ is an orbit of the two-sided Poincaré map $P_\pm$.

and

$$\Sigma_- := \{x : \langle h, x - x_\Sigma \rangle < 0\}, \tag{2.5}$$

where $h \in \mathbb{R}^n$ is a vector normal to Σ, $x_\Sigma \in \mathbb{R}^n$ is some point lying on the hyperplane, and $\langle u, v \rangle := u^T v$ is the *inner product.* If Σ is chosen properly, then the trajectory under observation will repeatedly pass through Σ crossing from Σ_- to Σ_+ to Σ_-, etc., as in Fig. 2.3.

Given the hyperplane Σ, three different Poincaré maps can be defined:

$P_+ : \Sigma \to \Sigma$. $P_+(x)$ is the point where $\phi_t(x)$ first intersects Σ in a positive direction (i.e., $\langle h, f(\phi_t(x)) \rangle \geq 0$) for $t > 0$.

$P_- : \Sigma \to \Sigma$. $P_-(x)$ is the point where $\phi_t(x)$ first intersects Σ in a negative direction (i.e., $\langle h, f(\phi_t(x)) \rangle \leq 0$) for $t > 0$.

$P_\pm : \Sigma \to \Sigma$. $P_\pm(x)$ is the first point where $\phi_t(x)$ intersects Σ in either direction for $t > 0$.

P_+ and P_- are called *one-sided Poincaré maps* while $P_\pm$ is called the *two-sided Poincaré map.* Note that a point of tangent intersection (i.e., an $x \in \Sigma$ for which $\langle h, f(x) \rangle = 0$) satisfies the direction criterion of each of the three maps.

Remark: $P_\pm(x)$ is the point where the trajectory emanating from x first hits Σ so $P_\pm$ is called a *first-return map.*

There is no guarantee that any of the maps are well-defined since $\phi_t(x)$ may never intersect Σ for $t > 0$. For a system in a Euclidean state space, with bounded behavior that does not approach an equilibrium point, there is always some choice of Σ for which all three maps are well-defined. This is not true for a system with a non-Euclidean state space. As an example, consider the Poincaré map of a non-autonomous system. Since the trajectory always crosses Σ in the same direction, one of the one-sided Poincaré maps is not defined; whether it is P_+ or P_- depends on the choice of the normal vector h.

If one of the maps is well-defined, continuity and, hence, differentiability are not guaranteed; however, if f is transversal to Σ at x and at $P(x)$, then the map is locally differentiable. A proof of this fact is presented in Appendix D.

The map P_A is related to the three maps just defined as follows. In a Euclidean state space, the trajectory emanating from the fixed point x^* intersects Σ more than once before it returns to x^*. Let there be k intersections including the final one at x^*, and assume that all the intersections are transversal. Then P_A is equivalent to k applications of $P_\pm$, that is, $P_A(x) = P_\pm^k(x)$. Note that in a Euclidean space, k is always even and, therefore, P_A is also equivalent to $k/2$ applications of P_+ or P_-; whether P_+ or P_- is used depends on whether $f(x^*)$ points into Σ_+ or Σ_-.

2.2 Limit Sets of Poincaré Maps

In this section we present the relationship between limit sets of the Poincaré map and limit sets of the underlying flow. Except where otherwise noted, the discussion is limited to structurally stable limit sets of systems in a Euclidean state space.

2.2.1 Equilibrium points

There is no Poincaré limit set that corresponds to an equilibrium point.[1]

[1] If Σ contains an equilibrium point x_{eq} and the Poincaré map is well-defined in a neighborhood of x_{eq}, then the limit set of the Poincaré map is x_{eq}; however, this limit set is not structurally stable since the equilibrium point does not lie on Σ for almost any perturbation of Σ.

2.2.2 Periodic solutions

We discuss the non-autonomous and autonomous cases separately but first, we present two definitions.

x^* is a *fixed point* of the map P if $x^* = P(x^*)$.

The set $\{x_1^*, \ldots, x_K^*\}$ is a *period-K closed orbit* of the map P if $x_{k+1}^* = P(x_k^*)$ for $k = 1, \ldots, K-1$ and $x_1^* = P(x_K^*)$.

Non-autonomous systems

A period-one solution of a continuous-time system corresponds to a *fixed point* x^* of the Poincaré map P_N.

A Kth-order subharmonic corresponds to a period-K closed orbit $\{x_1^*, \ldots, x_K^*\}$ of the Poincaré map P_N.

Remark: The Poincaré map freezes any periodic component of the solution that is commensurate with the forcing frequency. This is the same action as a periodically flashing stroboscope such as that commonly found on audio turntables.

Autonomous systems

$\mathbf{P_A}$: A limit cycle of ϕ_t corresponds to a fixed point x^* of P_A.

A period-K closed orbit of P_A indicates a subharmonic solution (relative to Γ) of the underlying flow. Remember that one must be careful in using the term subharmonic with autonomous systems. In particular, if the minimum period of Γ is T, then the minimum period of a Kth-order subharmonic will be close to but not usually equal to KT because, unlike the non-autonomous Poincaré map, P_A is defined in terms of a cross-section and not in terms of time. Thus, the return time for x^* is T, but the return time for a point near x^* is close to, but not usually equal to T (see Appendix D).

$\mathbf{P_+}$, $\mathbf{P_-}$, and $\mathbf{P_\pm}$: For these maps, the classification of limit cycles is ambiguous because the limit set of the Poincaré map depends on the position of Σ. Specifically, given a limit cycle of the underlying flow, different choices of Σ can lead to closed orbits of different orders (Fig. 2.4). The most general statement that can be made is that a closed orbit of one of these Poincaré maps corresponds to a limit cycle of the underlying flow.

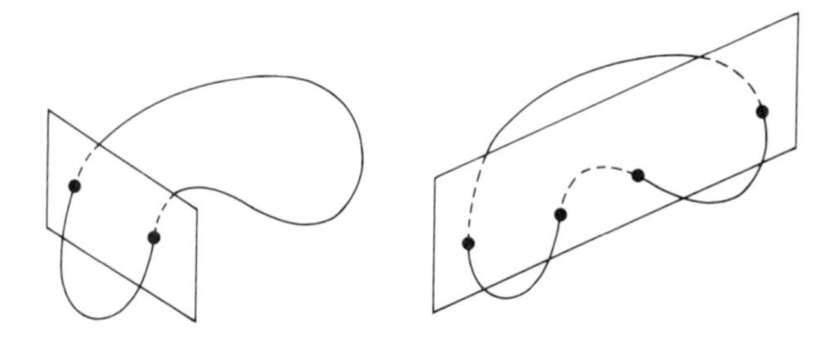

Figure 2.4: The limit sets of the one- and two-sided Poincaré maps depend on the choice of cross-section Σ.

In a Euclidean state space, if a limit cycle intersects Σ transversally at all its crossings, then the order of the corresponding closed orbit of P_+ equals that of the corresponding closed orbit of P_- and is equal to half the order of the corresponding closed orbit of $P_\pm$. Almost any perturbation of Σ eliminates a non-transversal crossing so, generically, all closed orbits of $P_\pm$ have even order.

If one is careful, subharmonics can be defined with these maps. Consider a period-m closed orbit (with transversal crossings) that corresponds to a limit cycle Γ with period T. If there is an (mK)th-order closed orbit nearby, it is a Kth-order subharmonic with respect to Γ and the period of its underlying limit cycle is approximately KT. The key word here is "nearby" since if the two orbits are not near one another, they might be caused by two completely unrelated limit cycles. More precisely, there exists a neighborhood of the period-m orbit in which P_+ is a diffeomorphism; "nearby" means within that neighborhood.

2.2.3 Quasi-periodic solutions

Again, the non-autonomous and autonomous cases are discussed separately.

The non-autonomous case

Consider a two-periodic solution $\phi_t(x)$ of a non-autonomous system with frequency base $\{f_1, f_f\}$ where $f_f := 1/T_f$ is the forcing period. Using coordinates (θ_1, θ_2) on the torus, $\phi_t(x)$ can be written

$$x(t) = F(\theta_1(t), \theta_2(t)) \tag{2.6}$$

where $F: S^1 \times S^1 \to \mathbb{R}^n$ and

$$\begin{bmatrix} \theta_1(t) \\ \theta_2(t) \end{bmatrix} = \begin{bmatrix} 2\pi f_1 t \bmod 2\pi \\ 2\pi f_f t \bmod 2\pi \end{bmatrix}. \tag{2.7}$$

The action of P_N is to sample $\phi_t(x)$ every $1/f_f$ seconds. Hence,

$$\begin{bmatrix} \theta_1(k/f_f) \\ \theta_2(k/f_f) \end{bmatrix} = \begin{bmatrix} 2\pi k f_1/f_f \bmod 2\pi \\ 2\pi k \bmod 2\pi \end{bmatrix} \tag{2.8}$$

$$= \begin{bmatrix} 2\pi k f_1/f_f \bmod 2\pi \\ 0 \end{bmatrix}, \quad k = 0, 1, \ldots. \tag{2.9}$$

Since f_1 and f_f are incommensurate, $\{\theta_1(k/f_f)\}_{k=0}^{\infty}$ is not periodic and repeatedly comes arbitrarily close to every point in $[0, 2\pi)$. Thus, in the (θ_1, θ_2) coordinate system, the limit set of the Poincaré map is the circle defined by $\theta_2 = 0$. In the original Euclidean coordinates, the limit set is a diffeomorphic copy of a circle, that is, an embedded circle. The Poincaré limit set for the forced van der Pol equation (1.23) is shown in Fig. 2.5.

We have seen that the limit set of a Kth-order subharmonic solution is a period-K closed orbit. The analogous behavior in the two-periodic case is a *Kth-order two-subharmonic* with frequency base $\{f_1, f_f/K\}$. In this case, the limit set of the Poincaré map consists of K embedded circles.

Again, we see the similarity between P_N and a stroboscope: P_N freezes the dynamics that are commensurate with the sampling frequency.

In general, the Poincaré limit set of a K-periodic trajectory is one or more embedded $(K-1)$-tori.

The autonomous case

We have not defined P_A in the quasi-periodic case. Though this can be done, it is of little interest to us and we will not discuss it further.

Recall that the limit set of a two-periodic solution of an autonomous system is a two-torus. It follows that the possible limit sets of the one- and two-sided Poincaré maps are the intersections of an $(n-1)$-dimensional hyperplane with an embedded two-torus.

Consider the two different cross-sections shown in Fig. 2.6. In

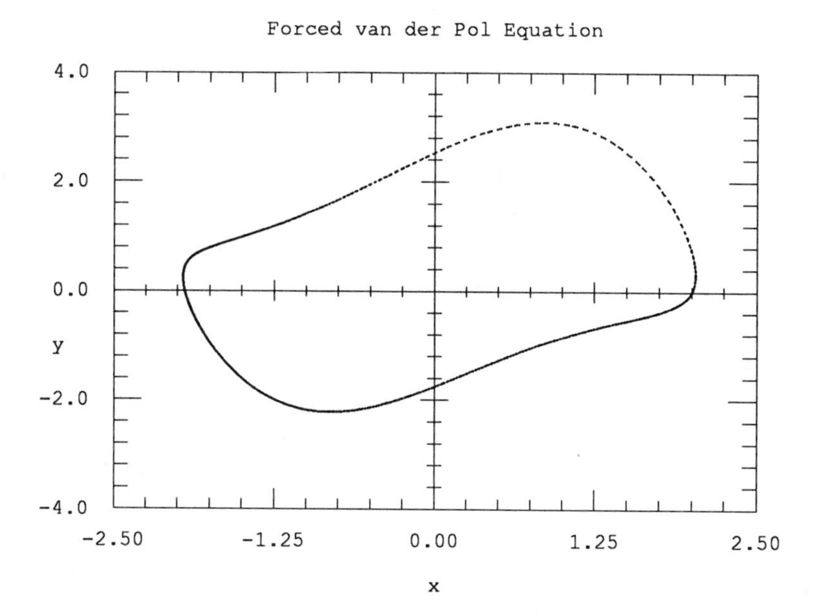

Figure 2.5: The limit set of the Poincaré map of the forced van der Pol equation for $t_0 = 0$. The parameters are the same as in Fig. 1.5.

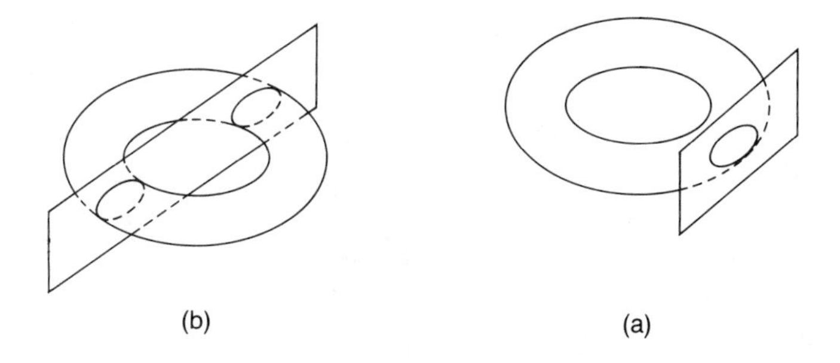

Figure 2.6: Limit sets of the Poincaré map in the autonomous two-periodic case. (a) The limit set of $P_{\pm}$ is two embedded circles; (b) the limit set is a single embedded circle.

Fig. 2.6(a), Σ intersects the hole of the torus and the limit set of the two-sided map $P_{\pm}$ is two embedded circles. In this case, the P_{+} limit set is one of the circles and the P_{-} limit set is the other. If Σ is shifted such that it misses the hole but still intersects the torus, as in Fig. 2.6(b)), then the two-sided limit set is an embedded circle, the P_{+} limit set is a portion of arc of this circle, and the P_{-} limit set is the remaining portion. Since both P_{+} and P_{-} include points of tangent intersection, the endpoints of the arcs are common to both one-sided limit sets.

If the two-torus is not as well-behaved as the one in Fig. 2.6, but bends and twists in state space like a tangled inner tube, the limit set can contain more elements than just described. For example, the limit set of $P_{\pm}$ can consist of four circles and the limit set of P_{+} can be two circles.

In summary, for two-periodic behavior, the limit set of $P_{\pm}$ is composed of embedded circles. The limit sets of the one-sided Poincaré maps are composed of embedded circles and arcs of embedded circles. For a given cross-section, the $P_{\pm}$ limit set is the union of the P_{+} and P_{-} limit sets. Furthermore, the only points common to the P_{+} and P_{-} limit sets are the endpoints of the arcs.

For K-periodic trajectories, the $P_{\pm}$ limit set consists of one or more embedded $(K-1)$-tori. The limit sets for P_{+} and P_{-} are composed of embedded $(K-1)$-tori and, perhaps, portions of embedded $(K-1)$-tori.

2.2.4 Chaos

The steady-state Poincaré orbits for chaotic systems are distinctive and often quite beautiful. Fig. 2.7 shows two typical orbits. Comparison of these chaotic orbits with the corresponding trajectories in Figs. 1.8 and 1.9 demonstrates the power of the Poincaré map in exposing the underlying structure of chaotic flows.

Looking at these orbits, two points are immediately clear. First, the steady-state orbits do not lie on a simple geometrical object as is the case with periodic and quasi-periodic behavior. Second, the attractor has a fine structure. As the magnifications in Figs. 2.7(b,c) show, there appear to be layers within layers, much like baklava or a French pastry. This fine structure, reminiscent of Cantor sets, is typical of chaotic systems and is characterized by a fractional dimension (see Chapter 7).

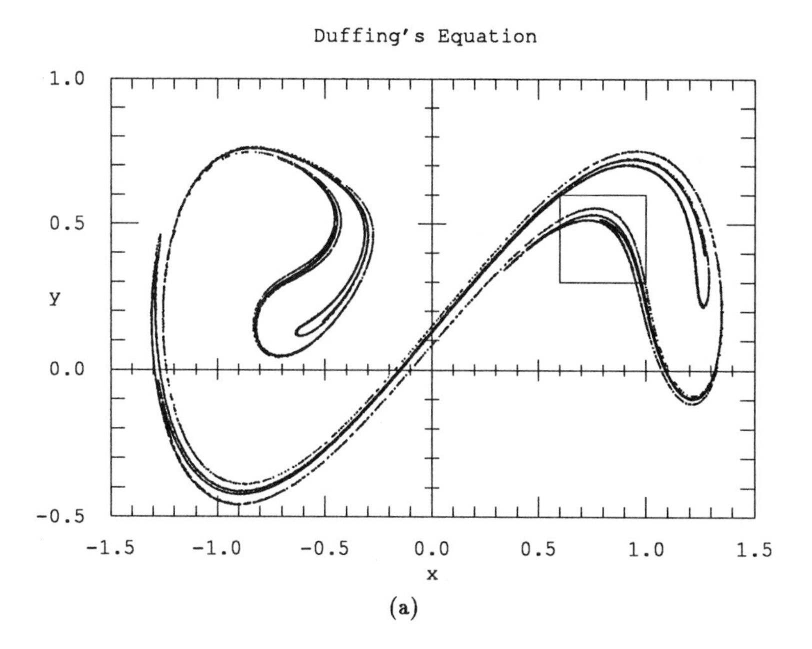

(a)

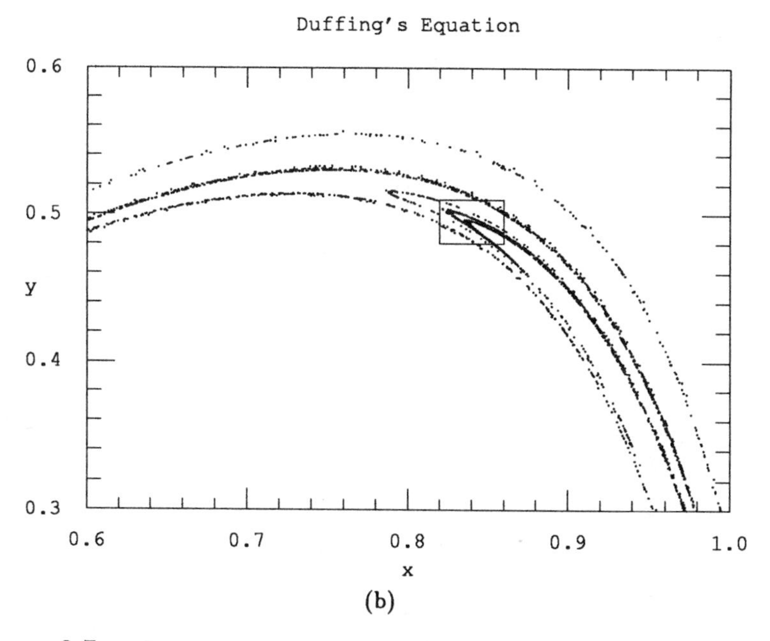

(b)

Figure 2.7: Strange attractors of Poincaré maps. (a) A chaotic orbit for Duffing's equation with $t_0 = 0$ and the same parameters as in Fig. 1.8; (b) and (c) are successive magnifications of (a) displaying the fine structure of the attractor; (d) a chaotic orbit for the double scroll equation with $h = [1\ 0\ 0]^T$, $x_\Sigma = [1.75\ 0\ 0]^T$ using the same parameters as Fig. 1.9.

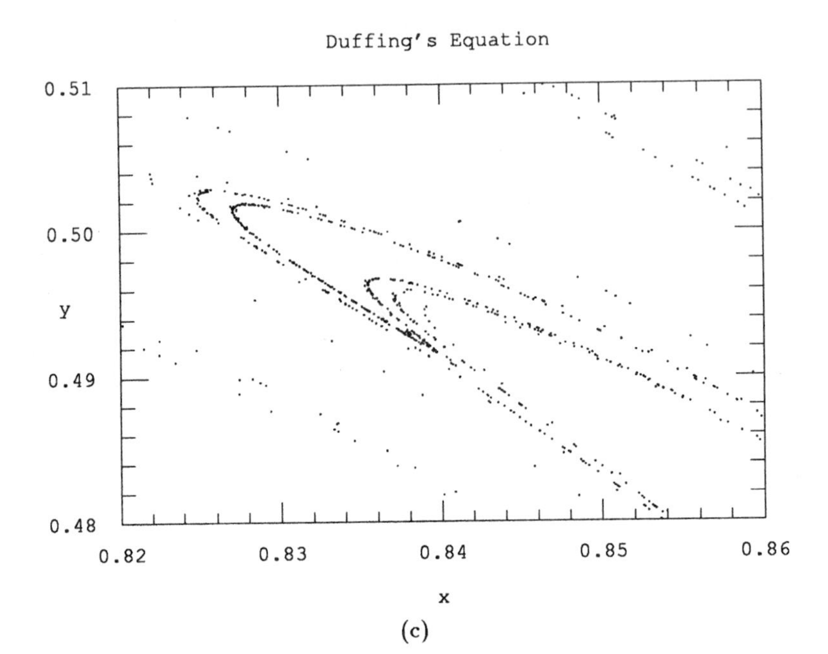

(c)

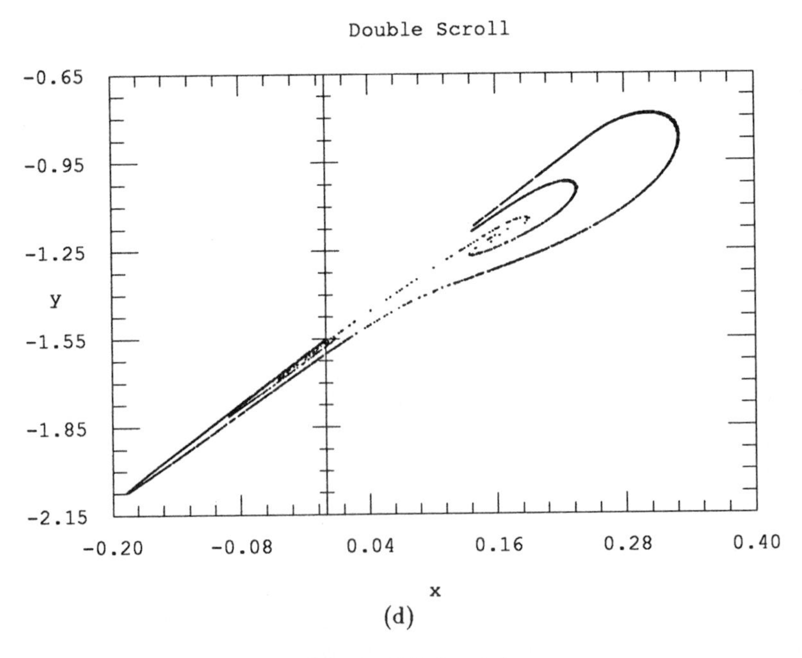

(d)

Figure 2.7 (cont.)

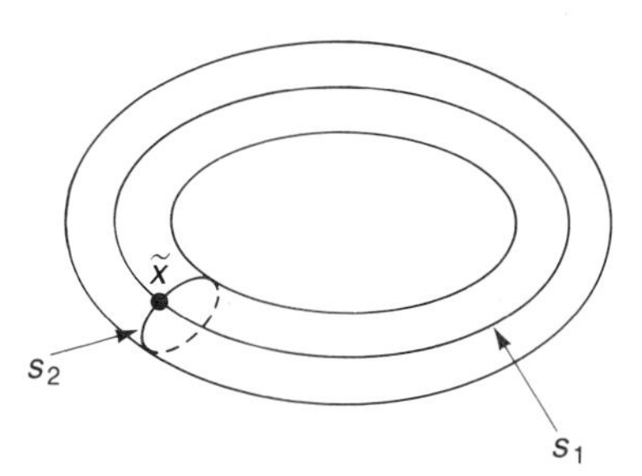

Figure 2.8: Two-periodic behavior on a torus. S_1 is the limit set when the two-periodic trajectory is sampled at the base frequency $1/T_1$, and S_2 is the limit set when the sampling rate is $1/T_2$. The point of intersection $\tilde{x}$ of S_1 and S_2 is the limit set of the second-order Poincaré map as $\epsilon \to 0$.

2.3 Higher-order Poincaré maps

The sampling action of the Poincaré map reduces the dimension of a limit set by one—a limit cycle becomes a point, a torus becomes a circle, and a K-torus becomes a $(K-1)$-torus. Another sampling should reduce the dimension of the limit set further. This idea leads to higher-order Poincaré maps.

The non-autonomous case Consider a continuous-time system that has two periodic forcing terms with incommensurate periods, T_1 and T_2, and suppose that the system exhibits two-periodic behavior with frequency base $\{1/T_1, 1/T_2\}$. As we have seen, the limit set of two-periodic behavior is a two-torus $S^1 \times S^1$ where each S^1 corresponds to one of the frequencies in the base set. Let the coordinates on the torus be (θ_1, θ_2) where θ_1 corresponds to the component with period T_1 and θ_2 corresponds to the component with period T_2. Sampling the two-periodic trajectory at a rate of $1/T_1$ freezes the motion along θ_1. The limit set of the resulting orbit is $S_1 := \{(\theta_1, \theta_2) : \theta_1 = \theta_0\}$ where θ_0 corresponds to the time origin (see Fig. 2.8). Likewise, sampling the trajectory at a rate of $1/T_2$ produces an orbit whose limit set is the circle S_2 defined by $\theta_2 = \theta_0$. If the trajectory could be sampled at $1/T_1$ and $1/T_2$ simultaneously, the motion on both θ_1 and θ_2 would be frozen, and the limit set would be the single point, $\tilde{x} := (\theta_0, \theta_0)$, where S_1 intersects S_2.

The objective of the *second-order Poincaré map* is to perform this double-sampling. For simplicity, assume $t_0 = 0$. The first sampling, at a rate of $1/T_1$, creates a set $\{x_k\}_{k=0}^{\infty}$ of points at times kT_1. The

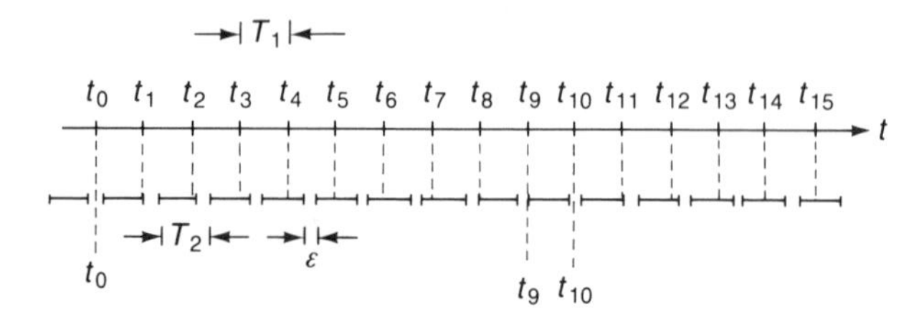

Figure 2.9: A second-order Poincaré map of a non-autonomous system. The first sampling selects time-points that are evenly spaced with period T_1. The second sampling selects those time-points $\{t_0, t_9, t_{10}\}$ from the first sampling that fall through ϵ-windows which are evenly spaced with period T_2.

second sampling, at a rate of $1/T_2$, selects points from this set that lie at times that are integer multiples of T_2, that is, x_k is selected if there exists an integer m such that

$$kT_1 = mT_2. \tag{2.10}$$

Of course, since T_1 and T_2 are incommensurate, condition (2.10) is never satisfied. Instead, choose $\epsilon > 0$. The point x_k is selected if its time kT_1 lies within ϵ of an integer multiple of T_2 (see Fig. 2.9). Specifically, the orbit of the second-order Poincaré map is composed of those points $x_k := \phi_{kT_1}(x_0, 0)$ where k satisfies

$$(kT_1 + \epsilon) \bmod T_2 < 2\epsilon. \tag{2.11}$$

Due to ϵ, the limit set of the second-order Poincaré map is not the single point $\tilde{x}$; it is a small segment of S^1_2 that contains $\tilde{x}$. As $\epsilon \to 0$, (2.11) approaches (2.10) and the limit set approaches $\tilde{x}$.

The extension to third- and higher-order Poincaré maps is obvious. Every additional sampling reduces the dimension of the limit set by one.

The autonomous case A second-order Poincaré map for an autonomous system is similar to the non-autonomous case in that the trajectory is sampled twice. The difference is that in the autonomous case, the trajectory is state-space sampled instead of time sampled.

The first sampling is done in the usual manner using an $(n-1)$-dimensional hyperplane Σ_1. The points where the trajectory intersects Σ form an orbit $\{x_k\}_{k=0}^{\infty}$ of the first-order Poincaré map. The second sampling uses an $(n-2)$-dimensional hyperplane $\Sigma_2 \subset \Sigma_1$.

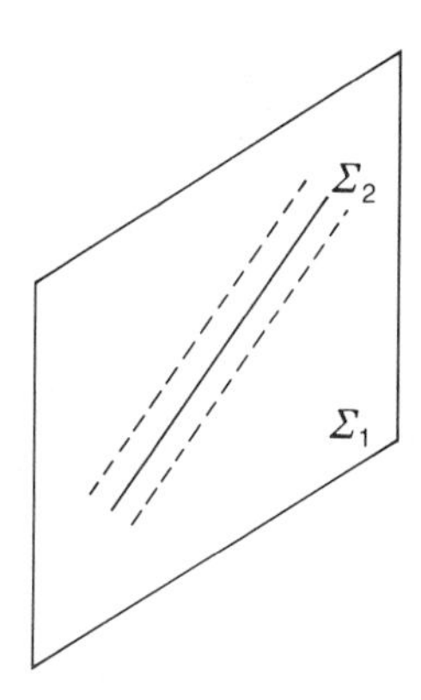

Figure 2.10: The orbit of the second-order Poincaré map of an autonomous system consists of those points of the first-order map (defined by the plane Σ_1) that lie within ϵ of the line Σ_2. The trajectory defining the orbit is not shown.

The x_k that lie on Σ_2 make up the orbit of the second-order Poincaré map. In practice, none of the x_k lies exactly on Σ_2, so those x_k within ϵ of Σ_2 are selected (see Fig. 2.10).

The extension to third- and higher-order Poincaré maps is obvious.

Hybrid maps Time and state-space sampling can be mixed to form hybrid higher-order Poincaré maps. For instance, the trajectory of a non-autonomous system with two forcing terms can be time sampled at each of the forcing periods, and the resulting orbit state-space sampled using cross-sections. Each sampling reduces the dimension of the limit set by one.

Discussion The advantage of a higher-order Poincaré map is that it allows easier identification of K-periodic behavior. For example, consider a flow with a limit set that is a three-torus. The limit set of the first-order Poincaré map is a two-torus. As Fig. 2.11(a) demonstrates, it is difficult to tell that a sequence of points lies on a two-torus. However, the limit set of the second-order Poincaré map is a closed curve (Fig. 2.11(b)) and can be identified easily.

The main drawback of the higher-order Poincaré map is that it requires an enormous amount of input data to produce a reasonable amount of output. If one assumes that each sampling reduces the amount of data by a factor of 100, then a third-order Poincaré map

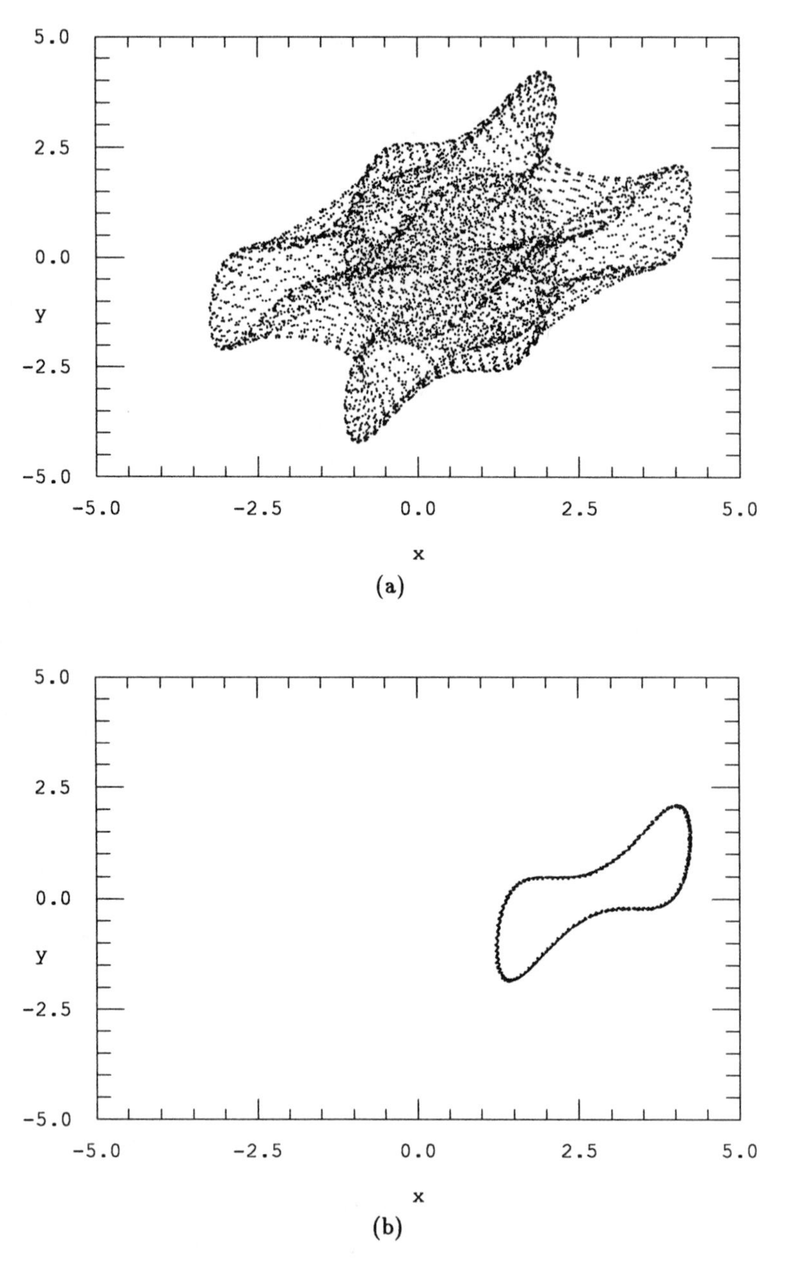

Figure 2.11: Higher-order Poincaré map of a three-periodic trajectory with frequency base $\{1, \sqrt{2}, \sqrt{3}\}$. (a) First-order Poincaré map sampled at rate 1; (b) the second-order Poincaré map sampled at rates 1 and $\sqrt{2}$.

requires 10000 more data points than a first-order map. Another disadvantage is that, owing to the use of ϵ-neighborhoods, the limit sets are fuzzy—a point becomes a segment and a circle becomes an annulus. This effect can be lessened by choosing a smaller value of ϵ, but this will increase the simulation time since more data points will be required.

2.4 Algorithms for calculating the Poincaré map

2.4.1 Choosing the hyperplane Σ

In a Poincaré map program, an $(n-1)$-dimensional hyperplane Σ can be chosen in several ways.

1. The user can specify n linearly independent points.
2. The user can specify a normal vector h and a point x_Σ on the hyperplane.
3. The user can specify x_Σ and let the program choose $h = f(x_\Sigma)$ where f is the vector field.
4. If the program already knows the position of a limit cycle, it can choose $x_\Sigma = x^*$ and $h = f(x^*)$ where x^* is any point on the limit cycle.

Methods 1 and 2 give the user full control over the position of the hyperplane and one of them should be included in any Poincaré map program. Of these two methods, Method 1 is perhaps a more natural way of specifying a hyperplane, but Method 2 requires the user to enter less data.

Method 3 is less flexible than the first two, but has two advantages. First, the user has to enter less data. Second, the choice of h implies that the vector field is orthogonal to Σ at x_Σ. This is advantageous for the algorithms that find hyperplane crossings and that locate limit sets because even though these algorithms require only that the crossings be transversal, they are more accurate and more efficient when the crossings are near orthogonal.

A good compromise is a combination of methods 2 and 3. If the user specifies h and x_Σ, those values are used. If h is not specified, it defaults to $f(x_\Sigma)$ unless that is 0, in which case h is set to (0, 0, ..., 1). The default value for x_Σ is (0, ..., 0).

```
begin find_crossing(x0[], t0, tf)
      set x2[] = x0[]
      set t2 = t0
      set α2 = H(x0[])
      while (t2 < tf) do
            set x1[] = x2[]
            set t1 = t2
            set α1 = α2
            set t2 = t1 + step-size
            if t2 > tf then
               set t2 = tf
            endif
            set x2[] = φt2(x[])
            set α2 = H(x2[])
            if (α1α2 < 0.0) then
               call a routine to calculate x̂[] and t̂
               return x̂[] and t̂
            endif
      endwhile
      return x2[] and t2
end find_crossing
```

Figure 2.12: Pseudo-code for `find_crossing`. Return values are the position $\hat{x}[]$ and time $\hat{t}$ of the first hyperplane crossing or the state at time t_f, whichever occurs earlier.

2.4.2 Locating hyperplane crossings

Calculation of hyperplane crossings is required for the computation of Poincaré maps of autonomous systems and is also useful in the simulation of piecewise-defined systems.

Suppose the hyperplane is represented by the equation

$$H(x) := \langle h, x - x_\Sigma \rangle = 0. \tag{2.12}$$

To locate the first hyperplane crossing of a trajectory $\phi_t(x)$, integrate the trajectory, stopping to calculate $H(\phi_t(x))$ at every time-step. Keep integrating until two consecutive points, $x_1 := \phi_{t_1}(x)$ and $x_2 := \phi_{t_2}(x)$, lie on different sides of Σ, that is, until $H(x_1)$ and $H(x_2)$ are of opposite sign. Once x_1 and x_2 are found, the exact crossing is known to be some point $\hat{x} := \phi_{\hat{t}}(x)$ with $t_1 < \hat{t} < t_2$.

Pseudo-code for a function `find_crossing` is shown in Fig. 2.12. `find_crossing` implements the algorithm just described with one additional feature: it returns if a crossing is not located before some

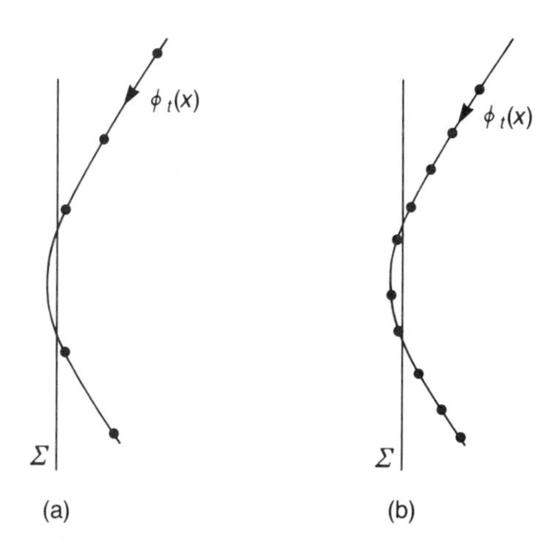

Figure 2.13: Near-tangent crossings are difficult to detect. (a) If all calculated points of the trajectory lie on one side of Σ, the crossing will not be detected; (b) the problem can be alleviated if the integration step-size is decreased by tightening the integration tolerance.

final time t_f. This *time-out* is a useful feature when one does not know beforehand whether a crossing exists.

Before we discuss how to calculate $\hat{x}$ given x_1 and x_2, observe that, besides the inaccuracy of the integration, there is another source of error in `find_crossing`. Consider the trajectory shown in Fig. 2.13 that crosses Σ into Σ_- and quickly returns to Σ_+. It is possible that none of the points calculated by the integration routine lie in Σ_-. If this occurs, the crossing will not be detected. The chance of missed detection is lessened by reducing the integration tolerance thereby reducing the step-size, but the possibility can never be eliminated totally.

We now present four methods for approximating $\hat{x}$ assuming x_1, x_2, t_1, and t_2 are known.

Interpolation

If efficiency is more important than accuracy, a simple linear interpolation scheme can be used. Let $\alpha_1 := H(x_1)$ and $\alpha_2 := H(x_2)$.

Then

$$\hat{x} \approx \frac{\alpha_2}{\alpha_2 - \alpha_1} x_1 + \frac{\alpha_1}{\alpha_1 - \alpha_2} x_2 \tag{2.13}$$

and

$$\hat{t} \approx \frac{\alpha_2}{\alpha_2 - \alpha_1} t_1 + \frac{\alpha_1}{\alpha_1 - \alpha_2} t_2. \tag{2.14}$$

```
begin time_half(t_1, x_1[], t_2)
      set α_1 = H(x_1[])
      set t̂ = t_1
      set x̂[] = x_1[]
      while (not converged) do
            set t̂_old = t̂
            set x̂_old[] = x̂[]
            set t̂ = (t_1 + t_2)/2.0
            set x̂[] = φ_t̂(x[])
            set α̂ = H(x̂[])
            if (α̂α_1 > 0.0) then
               set t_1 = t̂
               set α_1 = α̂
            else
               set t_2 = t̂
            endif
      endwhile
      return t̂ and x̂[]
end time_half
```

Figure 2.14: Pseudo-code for the `time_half` algorithm.

The accuracy of the interpolation is improved by decreasing the distance between x_1 and x_2, that is, by decreasing the integration tolerance.

Time-step halving

This is an iterative scheme. Since $\hat{t}$ must lie between t_1 and t_2, keep halving the interval $[t_1, t_2]$ until $\hat{t}$ is known as precisely as required. Pseudo-code for `time_half`, a routine which implements this algorithm, is presented in Fig. 2.14.

The relative/absolute convergence tests (see Appendix A)

$$\|\hat{x} - \hat{x}_{old}\| < E_r \|\hat{x}\| + E_a \tag{2.15}$$

and

$$|\hat{t} - \hat{t}_{old}| < E_r |\hat{t}| + E_a \tag{2.16}$$

are used to test for convergence of the iterations, and the inequality

$$\|H(\hat{x})\| < E_r \|H(0)\| + E_a \tag{2.17}$$

is used to test whether $\hat{x}$ lies close to Σ. Of course, different values of E_r and E_a can be used in the different tests.

There are a few considerations when calculating $\hat{x} = \phi_{\hat{t}}(x)$ in `time_half`. If an integration routine that incorporates interpolation is used, the integration routine interpolates to find $\hat{x}$ quickly and efficiently. If a non-interpolating routine is used, however, the routine must perform additional integration to calculate $\hat{x}$. The straightforward approach is to integrate $\phi_{\hat{t}}(x)$ always starting from the most recently calculated point on the trajectory. This is undesirable, however, since integration error accumulates even though the trajectory is not moving past t_2. Also, when integrating from t_2 to $\hat{t}$, the integration occurs in reverse time. This is inefficient for routines that alter step-size and order because the dynamics of the reverse-time problem are often quite different from the forward dynamics.

A better method is to have `find_crossing` pass a copy of the state of the integration at time t_1 to `time_half`. Subsequent integrations within `time_half` always start from this stored integration state. When `time_half` exits, `find_crossing` resumes integration from time t_2.

This approach has several advantages:

1. The integration is always in forward time.

2. The integration from t_1 to $\hat{t}$ is always accomplished in one step because $t_1 < \hat{t} < t_2$ and because the algorithm used by `find_crossing` ensures that the integration from t_1 to t_2 can be accomplished in one step.

3. The integration in `time_half` always restarts from the same state. Hence, there is no accumulated error in the calculation of $\hat{x}$.

4. The integration occurring inside `time_half` has no effect on the integration occurring inside `find_crossing`.

Note that since `find_crossing` may detect a crossing at any integration step, it must store the current state of the integration before each time-step. Storing the integration state—a process that consists of several memory read and write operations—is typically much faster than the computations involved in an integration step and the overhead involved is not substantial.

The Newton-Raphson method

This is an iterative approach similar in spirit to time-step halving.

```
begin NR_crossing(t_1, x_1[], t_2)
      set α_1 = H(x_1[])
      set t̂ = (t_1 + t_2)/2.0
      set x̂[] = x_1[]
      while (not converged) do
            set t̂ = t̂ − H(x̂[])/⟨h, f(x̂[])⟩
            if (t̂ < t_1 or t̂ > t_2) then
               set t̂ = (t_1 + t_2)/2.0
            endif
            set x̂[] = φ_{t̂−t_1}(x_1[])
            set α = H(x̂[])
            if (αα_1 < 0.0) then
               set t_2 = t̂
            else
               set t_1 = t̂
               set x_1[] = x̂[]
            endif
      endwhile
      return t̂ and x̂[]
end NR_crossing
```

Figure 2.15: Pseudo-code for the **NR_crossing** algorithm.

Define $F: \mathbb{R} \to \mathbb{R}$ by

$$F(t) := H(\phi_t(x)) = \langle h, \phi_t(x) - x_\Sigma \rangle. \tag{2.18}$$

The crossing times of $\phi_t(x)$ are the zeros of F, and the Newton-Raphson algorithm can be applied to find them. The Newton-Raphson iteration is easily found to be

$$\hat{t}^{(i+1)} = \hat{t}^{(i)} - \frac{H(\hat{x}^{(i)})}{\langle h, f(\hat{x}^{(i)}) \rangle} \tag{2.19}$$

where $\hat{x}^{(i)} := \phi_{\hat{t}^{(i)}}(x)$. Pseudo-code for NR_crossing, a routine which implements the Newton-Raphson procedure, is presented in Fig. 2.15.

The Newton-Raphson algorithm is ideally suited for this application. Since $\hat{t}$ lies between t_1 and t_2, the initial guess is always close to $\hat{t}$, and there is a high likelihood of convergence. In addition, if $\hat{t}$ does stray out of the interval $[t_1, t_2]$, NR_crossing further increases the chance of convergence by setting $\hat{t}$ to a value midway between t_1 and t_2.

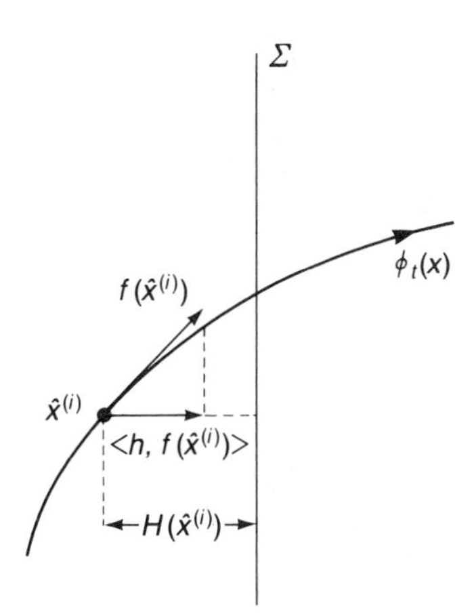

Figure 2.16: A geometrical interpretation of (2.19). $H(\hat{x}^{(i)})$ is the distance from $\hat{x}^{(i)}$ to the hyperplane Σ. $\langle h, f(\hat{x}^{(i)})\rangle$ is the component of the vector field that points toward Σ.

See Section 2.4.2 for comments on the efficient and accurate calculation of $\phi_{\hat{t}}(x)$.

It is always pleasing when the blind application of a technique like Newton-Raphson results in a physically meaningful equation and such is the case with (2.19). The distance from $\hat{x}^{(i)}$ to Σ is $H(\hat{x}^{(i)})$. The term $\langle h, f(\hat{x}^{(i)})\rangle$ is the component of the vector field at $\hat{x}^{(i)}$ that lies in the direction of Σ (see Fig. 2.16). Thus, the correction term, $H(\hat{x}^{(i)})/\langle h, f(\hat{x}^{(i)})\rangle$, is the time it would take the trajectory to travel from $\hat{x}^{(i)}$ to Σ if the the vector field were (locally) constant with value $f(\hat{x}^{(i)})$.

Henon's method

This clever method is due to Henon [1982]. The technique hinges on his observation that if one wants to know the state of a system at time t, one simply integrates the system to time t. It follows that if the independent variable of the system is x_n, the next crossing with the hyperplane $x_n = a$ is calculated by integrating to a; there is no interpolation or time-step halving.

To implement this idea, let the system of n differential equations be

$$\begin{aligned} dx_1/dt &= f_1(x_1,\ldots,x_n), \\ &\vdots \\ dx_n/dt &= f_n(x_1,\ldots,x_n). \end{aligned} \tag{2.20}$$

To make x_n the independent variable, divide the first $(n-1)$ equations by f_n and invert the nth equation:

$$\begin{aligned} dx_1/dx_n &= f_1(x_1,\ldots,x_n)/f_n(x_1,\ldots,x_n), \\ &\vdots \\ dx_{n-1}/dx_n &= f_{n-1}(x_1,\ldots,x_n)/f_n(x_1,\ldots,x_n), \\ dt/dx_n &= 1/f_n(x_1,\ldots,x_n). \end{aligned} \tag{2.21}$$

Remarks:

1. The intersection with an arbitrary hyperplane Σ can be found by rotating and translating the axes such that Σ corresponds to the hyperplane $x_n = a$.
2. (2.21) is singular whenever $f_n = 0$, that is, at every local minimum and maximum of x_n. Thus, system (2.21) should be used only when a hyperplane crossing has been located by a routine like `find_crossing`.
3. Since (2.21) is being integrated over short intervals, usually for just one time-step, Henon's method is feasible only with single-step integration algorithms; start-up costs prohibit its use with multi-step integration routines.
4. (2.20) and (2.21) can be combined into one $(n+1)$-dimensional system

$$\begin{aligned} dx_1/d\tau &= Kf_1(x_1,\ldots,x_n), \\ &\vdots \\ dx_n/d\tau &= Kf_n(x_1,\ldots,x_n), \\ dt/d\tau &= K. \end{aligned} \tag{2.22}$$

This system reduces to (2.20) for $K = 1$, and to (2.21) for $K = 1/f_n(x_1, \ldots, x_n)$. This particular formulation is cumbersome for a Poincaré map program since the position of Σ must be known beforehand. It is useful, however, in the simulation of piecewise-linear systems.

2.5 Summary

- *Poincaré map*: A technique that transforms a continuous-time system to a discrete-time system. The limit sets of the Poincaré map correspond to limit sets of the underlying continuous-time system.

- $P_N: \mathbb{R}^n \to \mathbb{R}^n$: The Poincaré map for time-periodic nonautonomous systems. P_N is equivalent to sampling the trajectory at a rate equal to the forcing frequency. P_N is a diffeomorphism.

- $P_A: \Sigma \to \Sigma$: The classical Poincaré map for autonomous systems. It is defined by an $(n-1)$-dimensional hyperplane Σ that intersects a limit cycle transversally. P_A is a diffeomorphism in a neighborhood of the intersection. P_A is useful in analytical studies, but is of little interest in simulations and experiments.

- $P_\pm: \Sigma \to \Sigma$: The two-sided Poincaré map. $P_\pm(x)$ is the next point where $\phi_t(x)$ intersects the $(n-1)$-dimensional hyperplane Σ. It is not well-defined for all Σ. When well-defined, $P_\pm$ is a local diffeomorphism at every point x such that $f(x)$ and $f(P(x))$ are transversal to Σ.

- $P_+, P_-: \Sigma \to \Sigma$: The one-sided Poincaré maps. $P_+(x)$ ($P_-(x)$) is the next point where $\phi_t(x)$ intersects the $(n-1)$-dimensional hyperplane Σ in the positive (negative) direction. They are not well-defined for all Σ. When well-defined, P_+ and P_- are local diffeomorphisms at every point x such that $f(x)$ and $f(P(x))$ are transversal to Σ.

- *Fixed point*: x^* is a fixed point of a map P if $x^* = P(x^*)$.

- *Closed orbit*: The sequence $\{x_1^*, \ldots, x_K^*\}$ is a period-K closed orbit of the map P if $x_{k+1}^* = P(x_k^*)$ for $k = 1, \ldots, K-1$, and if $x_1^* = P(x_K^*)$.

- *Periodic behavior*: Fixed points and closed orbits of any of the Poincaré maps indicate a periodic solution. A fixed point of P_N corresponds to a period-one solution of the underlying flow, and a K-periodic closed orbit corresponds to a Kth-order subharmonic.

- *Two-periodic behavior*: If the underlying flow has two-periodic behavior, the limit set of P_N or $P_\pm$ consists of (diffeomorphic copies of) one or more circles; the limit set of P_+ or P_- consists of half as many circles as are in the limit set of the corresponding $P_\pm$. If the limit set of $P_\pm$ contains an odd number of circles, then the limit sets of P_+ and P_- contain a portion of a circle.

- *K-periodic behavior*: If the underlying flow exhibits K-periodic behavior, the limit set of the Poincaré map consists of one or more (diffeomorphic copies of) a $(K-1)$-torus. The limit sets of the one-sided maps can contain portions of one or more $(K-1)$-tori.

- *Chaos*: The limit set, or strange attractor, of the Poincaré map of a chaotic flow is not a simple, geometrical object. Unlike the other limit sets, a strange attractor exhibits a fine structure, that is, successive magnifications uncover more and more details.

- *Higher-order Poincaré maps*: Higher-order Poincaré maps are based on the idea of multi-sampling. Each sampling reduces the dimension of the limit set by one. The sampling can be either in time (for non-autonomous systems) or in state space (for autonomous systems). Time and state-space sampling can be mixed to create hybrid higher-order Poincaré maps. Higher-order Poincaré maps are useful for identifying K-periodic behavior for $K > 2$.

Chapter 3

Stability of Limit Sets

Stable limit sets are of supreme importance in experimental and numerical settings because they are the only kind of limit set that can be observed naturally, that is, by simply letting the system run. In this chapter, we examine the conditions for a limit set to be stable.

We distinguish between four different types of stability. Examples of each type are shown in Fig. 3.1.

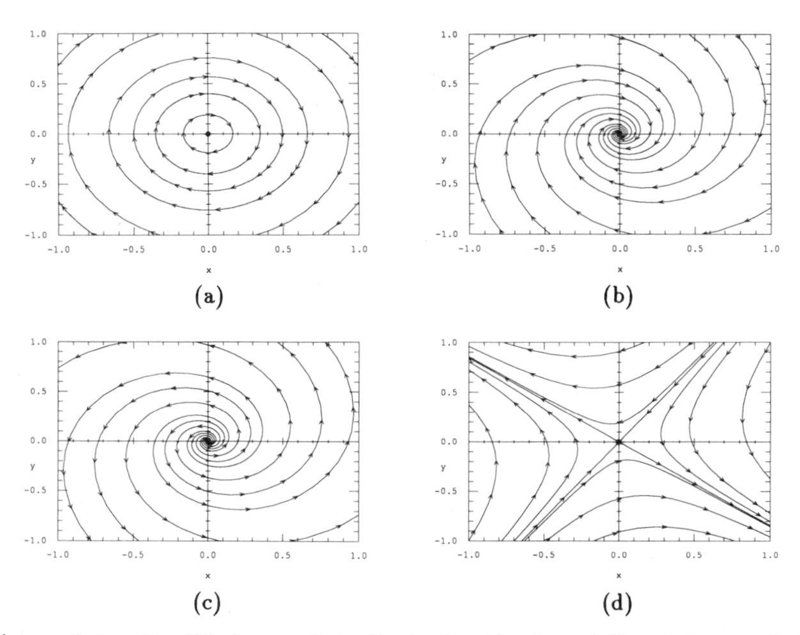

Figure 3.1: Equilibrium points displaying the four different types of stability. (a) Stable but not asymptotically stable; (b) asymptotically stable; (c) unstable; (d) non-stable.

1. A limit set L is *stable* if, for each open neighborhood U of L, there exists an open neighborhood V of L such that, for all $x \in V$ and for all $t > 0$, $\phi_t(x) \in U$. Equivalently, $\phi_t(V) \subset U$ for all $t > 0$.

2. A limit set is *asymptotically stable* if there exists an open neighborhood V of L such that the ω-limit set of every point in V is L.

3. A limit set is *unstable* if there exists an open neighborhood V of L such that the α-limit set of every point in V is L.

4. A limit set is *non-stable* if every neighborhood V of L contains at least one point not in L whose ω-limit set is L and at least one point not in L whose α-limit set is L.

Informally, L is stable if all nearby trajectories stay nearby. L is asymptotically stable if all nearby trajectories are attracted. L is unstable if all nearby trajectories are repelled (except of course trajectories in L). L is non-stable if at least one trajectory not in L is attracted and if at least one nearby trajectory is repelled.

For another interpretation of these definitions, define two sets A and R as follows. A (for attracting) is the set of all $x \notin L$ whose ω-limit set is L. R (for repelling) is the set of all $x \notin L$ whose α-limit set is L. Then

1. L is stable implies R is empty.

2. L is asymptotically stable implies R is empty and A is not.

3. L is unstable implies A is empty and R is not.

4. L is non-stable implies that neither A nor R is empty.

Remarks:

1. An asymptotically stable limit set is also stable. All other pairings of stability types are mutually exclusive.

2. It is common to see "unstable" defined as "not stable" and "non-stable" not defined at all. The motivation for our definitions is clear when the reverse-time dynamics of the system are considered. An unstable limit set is asymptotically stable in reverse time, and any algorithm or theorem pertaining to

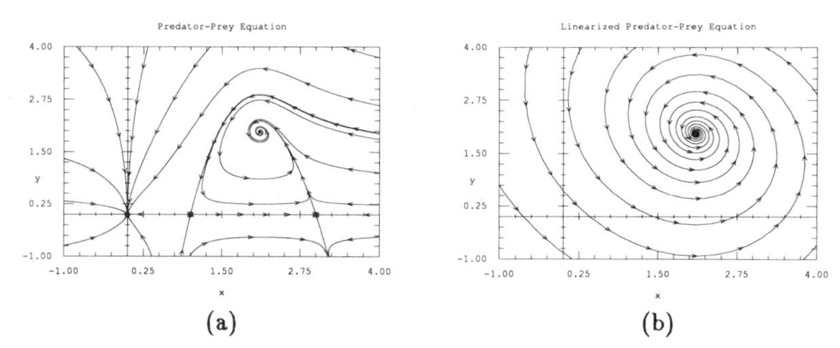

(a) (b)

Figure 3.2: The linearization of a vector field f at an equilibrium point x_{eq} is composed of two operations: differentiation of f and translation of the origin to x_{eq}. The linearized system governs the evolution of infinitesimal perturbations near x_{eq}.

an asymptotically stable system can be applied to an unstable system by running the system backwards in time. This is not true, however, for a non-stable system because a non-stable limit set remains non-stable in reverse time.

We now present techniques for determining the stability type of arbitrary limit sets. We use the familiar case of the eigenvalues at an equilibrium point as a starting point, then introduce the characteristic multipliers of a periodic solution, and finish with Lyapunov exponents. Each stage in this succession is presented as a generalization of the previous step.

3.1 Equilibrium points—eigenvalues

The linearized system Consider an equilibrium point x_{eq} of an autonomous system described by a vector field f. The local behavior of the flow near x_{eq} is determined by linearizing the vector field at x_{eq}. In particular, the linear vector field

$$\delta x = Df(x_{eq})\,\delta x, \qquad \delta x(0) = \delta x_0 \tag{3.1}$$

governs the time evolution of a perturbation δx_0 in a neighborhood of the equilibrium point x_{eq}. As Fig. 3.2 shows, linearization is composed of two operations: differentiation and translation.

Let the eigenvalues of $Df(x_{eq})$ be $\lambda_i \in \mathbb{C}$, with corresponding eigenvectors $\eta_i \in \mathbb{C}^n$, for $i = 1, \ldots, n$. It is a standard result from linear system theory (e.g., Hirsch and Smale [1974]) that, assuming the eigenvalues are distinct, the trajectory with initial condition $x_{eq}+\delta x_0$ is, to the first order,

$$\begin{aligned} \phi_t(x_{eq}+\delta x_0) &= x_{eq}+\delta x(t) \\ &= x_{eq}+e^{Df(x_{eq})t}\delta x_0 \\ &= x_{eq}+c_1 e^{\lambda_1 t}\eta_1+\cdots+c_n e^{\lambda_n t}\eta_n \end{aligned} \tag{3.2}$$

where $c_i \in \mathbf{C}$ are constants chosen to achieve the correct initial condition.

If λ_i is real, η_i and c_i are also real, and it is clear from (3.2) that the eigenvalue is the rate of contraction (if $\lambda_i < 0$) or expansion (if $\lambda_i > 0$) near x_{eq} in the direction of η_i.

Next, consider the case of complex eigenvalues. Since $Df(x_{eq})$ is a real matrix, complex eigenvalues occur in complex-conjugate pairs, λ_i and $\lambda_{i+1}=\bar{\lambda}_i$, as do the corresponding eigenvectors, η_i and $\eta_{i+1}=\bar{\eta}_i$, and constants, c_i and $c_{i+1}=\bar{c}_i$. It follows that

$$\begin{aligned} c_i\, e^{\lambda_i t}\,\eta_i + c_{i+1}\, e^{\lambda_{i+1} t}\,\eta_{i+1} &= c_i\, e^{\lambda_i t}\eta_i + \bar{c}_i\, e^{\bar{\lambda}_i t}\,\bar{\eta}_i \\ &= 2\,\mathrm{Re}[c_i\, e^{\lambda_i t}\,\eta_i]. \end{aligned} \tag{3.3}$$

Since

$$e^{\alpha+j\beta}=e^{\alpha}(\cos(\beta)+j\sin(\beta)), \tag{3.4}$$

after some algebra, (3.3) becomes

$$2\,|c_i|\,e^{\mathrm{Re}[\lambda_i]t}\left\{\cos(\mathrm{Im}[\lambda_i]t+\theta_i)\,\mathrm{Re}[\eta_i]-\sin(\mathrm{Im}[\lambda_i]t+\theta_i)\,\mathrm{Im}[\eta_i]\right\} \tag{3.5}$$

where $\theta_i := \arg(c_i)$.

Equation (3.5) is a parametric equation of a spiral on the plane spanned by $\mathrm{Re}[\eta_i]$ and $\mathrm{Im}[\eta_i]$. The real part of λ_i gives the rate of expansion (if $\mathrm{Re}[\lambda_i] > 0$) or contraction (if $\mathrm{Re}[\lambda_i] < 0$) of the spiral; the imaginary part of the eigenvalue is the frequency of rotation.

Stability Combining the results for real and complex eigenvalues, we reach the following conclusions. If $\mathrm{Re}[\lambda_i] < 0$ for all λ_i, then all sufficiently small perturbations tend toward 0 as $t \to \infty$, and x_{eq} is asymptotically stable (see Fig. 3.3). If $\mathrm{Re}[\lambda_i] > 0$ for all λ_i, then any perturbation grows with time, and x_{eq} is unstable. If there exists i and j such that $\mathrm{Re}[\lambda_i] < 0$ and $\mathrm{Re}[\lambda_j] > 0$, then x_{eq} is non-stable. A non-stable equilibrium point is often called a *saddle point*. A stable

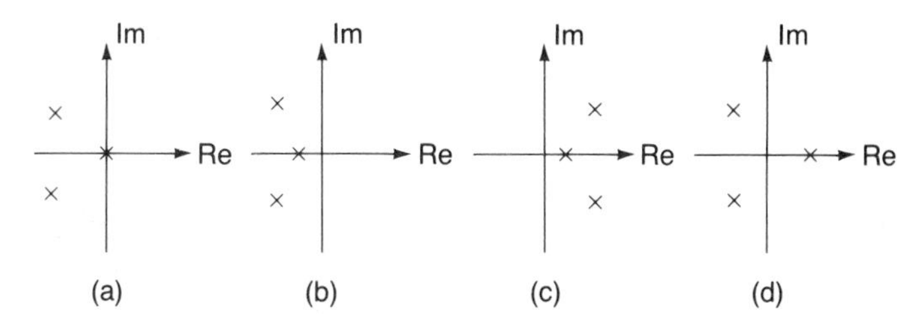

Figure 3.3: The position in the complex plane of the eigenvalues at a hyperbolic equilibrium point determines the stability of the equilibrium point. (a) Non-hyperbolic; (b) asymptotically stable; (c) unstable; (d) non-stable.

or unstable equilibrium point with no complex eigenvalues is often called a *node*.

Though (3.2) is not always valid when two or more of the eigenvalues are equal, the classification scheme just presented remains valid as long as none of the eigenvalues has zero real part (see Hirsch and Smale [1974]). An equilibrium point whose eigenvalues all have a non-zero real part is called *hyperbolic*. The stability of a non-hyperbolic equilibrium point cannot be determined from the eigenvalues alone unless one eigenvalue has real part greater than zero and another eigenvalue has real part less than zero, in which case the equilibrium point is non-stable.

Hyperbolic equilibrium points are generic. They are also structurally stable, that is, they still exist under small perturbations of the vector field. Furthermore, the perturbed equilibrium point has the same stability type. Typically, non-hyperbolic equilibrium points are not structurally stable, and cannot be observed experimentally or in simulations.

3.2 Periodic Solutions—Characteristic Multipliers

3.2.1 Characteristic multipliers

The stability of a periodic solution is determined by its characteristic multipliers, also called Floquet multipliers. Characteristic multipliers are a generalization of the eigenvalues at an equilibrium point.

A periodic solution corresponds to a fixed point[1] x^* of the Poin-

[1]If the periodic solution corresponds to a period-K closed orbit of P, any point

caré map P. Not surprisingly, the stability of the periodic solution is the same as the stability of the fixed point.

A fixed point is the discrete-time analog of an equilibrium point. To highlight the similarities (and the differences), we use, whenever appropriate, the same wording as in the previous section.

The linearized system Consider a fixed point x^* of a map P. The local behavior of the map near x^* is determined by linearizing the map at x^*. In particular, the linear map

$$\delta x_{k+1} = DP(x^*)\,\delta x_k \tag{3.6}$$

governs the evolution of a perturbation δx_0 in a neighborhood of the fixed point x^*.

Let p be the dimension of the Poincaré map; $p = n$ for non-autonomous systems and $p = n - 1$ for autonomous systems. Let the eigenvalues of $DP(x^*)$ be $m_i \in \mathbf{C}$, with corresponding eigenvectors $\eta_i \in \mathbf{C}^n$, for $i = 1, \ldots, p$. Assuming that the eigenvalues are distinct, the orbit of P with initial condition $x^* + \delta x_0$ is, to first order,

$$\begin{aligned} x_k &= x^* + \delta x_k \\ &= x^* + DP(x^*)^k\,\delta x_0 \\ &= x^* + c_1 m_1^k \eta_1 + \cdots + c_p m_p^k \eta_p \end{aligned} \tag{3.7}$$

where $c_i \in \mathbf{C}$ are constants chosen to achieve the correct initial condition.

The eigenvalues $\{m_i\}$ are called the *characteristic multipliers* of the periodic solution. Like eigenvalues at an equilibrium point, the characteristic multipliers' position in the complex plane determines the stability of the fixed point.

If m_i is real, then η_i and c_i are also real and it is clear from (3.7) that the characteristic multiplier is the amount of contraction (if $m_i < 0$) or expansion (if $m_i > 0$) near x^* in the direction of η_i for one iteration of the map.

Next, consider the case of complex eigenvalues. Since $DP(x^*)$ is a real matrix, complex eigenvalues occur in complex-conjugate pairs, m_i and $m_{i+1} = \bar{m}_i$, as do the corresponding eigenvectors, η_i and

---

of the orbit is a fixed point of the K-iterated map P^K and the method presented here can be applied to P^K.

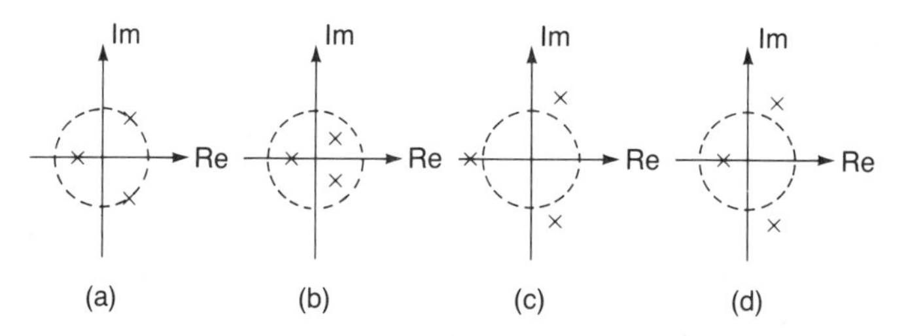

Figure 3.4: The position in the complex plane of the characteristic multipliers at a hyperbolic fixed point determines the stability of the fixed point. (a) Non-hyperbolic; (b) asymptotically stable; (c) unstable; (d) non-stable.

$\eta_{i+1} = \bar{\eta}_i$, and constants, c_i and $c_{i+1} = \bar{c}_i$. It follows that

$$\begin{aligned} c_i\, m_i^k\, \eta_i + c_{i+1}\, m_i^k\, \eta_{i+1} &= c_i\, m_i^k\, \eta_i + \bar{c}_i\, \bar{m}_i^k\, \bar{\eta}_i \\ &= 2\,\mathrm{Re}[c_i\, m_i^k\, \eta_i] \end{aligned} \tag{3.8}$$

which, after some algebra, becomes

$$2|c_i||m_i|^k \left\{\cos(k\phi_i + \theta_i)\,\mathrm{Re}[\eta_i] - \sin(k\phi_i + \theta_i)\,\mathrm{Im}[\eta_i]\right\} \tag{3.9}$$

where $\phi_i := \arg(m_i)$, and $\theta_i := \arg(c_i)$.

Equation (3.9) is a parametric equation of a (sampled) spiral on the plane spanned by $\mathrm{Re}[\eta_i]$ and $\mathrm{Im}[\eta_i]$. The magnitude of m_i gives the amount of expansion (if $|m_i| > 1$) or contraction (if $|m_i| < 1$) of the spiral for one iteration of the map; the angle of the characteristic multiplier is the frequency of rotation.

Stability Combining the results for real and complex characteristic multipliers, we reach the following conclusions. If $|m_i| < 1$ for all m_i, then all sufficiently small perturbations tend toward 0 as $k \to \infty$, and x^* is asymptotically stable (see Fig. 3.4). If $|m_i| > 1$ for all m_i, then any perturbation grows with time and x^* is unstable. If there exists i and j such that $|m_i| < 1$ and $|m_j| > 1$, then x^* is non-stable. A non-stable fixed point is often called a *saddle point.*

Though (3.7) is not always valid when two or more of the characteristic multipliers are equal, the classification scheme just presented remains valid as long as none of the characteristic multipliers lies on the unit circle. A fixed point with no characteristic multiplier on the unit circle is called *hyperbolic.* The stability of a non-hyperbolic fixed point cannot be determined from the characteristic multipliers alone unless one characteristic multiplier has magnitude greater than one and another characteristic multiplier has magnitude less than one, in which case the fixed point is non-stable.

Hyperbolic fixed points are generic. They are also structurally stable, that is, they still exist under small perturbations of the map; furthermore, the perturbed fixed point has the same stability type. Typically non-hyperbolic fixed points are not structurally stable, and cannot be observed experimentally or in simulations.

Remarks:

1. The stability type of a fixed point of a Poincaré map corresponds to the stability type of the underlying periodic solution. In the preceding discussion of stability, "fixed point" may be replaced everywhere by "periodic solution."
2. A periodic solution of an nth-order non-autonomous system has n characteristic multipliers; a limit cycle has only $(n-1)$. This is not surprising since a non-autonomous system is equivalent to an $(n+1)$th-order autonomous system.
3. In the autonomous case, the characteristic multipliers are independent of the position of the cross-section Σ. It makes sense, therefore, to speak of *the* characteristic multipliers of a limit cycle. This result is proved in Appendix D.
4. It follows from transformation (1.3) and the preceding remark, that, in the non-autonomous case, the characteristic multipliers are independent of the choice of t_0.
5. The eigenvectors, however, do depend on the position of Σ in the autonomous case and on t_0 in the non-autonomous case.

3.2.2 Characteristic multipliers and the variational equation

In this section, we discuss the relationship between the characteristic multipliers and the eigenvalues of $\Phi_T(x^*, t_0)$, the solution of the variational equation (B.4).

The non-autonomous case In the non-autonomous case,

$$P_N(x) := \phi_T(x, t_0), \tag{3.10}$$

and, therefore,

$$\begin{aligned} DP_N(x^*) &= D_x\phi_T(x^*, t_0) \\ &= \Phi_T(x^*, t_0). \end{aligned} \tag{3.11}$$

Thus, for non-autonomous systems, the characteristic multipliers are equal to the eigenvalues of $\Phi_T(x^*, t_0)$.

The autonomous case Calculating $DP(x^*)$ in the autonomous case is more complicated than in the non-autonomous case.[2] It is shown in Appendix D that for a special cross-section Σ^*,

$$DP(x^*) = \Phi_T(x^*) \tag{3.12}$$

where T is the minimum period of the limit cycle associated with x^*.

In light of (3.11) and the fact that non-autonomous systems can be converted to autonomous ones, (3.12) is not surprising. Since the characteristic multipliers are independent of the choice of Σ, it follows that the characteristic multipliers are a subset of the eigenvalues of $\Phi_T(x^*)$. They are a subset because $DP(x^*)$ has $n-1$ eigenvalues—it maps an $(n-1)$-dimensional hyperplane to itself—whereas $\Phi_T(x^*): \mathbb{R}^n \to \mathbb{R}^n$ has n eigenvalues.

It is proved in Appendix D that one eigenvalue of $\Phi_T(x^*)$ is always 1 and that the corresponding eigenvector is $f(x^*)$. It is also shown that the remaining eigenvalues of $\Phi_T(x^*)$ are the characteristic multipliers, that is, the eigenvalues of $\Phi_T(x^*)$ are $\{m_1, \ldots, m_{n-1}, 1\}$.

There is an intuitive explanation why one eigenvalue of $\Phi_T(x^*)$ is always 1. Consider a limit cycle Γ with period T. Let x^* be a point on the limit cycle. Perturb x^* slightly in the direction of $f(x^*)$; that is, for some small $\epsilon > 0$, choose $\delta x = \epsilon f(x^*)$. To first order, $x^* + \delta x$ lies on Γ and, therefore, to first order,

$$x^* + \delta x = \phi_T(x^* + \delta x). \tag{3.13}$$

Since infinitesimal perturbations evolve according to $\Phi_t(x^*)$, (3.13) implies $\delta x = \Phi_T(x^*)\delta x$, from which it follows that $f(x^*)$ is an eigenvector of $\Phi_T(x^*)$ with eigenvalue 1.

[2] Here, P stands for P_A, $P_\pm$, P_+, P_-, or perhaps the Kth iterate of any of these maps.

3.2.3 Characteristic multipliers and equilibrium points

Characteristic multipliers can be considered a generalization of the eigenvalues at an equilibrium point for, as we show in this section, they can be used to find the stability of equilibrium points as well as the stability of limit cycles.

Let x_{eq} be an equilibrium point of an autonomous system. Choose any $T > 0$. Since the vector field of an autonomous system is independent of time, the autonomous system can be considered as a time-periodic, non-autonomous system with (non-minimum) period T. The Poincaré map, $P_N(x) = \phi_T(x)$, is well-defined, and since $x_{eq} = P(x_{eq})$, x_{eq} may be interpreted as a periodic solution. To find the characteristic multipliers of this "periodic solution," we need to calculate $DP_N(x_{eq}) = \Phi_T(x_{eq})$.

Since $\phi_t(x_{eq}) = x_{eq}$, the variational equation is

$$\dot{\Phi} = Df(x_{eq})\,\Phi, \qquad \Phi_0 = I \tag{3.14}$$

which is a linear time-invariant system with state transition matrix

$$\Phi_t(x_{eq}) = e^{Df(x_{eq})t}. \tag{3.15}$$

Let the eigenvalues of $Df(x_{eq})$ be $\lambda_1, \ldots, \lambda_n$. The characteristic multipliers are the eigenvalues of $e^{Df(x_{eq})T}$, and by the spectral mapping theorem (Desoer [1970]), λ_i and m_i are related by

$$m_i = e^{\lambda_i T}, \qquad i = 1, \ldots, n. \tag{3.16}$$

λ_i is the *rate* of contraction or expansion; m_i is the *amount* of contraction or expansion in T seconds. Since $|m_i| < 1$ if and only if $\mathrm{Re}[\lambda_i] < 0$ and $|m_i| > 1$ if and only if $\mathrm{Re}[\lambda_i] > 0$, λ_i and m_i provide the same stability information.

As an aside, note that x_{eq} may also be considered as a period-T limit cycle of the original autonomous system. The difficulty with this interpretation is that since $f(x_{eq}) = 0$, $f(x_{eq})$ is not an eigenvector of $\Phi_T(x_{eq})$, and it is no longer true that one eigenvalue of $\Phi_T(x_{eq})$ must be 1. Thus the "limit cycle" x_{eq} has n characteristic multipliers, not $n-1$.

3.3 Lyapunov exponents

Lyapunov exponents are a generalization of the eigenvalues at an equilibrium point and of characteristic multipliers. They can be used to determine the stability of quasi-periodic and chaotic behavior as well as that of equilibrium points and periodic solutions.

3.3.1 Definition

Continuous-time systems Pick any initial condition $x_0 \in \mathbb{R}^n$. Let $m_1(t)$, ..., $m_n(t)$ be the eigenvalues of $\Phi_t(x_0)$. The *Lyapunov exponents* of x_0 are

$$\lambda_i := \lim_{t\to\infty} \frac{1}{t} \ln |m_i(t)|, \qquad i = 1, \ldots, n \tag{3.17}$$

whenever the limit exists.[3]

Discrete-time systems Pick any initial condition x_0, and let $\{x_k\}_{k=0}^{\infty}$ be the corresponding orbit of a p-dimensional, discrete-time system P. Let $m_1(k)$, ..., $m_p(k)$ be the eigenvalues of $DP^k(x_0)$. The *Lyapunov numbers* of x_0 are

$$m_i := \lim_{k\to\infty} |m_i(k)|^{1/k}, \qquad i = 1, \ldots, p \tag{3.18}$$

whenever the limit exists.

3.3.2 Lyapunov exponents of an equilibrium point

To gain some understanding of Lyapunov exponents, we now find the Lyapunov exponents of an equilibrium point x_{eq}. It is common to use λ as a symbol for the eigenvalues at an equilibrium point as well as for Lyapunov exponents. In this section, to distinguish between the two, we use $\hat{\lambda}$ to denote the eigenvalues at an equilibrium point.

Let $\hat{\lambda}_1$, ..., $\hat{\lambda}_n$, be the eigenvalues of $Df(x_{eq})$. It was shown in the previous section that $\Phi_t(x_{eq}) = e^{Df(x_{eq})t}$. It follows that $m_i(t) = e^{\hat{\lambda}_i t}$, and

$$\lambda_i = \lim_{t\to\infty} \frac{1}{t} \ln |e^{\hat{\lambda}_i t}| = \lim_{t\to\infty} \frac{1}{t} \mathrm{Re}[\hat{\lambda}_i]t = \mathrm{Re}[\hat{\lambda}_i]. \tag{3.19}$$

Hence, the Lyapunov exponents are equal to the real parts of the eigenvalues at the equilibrium point. They indicate the rate of contraction (if $\lambda_i < 0$) or expansion (if $\lambda_i > 0$) near the equilibrium

[3] *lim* can be replaced by *lim sup* to guarantee existence of the Lyapunov exponents; however, the interpretation of the Lyapunov exponents presented shortly is valid only when the limit in (3.17) exists.

point. Furthermore, the subspace in which the expansion or contraction occurs is determined by the eigenvectors of $Df(x_{eq})$ (see (3.2)).

As defined, the Lyapunov exponents depend on the initial condition. Suppose that $x_0 \neq x_{eq}$, but $\phi_t(x_0) \to x_{eq}$ as $t \to \infty$, that is, x_0 lies in the basin of attraction of the equilibrium point. Since the Lyapunov exponents are defined in the limit as $t \to \infty$, any transient can be ignored and, therefore, the Lyapunov exponents of x_{eq} and x_0 are the same. In general, every point[4] in the basin of attraction of an attractor has the same Lyapunov exponents as the attractor, and it is convenient to speak of *the* Lyapunov exponents of an attractor.

3.3.3 Lyapunov numbers of a fixed point

Consider a fixed point x^* of a p-dimensional map P. Let $\hat{m}_1, \ldots, \hat{m}_p$ be the characteristic multipliers of P at x^*, that is, the $\hat{m}_i$ are the eigenvalues of $DP(x^*)$. Since $DP^k(x^*) = DP(x^*)^k$, the eigenvalues of $DP^k(x^*)$ are $\hat{m}_1^k, \ldots, \hat{m}_p^k$. The Lyapunov numbers of the fixed point are

$$m_i = \lim_{k\to\infty} |\hat{m}_i^k|^{1/k} = |\hat{m}_i|. \tag{3.20}$$

Hence, the Lyapunov numbers are equal to the magnitudes of the characteristic multipliers at the fixed point. They are the amount of contraction (if $m_i < 1$) or expansion (if $m_i > 1$) per iteration near x^*. Furthermore, the subspace in which the expansion or contraction occurs is determined by the eigenvectors of $DP(x^*)$ (see (3.7)).

Assuming P is a Poincaré map, it is instructive to find the Lyapunov exponents of the underlying periodic solution Γ that corresponds to x^*. Let T be the period of Γ, and let $m_1, \ldots, m_n$ be the eigenvalues of $\Phi_T(x^*)$. Due to periodicity, $\Phi_{kT}(x^*) = \Phi_T(x^*)^k$, so $\{m_1^k, \ldots, m_n^k\}$ are the eigenvalues of $\Phi_{kT}(x^*)$. The Lyapunov exponents of Γ are[5]

$$\lambda_i = \lim_{k\to\infty} \frac{1}{kT} \ln |m_i^k| = \frac{1}{T} \ln |m_i|. \tag{3.21}$$

Equation (3.21) may be solved for m_i to obtain

$$|m_i| = e^{\lambda_i T}. \tag{3.22}$$

[4]This wording may need to be changed to *almost every point* for some definitions of strange attractor, but it is always valid for non-strange attractors.

[5]Equation (3.21) is (3.17) sampled at uniform intervals of T. By itself, the convergence of this sequence does not guarantee the existence of the continuous-time limit in (3.17); however, the additional fact that the m_i are independent of the choice of x^* does.

Since $|m_i|$ is the amount of contraction or expansion per T seconds, the Lyapunov exponents give the *average* rate of contraction or expansion near the periodic solution. The subspace in which the expansion or contraction occurs is determined by the eigenvectors of $\Phi_T(x^*)$.

In the autonomous case, one eigenvalue of $\Phi_T(x^*)$ is always 1 ($m_n = 1$ in our indexing scheme) and, therefore, one Lyapunov exponent of a limit cycle is always 0. In fact, it is shown in Appendix E that, for any bounded attractor of an autonomous system except an equilibrium point, one Lyapunov exponent is always 0.

3.3.4 Perturbation subspaces

As the previous examples demonstrate, λ_i is the average rate of contraction (if $\lambda_i < 0$) or expansion (if $\lambda_i > 0$) in a particular subspace near a particular limit set.

What is meant by "in a particular subspace?" Order the λ_i such that $\lambda_1 \geq \cdots \geq \lambda_n$. Then, in the linearized system (B.4), there are n nested linear subspaces

$$W_1 \supset W_2 \supset \cdots \supset W_n, \tag{3.23}$$

with

$$\dim W_1 = n, \ \dim W_2 = n-1, \ \ldots, \ \dim W_n = 1, \tag{3.24}$$

such that almost all perturbations in W_i evolve, on the average, as $e^{\lambda_i t}$.

As an example, consider a third-order system with an equilibrium point x_{eq}. Assume the linearized system at x_{eq} has real eigenvalues, $\lambda_1 > \lambda_2 > \lambda_3$, with corresponding eigenvectors η_1, η_2, and η_3. Note that since the eigenvalues are real, they are equal to the Lyapunov exponents. A solution of the linearized system has the form

$$\delta x(t) = c_1 e^{\lambda_1 t}\,\eta_1 + c_2 e^{\lambda_2 t}\,\eta_2 + c_3 e^{\lambda_3 t}\,\eta_3 \tag{3.25}$$

where the $c_i \in \mathbb{R}$ depend on the initial condition.

For this system, $\delta x(t)$ grows as $e^{\lambda_3 t}$ only if $c_1 = c_2 = 0$; hence, W_3 is the line spanned by η_3. If $c_1 = 0$ but c_2 and c_3 are non-zero, $\delta x(t)$ has two terms,

$$\phi_t(x_0) = c_2 e^{\lambda_2 t}\,\eta_2 + c_3 e^{\lambda_3 t}\,\eta_3. \tag{3.26}$$

Since $\lambda_2 > \lambda_3$, the $e^{\lambda_3 t}$ term becomes negligible after a finite time, and as $t \to \infty$, the perturbation evolves as $e^{\lambda_2 t}$. Therefore, W_2 is the plane spanned by η_2 and η_3. Similarly, $W_1 = \text{span}\{\eta_1, \eta_2, \eta_3\} = \mathbb{R}^3$.

3.3.5 Lyapunov exponents of non-chaotic limit sets

Lyapunov exponents are convenient for categorizing steady-state behavior. For an attractor, contraction must outweigh expansion so $\sum_{i=1}^{n} \lambda_i < 0$. Non-chaotic attractors are classified as follows.[6]

For an asymptotically stable equilibrium point, $\lambda_i < 0$ for $i = 1, \ldots, n$.

For an asymptotically stable limit cycle, $\lambda_1 = 0$, and $\lambda_i < 0$ for $i = 2, \ldots, n$.

For an asymptotically stable two-torus, $\lambda_1 = \lambda_2 = 0$, and $\lambda_i < 0$ for $i = 3, \ldots, n$.

For an asymptotically stable K-torus, $\lambda_1 = \cdots = \lambda_K = 0$, and $\lambda_i < 0$ for $i = K + 1, \ldots, n$.

Remarks:

1. No Lyapunov exponent of a non-chaotic attracting set is positive.

2. Generically, the number of zero Lyapunov exponents of a non-chaotic hyperbolic attracting set indicates the (topological) dimension of the attractor: an equilibrium point has dimension 0, a limit cycle has dimension 1, a two-torus has dimension 2, and a K-torus has dimension K.

3. A non-chaotic limit set is *hyperbolic* if its dimension equals the number of zero Lyapunov exponents.

---

[6]This classification scheme is generic in that it ignores non-hyperbolic attractors. The stability of non-hyperbolic limit sets cannot be determined from the Lyapunov exponents alone.

3.3.6 Lyapunov exponents of chaotic attractors

One feature of chaos is sensitive dependence on initial conditions. As we saw in Section 1.2.5, sensitive dependence occurs in a flow with an expanding component. Since a positive Lyapunov exponent indicates expansion, what distinguishes strange attractors from non-chaotic attractors is the existence of a positive Lyapunov exponent.

From the facts that at least one Lyapunov exponent of a chaotic system must be positive, that one Lyapunov exponent of any limit set other than an equilibrium point must be 0, and that the sum of the Lyapunov exponents of an attractor must be negative, it follows that a strange attractor must have at least three Lyapunov exponents. Hence, chaos cannot occur in a first- or second-order autonomous continuous-time system or in a first-order non-autonomouscontinuous-time system.

In the three-dimensional case, the only possibility is $(+\,0\,-)$, that is, $\lambda_1 > 0$, $\lambda_2 = 0$, and $\lambda_3 < 0$. Since contraction must outweigh expansion, a further condition on stable three-dimensional chaos is $\lambda_3 < -\lambda_1$.

In fourth-order systems, there are three possibilities:

$(+\,0\,-\,-)$: $\lambda_1 > 0$, $\lambda_2 = 0$, and $\lambda_4 \leq \lambda_3 < 0$.

$(+\,+\,0\,-)$: $\lambda_1 \geq \lambda_2 > 0$, $\lambda_3 = 0$, and $\lambda_4 < 0$. This has been termed *hyper-chaos* by Rössler [1979].

$(+\,0\,0\,-)$: $\lambda_1 > 0$, $\lambda_2 = \lambda_3 = 0$, and $\lambda_4 < 0$. This corresponds to a chaotic two-torus. As far as we know, this case has not been observed.

The Lyapunov exponents of the different types of limit sets are summarized in Table 3.1.

3.4 Algorithms for calculating the stability of limit sets

3.4.1 Eigenvalues at an equilibrium point

We wish to find the eigenvalues of $Df(x_{eq})$. Since this matrix is required by the algorithm that locates equilibrium points (Section 5.2), we assume it has already been found.

Its eigenvalues can be calculated by applying the QR algorithm (see Dahlquist and Björck [1974]). Code for the QR algorithm is

Steady State	Attracting Set: Flow	Attracting Set: Poincaré Map	Lyapunov Exponents	Dimension
Equilibrium point	point		$0 > \lambda_1 \geq \cdots \geq \lambda_n$	0
Periodic	circle	one or more points	$\lambda_1 = 0$ $0 > \lambda_2 \geq \cdots \geq \lambda_n$	1
Two-periodic	torus	one or more closed curves	$\lambda_1 = \lambda_2 = 0$ $0 > \lambda_3 \geq \cdots \geq \lambda_n$	2
K-periodic	K-torus	one or more $(K-1)$-tori	$\lambda_1 = \cdots = \lambda_K = 0$ $0 > \lambda_{K+1} \geq \cdots \geq \lambda_n$	K
Chaotic	Cantor-like	Cantor-like	$\lambda_1 > 0$ $\sum \lambda_i < 0$	non-integer

Table 3.1

widely available in numerical libraries such as EISPACK (see Garbow *et al.* [1977] and Smith *et al.* [1976]).

3.4.2 Characteristic multipliers

The characteristic multipliers are the eigenvalues of $\Phi_T(x^*)$. All of the shooting methods require calculation of $\Phi_T(x^*)$, and its eigenvalues may be obtained using the QR algorithm.

3.4.3 Lyapunov exponents

In this section, we present an algorithm that uses integration to estimate simultaneously all n Lyapunov exponents of any asymptotically stable limit set. This technique may be used to find the Lyapunov exponents of an unstable limit set by running the system in reverse time, but estimating the Lyapunov exponents of a non-stable limit set is an open problem. The method requires the solution of the variational equation and is not useful in experimental settings. There are algorithms for calculating Lyapunov exponents from experimentally obtained data (see Wolf *et al.* [1985] and Eckmann and Ruelle [1985]), but these techniques can reliably find only the two largest positive Lyapunov exponents.

The direct approach

This technique uses the definition of λ_i directly. It is not a workable approach and is presented solely as motivation for the more sophisticated technique that is explained in the next section.

Choose an initial state x_0 that lies in the basin of attraction of the limit set. Integrate the variational equation (B.4) from initial condition $\Phi_0(x_0) = I$ for T seconds. If T is large enough, then

$$\lambda_i = \frac{1}{T} \ln |m_i(T)|, \qquad i = 1, \ldots, n \tag{3.27}$$

where the m_i are the eigenvalues of $\Phi_T(x_0)$.

This approach faces two difficulties.

1. For a chaotic system, at least one Lyapunov exponent is positive which implies that $\Phi_t(x)$ is unbounded as $t \to \infty$. Thus, for chaotic systems, serious numerical problems can arise in the integration of the variational equation.

2. Consider the third-order linearized system with solution given by (3.25). It is almost certain that each column of $\Phi_0(x)$ $(= I)$ lies in the largest perturbation subspace W_1, that is, for each column of $\Phi_0(x)$, $c_1 \neq 0$. It follows that all the columns of $\Phi_t(x)$ tend to line up with the eigenvector η_1. This alignment occurs in nonlinear systems as well, and implies that for large T, the matrix $\Phi_t(x)$ is ill-conditioned which, in turn, implies that the $m_i(t)$ cannot be calculated reliably.

The obvious modification is to perform the integration of the variational equation in stages. Choose $T > 0$, an initial state x_0, and an integer $K > 0$. Let $x^{(0)} := x_0$, and let $x^{(k)} := \phi_T(x^{(k-1)})$ for $k = 1, \ldots, K$. Then

$$\Phi_{KT}(x) = \Phi_T(x^{(K-1)}) \cdots \Phi_T(x^{(0)}). \tag{3.28}$$

If T is not too big, then each of the $\Phi_T(x^{(k)})$ is well-behaved and can be integrated accurately.

Unfortunately, there is no formula relating the eigenvalues of a set of matrices to the eigenvalues of their product. One cannot multiply the $\Phi_T(x^{(k)})$ together and then apply the QR algorithm to find the eigenvalues of $\Phi_{KT}(x)$ because the same two problems mentioned above will be encountered. Another approach must be taken.

A better approach

This approach takes advantage of the fact that almost every initial perturbation in W_i grows, on the average, as $e^{\lambda_i t}$.

Finding λ_1 Almost every perturbation lies in W_1, so following the evolution of a randomly chosen initial perturbation allows one to estimate λ_1.

Choose an initial condition x_0 and an initial perturbation δx_0. Set $\delta x^{(0)} = \delta x_0$, $u^{(0)} = \delta x_0 / \|\delta x_0\|$, and $x^{(0)} = x_0$. Integrate the variational equation from $u^{(0)}$ for T seconds to obtain

$$\delta x^{(1)} := \delta x(T; u^{(0)}, x^{(0)}) := \Phi_T(x^{(0)})\, u^{(0)}. \tag{3.29}$$

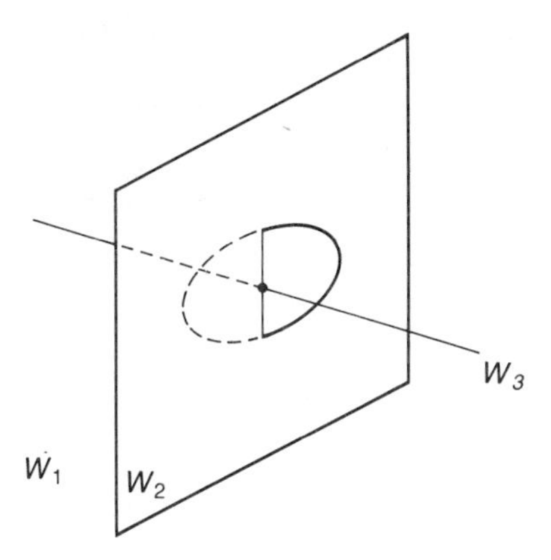

Figure 3.5: A disk in the perturbation space of a three-dimensional system. The disk lies in $W_1 = \mathbb{R}^3$ and it intersects W_2 along a diameter. The area of the disk evolves, on the average, as $e^{(\lambda_1+\lambda_2)t}$.

Let $u^{(1)} := \delta x^{(1)}/\|\delta x^{(1)}\|$ be the normalized version of $\delta x^{(1)}$. Integrate the variational equation from $u^{(1)}$ for T seconds to obtain

$$\delta x^{(2)} := \delta x(T; u^{(1)}, x^{(1)}) = \Phi_T(x^{(1)})\, u^{(1)} \tag{3.30}$$

where $x^{(1)} := \phi_T(x^{(0)})$.

Repeat this integration/normalization procedure K times. Then

$$\delta x(KT; \delta x_0, x_0) = \|\delta x^{(K)}\| \cdots \|\delta x^{(1)}\|\, u^{(K)}, \tag{3.31}$$

and for K large enough,

$$\begin{aligned} \lambda_1 &\approx \frac{1}{KT} \ln \|\delta x(KT; \delta x_0, x_0)\| \\ &= \frac{1}{KT} \ln \prod_{k=1}^{K} \|\delta x^{(k)}\| \\ &= \frac{1}{KT} \sum_{k=1}^{K} \ln \|\delta x^{(k)}\|. \end{aligned} \tag{3.32}$$

Finding λ_2 Consider a unit disk centered at the origin of the linearized system (B.4), and let the orientation of the disk be random (see Fig. 3.5). Since the disk is two-dimensional and W_2 is $(n-1)$-dimensional, it is almost certain that the intersection of the disk and W_2 has dimension one, that is, the intersection is a diameter of the disk.[7] The length of this diameter evolves with rate λ_2 while any other diameter of the disk evolves with rate λ_1. It follows that the area of the disk evolves as $e^{(\lambda_1+\lambda_2)t}$.

Since the variational equation is linear, the area of any area element that lies in the same plane as the disk evolves at the same rate. Thus $\lambda_1 + \lambda_2$ can be found by measuring the average rate of change of the area of almost any area element as it evolves in the linearized system. Since λ_1 can be calculated separately, this procedure yields λ_2.

Let δx_1 and δx_2 be two linearly independent vectors. For an area element, use the parallelogram with vertices at δx_1, δx_2, $\delta x_1 + \delta x_2$, and the origin.

One could follow the evolution of the area of the parallelogram by following separately the evolution of δx_1 and δx_2 using the integration/normalization method presented for λ_1, but there is the previously mentioned problem that $\delta x_1(t)$ and $\delta x_2(t)$ tend to line up as $t \to \infty$. This alignment makes the calculation of the area spanned by $\delta x_1(t)$ and $\delta x_2(t)$ unreliable.

The effect of the alignment can be lessened by orthogonalizing as well as normalizing δx_1 and δx_2 every T seconds. The only constraint is that the orthonormalized vectors must span the same subspace as the original vectors.

Define $\delta x_1^{(0)} := \delta x_1$ and $u_1^{(0)} := \delta x_1/\|\delta x_1\|$. Likewise, define $\delta x_2^{(0)}$ and $u_2^{(0)}$ with respect to δx_2. As before, $x^{(0)} := x_0$ and $x^{(k)} := \phi_T(x^{(k-1)})$ for $k = 1, \ldots, K$. At the kth step, the perturbation equations are

$$\begin{aligned} \delta x_1^{(k)} &= \delta x(kT; u_1^{(k-1)}, x^{(k-1)}) \\ \delta x_2^{(k)} &= \delta x(kT; u_2^{(k-1)}, x^{(k-1)}), \end{aligned} \qquad (3.33)$$

and the orthonormalization equations are

$$\begin{aligned} v_1^{(k)} &= \delta x_1^{(k)} \\ u_1^{(k)} &= v_1^{(k)}/\|v_1^{(k)}\| \\ v_2^{(k)} &= \delta x_2^{(k)} - \langle \delta x_2^{(k)}, u_1^{(k)} \rangle u_1^{(k)} \\ u_2^{(k)} &= v_2^{(k)}/\|v_2^{(k)}\|. \end{aligned} \qquad (3.34)$$

[7]The other, measure zero possibility is that the disk lies completely in W_2.

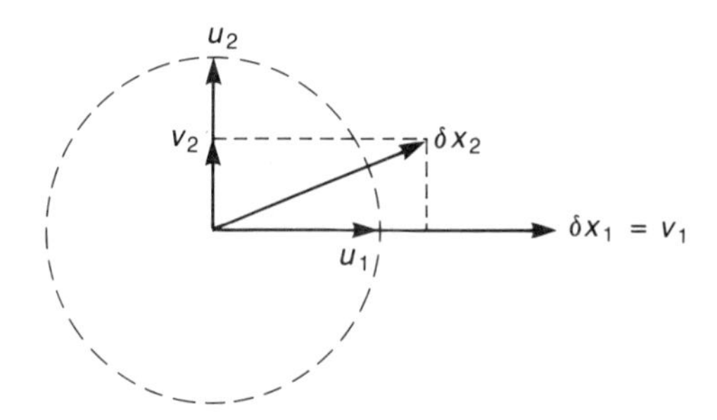

Figure 3.6: Orthonormalization of two vectors, δx_1 and δx_2. The product of the magnitudes of v_1 and v_2 is the area of the parallelogram defined by δx_1 and δx_2.

Since $\langle u_1^{(k)}, u_2^{(k)} \rangle = 0$, the vectors, $u_1^{(k)}$ and $u_2^{(k)}$, are orthogonal. It is clear that $u_1^{(k)}$ and $u_2^{(k)}$ span the same subspace as $\{\delta x_1^{(k)}, \delta x_2^{(k)}\}$ because they are linear combinations of $\delta x_1^{(k)}$ and $\delta x_2^{(k)}$.

Fig. 3.6 shows a geometrical interpretation of the orthonormalization formula (3.34). For simplicity, the superscripts have been dropped. The vector v_1 is equal to δx_1, and the vector u_1 is a normalized version of v_1. The factor $\langle \delta x_2, u_1 \rangle$ is the length of the orthogonal projection of δx_2 onto u_1. Subtract $\langle \delta x_2, u_1 \rangle u_1$ from δx_2 to produce a vector v_1 that is orthogonal to u_1. Normalization of v_2 yields the ortho-normal vector u_2. The area of the parallelogram spanned by δx_1 and δx_2 is $A = \|v_1\| \, \|v_2\|$.

For each iteration, the area is $\|v_1^{(k)}\| \, \|v_2^{(k)}\|$, so, for K large,

$$\begin{aligned}
\lambda_1 + \lambda_2 &\approx \frac{1}{KT} \ln \prod_{k=1}^{K} \|v_1^{(k)}\| \, \|v_2^{(k)}\| \\
&= \frac{1}{KT} \sum_{k=1}^{K} \ln \|v_1^{(k)}\| \, \|v_2^{(k)}\| \\
&= \frac{1}{KT} \sum_{k=1}^{K} \ln \|v_1^{(k)}\| + \frac{1}{KT} \sum_{k=1}^{K} \ln \|v_2^{(k)}\| \\
&\approx \lambda_1 + \frac{1}{KT} \sum_{k=1}^{K} \ln \|v_2^{(k)}\|,
\end{aligned} \tag{3.35}$$

and, therefore,

$$\begin{aligned}\lambda_1 &\approx \frac{1}{KT}\sum_{k=1}^{K}\ln\|v_1^{(k)}\| \\ \lambda_2 &\approx \frac{1}{KT}\sum_{k=1}^{K}\ln\|v_2^{(k)}\|.\end{aligned} \tag{3.36}$$

Gram-Schmidt Orthonormalization The technique for simultaneously estimating all n Lyapunov exponents relies on the orthonormalization of n vectors. We now present the Gram-Schmidt orthonormalization procedure.

Let $\{\delta x_1, \ldots, \delta x_n\}$ be a set of linearly independent n-vectors in $\mathbb{R}^n$. Gram-Schmidt orthonormalization generates an orthonormal set $\{u_1, \ldots, u_n\}$ of n-vectors with the property that $\{u_1, \ldots, u_i\}$ spans the same subspace as $\{\delta x_1, \ldots, \delta x_i\}$ for $i = 1, \ldots, n$. The formulae are

$$\begin{aligned}v_1 &= \delta x_1 \\ u_1 &= v_1/\|v_1\| \\ v_2 &= \delta x_2 - \langle \delta x_2, u_1\rangle u_1 \\ u_2 &= v_2/\|v_2\| \\ &\vdots \\ v_n &= \delta x_n - \langle \delta x_n, u_1\rangle u_1 - \cdots - \langle \delta x_n, u_{n-1}\rangle u_{n-1} \\ u_n &= v_n/\|v_n\|.\end{aligned} \tag{3.37}$$

A simple calculation shows that for any $m \leq n$, $\langle u_m, u_i\rangle = 0$ for $1 \leq i < m$. It follows that $\{u_1, \ldots, u_n\}$ is an orthogonal set. Furthermore, since u_i is a linear combination of $\delta x_1, \ldots, \delta x_{i-1}$, it is clear that $\{u_1, \ldots, u_i\}$ and $\{\delta x_1, \ldots, \delta x_i\}$ span the same subspace for $i = 1, \ldots, n$.

Just as in the two-dimensional case, the area of the parallelepiped spanned by $\{\delta x_1, \ldots, \delta x_i\}$ is

$$\text{Volume}\{\delta x_1, \ldots, \delta x_i\} = \|v_1\| \cdots \|v_i\| \tag{3.38}$$

for $i = 1, \ldots, n$.

Finding all the λ_k We have seen that the length of almost any perturbation evolves as $e^{\lambda_1 t}$, and the area of almost any disk evolves as $e^{(\lambda_1+\lambda_2)t}$. By the same reasoning it can be shown that the volume of almost any k-dimensional volume element evolves, on the average, in the linearized system, as $e^{(\lambda_1+\cdots+\lambda_k)t}$.

To find all n Lyapunov exponents simultaneously, a set of n linearly independent perturbation vectors is repeatedly integrated and orthonormalized. By direct analogy with (3.36), at the Kth stage, the orthonormalization produces n vectors, $\{v_1^{(K)}, \ldots, v_n^{(K)}\}$ and for K large enough,

$$\begin{aligned} \lambda_1 &\approx \frac{1}{KT} \sum_{k=1}^{K} \ln \|v_1^{(k)}\| \\ &\vdots \\ \lambda_n &\approx \frac{1}{KT} \sum_{k=1}^{K} \ln \|v_n^{(k)}\|. \end{aligned} \tag{3.39}$$

Fig. 3.7 shows pseudo-code for a routine `lyapunov_exponents` that finds all n Lyapunov exponents.

Remarks:

1. The vector $x[\,]$ is the current state of the trajectory (of the underlying nonlinear system). Initially, it should be chosen in the basin of attraction of the limit set under study. Of course, the closer $x[\,]$ lies to the limit set, the more quickly `lyapunov_exponent` converges.
2. Initially, the orthonormalized perturbation matrix $u[\,][\,]$ is chosen to be the identity matrix. If a pseudo-random number generator is available, it can be used to select a random initial perturbation matrix which can then be orthonormalized by Gram-Schmidt to obtain $u[\,][\,]$. It is not clear that there is any advantage to this randomization though.
3. The step with Φ_T and the step with ϕ_T require integration which may be performed simultaneously by integrating (B.8) from initial condition

$$\left\{ \begin{array}{c} x(0) \\ \Phi(0) \end{array} \right\} = \left\{ \begin{array}{c} x_0[\,] \\ u[\,][\,] \end{array} \right\}. \tag{3.40}$$

4. Due to its improved numerical stability, the modified Gram-Schmidt orthonormalization procedure is used (see Noble and

```
begin lyapunov_exponent(T, E_r, E_a, k_max, x[])
      set u[][] = I
      for i from 1 to n do
          set λ[i] = 0
          set sum[i] = 0
      endfor
      set k = 0
      repeat
             set k = k + 1
             if (k = k_max) then
                exit--no convergence
             endif
             set λ_old[] = λ[]
             set δx[][] = Φ_T(x[])u[][]
             set x[] = φ_T(x[])
             for i from 1 to n do
                 set v[][i] = δx[][i]
                 for j from 1 to (i − 1) do
                     set v[][i] = v[][i] − ⟨v[][i], u[][j]⟩u[][j]
                 endfor
                 set u[][i] = v[][i]/||v[][i]||
                 set sum[i] = sum[i] + ln ||v[][i]||
                 set λ[i] = sum[i]/kT
             endfor
      until (||λ_old[] − λ[]|| < E_r||λ[]|| + E_a)
      return λ[]
end lyapunov_exponent
```

Figure 3.7: Pseudo-code for a subroutine that estimates the Lyapunov exponents. Single brackets indicate that a variable is an n-vector. Double brackets indicate that a variable is an $n \times n$ matrix (i.e., a two-dimensional array). An index in the second pair of brackets indicates a column vector (e.g., $v[\,][i]$ is the ith column of $v[\,][\,]$ and corresponds to v_i in the text).

Daniel [1979]). In the unmodified procedure, v_i is calculated by subtracting all the inner product terms from x_i. In the modified procedure, v_i is calculated in $(i-1)$ stages—at each stage, only one inner product term is subtracted from v_i, and this new v_i is used in the next stage.

5. The calculation continues until the Lyapunov exponents converge or a maximum iteration count is reached. As usual, a relative/absolute convergence test is used (see Appendix A).

6. When $i = 1$, the body of the j `for` loop is never executed.

It remains to discuss the selection of T. Too small a value results in excessive orthonormalization and if the $\|v_i\|$ are all nearly 1, leads to numerical inaccuracies in the volume calculations. Too large a value results in an ill-conditioned system of vectors and could lead to numerical overflow in the integration. Typically, T is chosen to be ten or twenty times the natural period of the system.

3.5 Summary

- *Stable*: A limit set is stable if all nearby trajectories remain nearby. A stable limit set is unstable in reverse time.

- *Asymptotically stable*: A limit set is asymptotically stable if all nearby trajectories approach the limit set as $t \to \infty$.

- *Unstable*: A limit set is unstable if no nearby trajectories (except those lying on the limit set) remain nearby. An unstable limit set is asymptotically stable in reverse time.

- *Non-stable*: A limit set is non-stable if at least one trajectory not in the limit set is attracted to the limit set in forward time and if at least one trajectory not in the limit set is attracted to the limit set in reverse time. A non-stable limit set remains non-stable in reverse time.

- *Eigenvalues of an equilibrium point*: The eigenvalues of an equilibrium point x_{eq} are the eigenvalues of $Df(x_{eq})$. They give the rate of expansion (if $\mathrm{Re}[\lambda_i] > 0$) and contraction (if $\mathrm{Re}[\lambda_i] < 0$) near x_{eq}.

- *Hyperbolic equilibrium point*: An equilibrium point is hyperbolic if $\mathrm{Re}[\lambda_i] \neq 0$ for $i = 1, \ldots, n$. Hyperbolicity is a generic property, and hyperbolic equilibrium points are structurally stable.

- *Stability of an equilibrium point*: A hyperbolic equilibrium point is asymptotically stable if all its eigenvalues have negative real part. It is unstable if all the real parts are positive. It is non-stable if at least one real part is positive and one real part is negative.

- *Characteristic multipliers*: The characteristic multipliers of a periodic solution are the eigenvalues of $DP(x^*)$. They give the amount of expansion (if $|m_i| > 1$) and contraction (if $|m_i| < 1$) near the periodic solution during one period of the solution. A periodic solution of an nth-order non-autonomous system has n characteristic multipliers; a limit cycle has $(n-1)$.

- *Hyperbolic periodic solution*: A periodic solution is hyperbolic if none of its characteristic multipliers has magnitude 1. Hyperbolicity is a generic property, and hyperbolic periodic solutions are structurally stable.

- *Stability of a periodic solution*: A hyperbolic periodic solution is asymptotically stable if all its characteristic multipliers lie inside the unit circle. It is unstable if all the characteristic multipliers lie outside the unit circle. It is non-stable if at least one characteristic multiplier lies in and one lies outside the unit circle.

- *Lyapunov exponents*: The Lyapunov exponents are the average rates of expansion (if $\lambda_i > 0$) and contraction (if $\lambda_i < 0$) near a limit set. They are generalizations of the eigenvalues at an equilibrium point and of characteristic multipliers. A non-chaotic hyperbolic limit set is asymptotically stable if all its Lyapunov exponents are negative. For any limit set in an autonomous system except an equilibrium point, one Lyapunov exponent is always 0. In a dissipative system, it is required that the sum of the Lyapunov exponents be negative. At least one Lyapunov exponent of a chaotic attractor is positive; it accounts for the sensitive dependence on initial conditions.

Chapter 4

Integration of Trajectories

The most important numerical task in the simulation of continuous-time dynamical systems is the calculation of trajectories.

It is quite common for researchers to treat an integration routine as a black box—given the initial condition, the final time, and the error tolerance, out pops the final state. There is nothing wrong with this approach; indeed, the user should be insulated from the internal details of the coding of an integration routine.

Unfortunately, it is also common for researchers to be unaware of many of the pitfalls of numerical integration. The state of the art is not to the point where there is one best algorithm. There are several types of integration algorithms from which to choose, each with its own peculiarities. What are the different types? What are the advantages and disadvantages of each? What types of integration error occur? How does the actual integration error relate to the user-supplied error tolerance? What is a stable integration algorithm? How does it differ from a stiffly stable algorithm? These and related questions are addressed in this chapter.

The discussion in this chapter is descriptive in nature. Readers interested in a more detailed discussion are referred to Chua and Lin [1975] and Gear [1971].

4.1 Types of integration algorithms

The goal of an integration algorithm is to approximate the behavior of a continuous-time system on a digital computer. Since digital com-

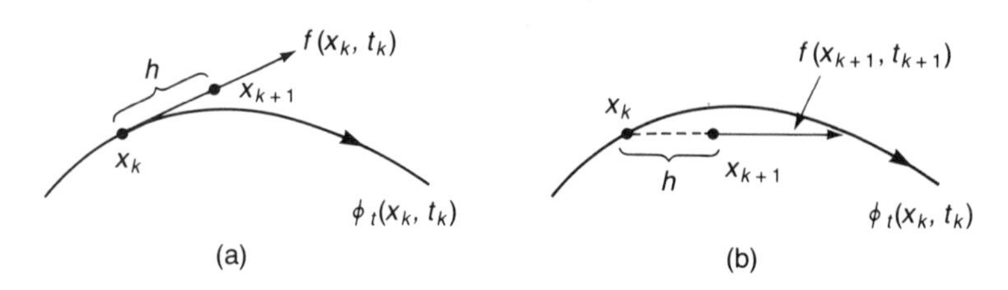

Figure 4.1: (a) The forward Euler algorithm; (b) the backward Euler algorithm.

puters are discrete-time in nature, an integration algorithm models a continuous-time system by a discrete-time system. This is a point worth repeating. The actual system being simulated is not a differential equation. It is a discrete-time system, often an extremely complex one. Integration algorithms differ in that they map the same differential equation into different discrete-time systems.

We now present several standard algorithms and at the same time, introduce the concepts used to classify them. All of these algorithms attempt to approximate the solution $\phi_t(x_0, t_0)$ of the continuous-time system

$$\dot{x} = f(x,t), \qquad x(t_0) = x_0 \tag{4.1}$$

by generating a sequence of points, $x_0, x_1, \ldots$, at times, $t_0, t_1, \ldots$, with the hope that $x_k \approx \phi_{t_k}(x_0, t_0)$. Unless otherwise noted, it is assumed that the time points are uniformly spaced with step-size $h > 0$, that is, $t_{k+1} = t_0 + kh$ for $k = 0, 1, \ldots$.

Forward Euler

Approximate the derivative of the state at time t_k by

$$\dot{x}(t_k) \approx \frac{x_{k+1} - x_k}{h}. \tag{4.2}$$

With this approximation, (4.1) becomes

$$x_{k+1} = x_k + hf(x_k, t_k). \tag{4.3}$$

The system (4.3) is known as the *forward Euler* algorithm.

Fig. 4.1(a) shows a graphical interpretation of the forward Euler algorithm. At the $(k+1)$th step, the vector field is assumed to be (locally) constant with value $f(x_k, t_k)$. A smaller value of the

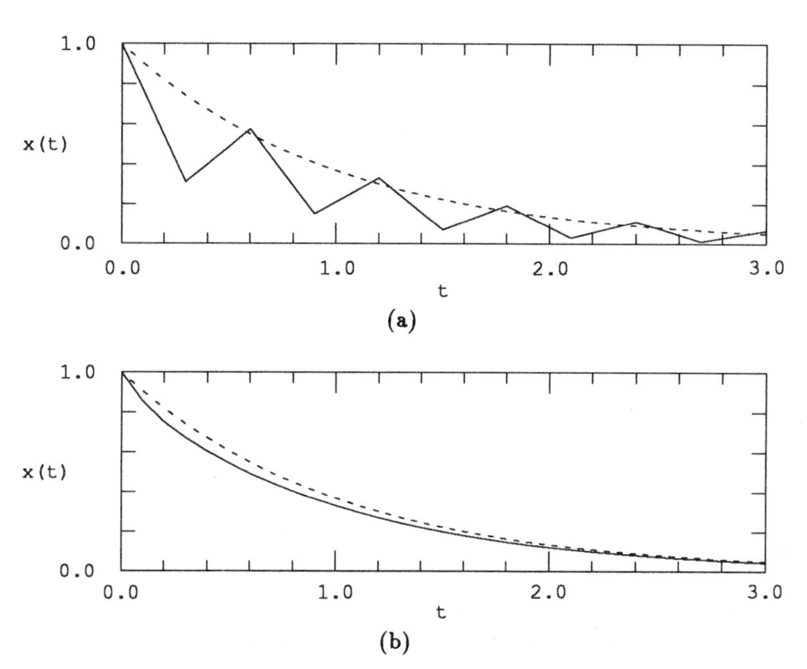

Figure 4.2: $\dot{x} = -6x + 5e^{-t}$ integrated from $x_0 = 1$ by the forward Euler algorithm with (a) $h = 0.3$; (b) $h = 0.1$. The exact solution is the dashed curve.

step-size h results in more frequent approximations and, as Fig. 4.2 demonstrates, leads to a more accurate integration. This makes sense mathematically because (4.2) approaches (4.1) as $h \to 0$.

Backward Euler

The backward Euler algorithm is similar to the forward Euler algorithm with the one difference that the approximation for the derivative is

$$\dot{x}(t_k) \approx \frac{x_k - x_{k-1}}{h}. \tag{4.4}$$

This approximation yields the *backward Euler* formula,

$$x_{k+1} = x_k + hf(x_{k+1}, t_{k+1}). \tag{4.5}$$

Fig. 4.1(b) shows a geometrical interpretation of the backward Euler formula. At the $(k+1)$th step, the vector field is assumed (locally) constant with value $f(x_{k+1}, t_{k+1})$.

The backward Euler algorithm is an example of an *implicit* integration algorithm, that is, x_{k+1} is a function of itself. In contrast, the

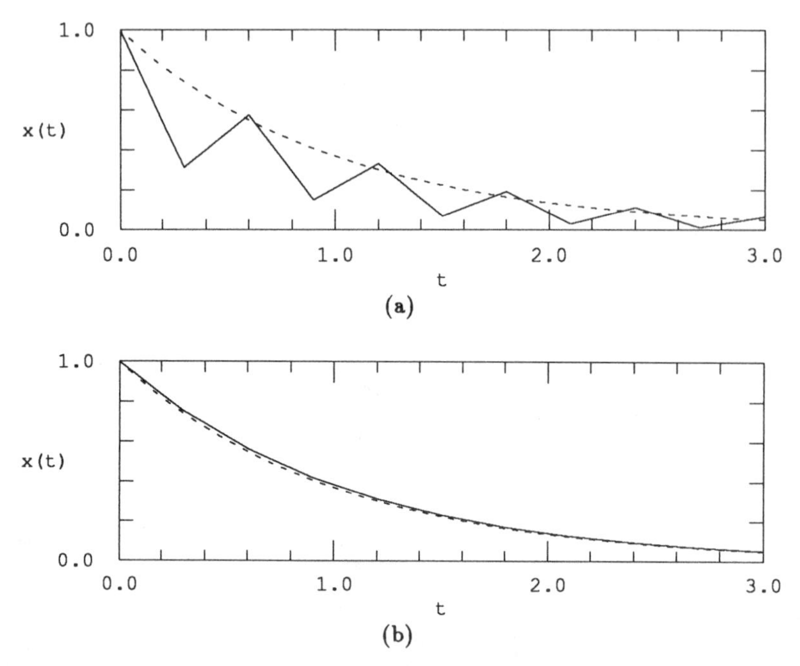

Figure 4.3: The same system as in Fig. 4.2 integrated from $x_0 = 1.0$ with $h = 0.3$ by (a) the forward Euler algorithm; (b) the backward Euler algorithm. The exact solution is the dashed curve.

forward Euler algorithm is an *explicit* algorithm. Implicit algorithms require additional computation to solve for x_{k+1}, but since the x_{k+1} term on the right-hand side may be viewed as a form of feedback, implicit algorithms offer improved stability properties and increased accuracy over their explicit counterparts (see Fig. 4.3).

Trapezoidal

The *trapezoidal* algorithm is

$$x_{k+1} = x_k + \frac{h}{2}\left\{f(x_k, t_k) + f(x_{k+1}, t_{k+1})\right\} \tag{4.6}$$

This is an implicit algorithm. It assumes that the vector field is (locally) constant with value equal to the average of $f(x_k, t_k)$ and $f(x_{k+1}, t_{k+1})$, and may be viewed as a merging of the forward and backward Euler algorithms.

Runge-Kutta

The Runge-Kutta family of algorithms has its origins in the idea of approximating $\phi_t(x_k)$ by its Taylor series expansion. We present second- and fourth-order algorithms. In this situation, the term *K th-order* means that K terms of the Taylor expansion are used in the approximation.

Second-order Runge-Kutta There is an entire family of second-order Runge-Kutta equations. We present one—the *modified Euler-Cauchy* algorithm—given by

$$x_{k+1} = x_k + hf(x_k + \frac{h}{2}f(x_k, t_k), t_k + \frac{h}{2}). \tag{4.7}$$

Close inspection of (4.7) reveals that the modified Euler-Cauchy algorithm is composed of two steps.

Step 1 moves a half-step forward to time $(t_k + h/2)$ using the forward Euler algorithm,

$$\tilde{x} = x_k + \frac{h}{2}f(x_k, t_k) \tag{4.8}$$

In step 2, this intermediate value is used to approximate the vector field, and a forward-type Euler iteration is used,

$$x_{k+1} = x_k + hf(\tilde{x}, t_k + \frac{h}{2}). \tag{4.9}$$

Unlike the trapezoidal algorithm, which uses the average value of the vector field at x_k and x_{k+1}, the modified Euler-Cauchy algorithm uses the value of the vector field midway between x_k and x_{k+1}. It may be viewed as an explicit algorithm that attempts to share some of the advantages of implicit algorithms by taking intermediate steps.

Fourth-order Runge-Kutta As in the case of the second-order algorithm, the fourth-order Runge-Kutta algorithm is an explicit algorithm that uses intermediate time points to calculate the state at time t_{k+1}. The formulae for the most common fourth-order Runge-Kutta algorithm are

$$\begin{aligned} k_1 &= f(x_k, t_k) \\ k_2 &= f(x_k + \frac{h}{2}k_1,\ t_k + \frac{h}{2}) \\ k_3 &= f(x_k + \frac{h}{2}k_2,\ t_k + \frac{h}{2}) \\ k_4 &= f(x_k + hk_3,\ t_k + h) \\ x_{k+1} &= x_k + \frac{h}{6}(k_1 + 2k_2 + 2k_3 + k_4). \end{aligned} \tag{4.10}$$

The mathematical derivation of (4.10) is tedious, but its interpretation is straightforward. Each of the four k_i is an approximate value of the vector field.

k_1 is the value of the vector field at x_k.

k_2 is the approximate value a half-step later, at time $t_k + h/2$. This is essentially a forward Euler step with time-step $h/2$.

k_3 is also the value at time $t_k + h/2$, but it is calculated using the slope k_2. This is similar to a backward Euler half-step, except that an explicit approximation to f at time $t_k + h/2$ is used instead of the implicit expression (4.5).

k_4 is the value of the vector field at t_{k+1}, and is calculated using the most recent value k_3. This is a modified Euler-Cauchy step.

Finally, these four values are averaged together to give the approximation to the vector field which is used to predict x_{k+1}.

Multi-step algorithms

All of the algorithms presented so far require only one input point x_k at each step. Such algorithms are called *single-step algorithms.*

Higher-order single-step algorithms are used because of their increased accuracy, but they tend to be computationally inefficient, especially if the cost per evaluation of f is high. For example, the fourth-order Runge-Kutta algorithm requires four evaluations of f per step. Furthermore, none of the function evaluations performed during the current step is used in any of the succeeding steps.

Unlike single-step algorithms, a *multi-step algorithm* reuses past information about the trajectory. An m-step algorithm uses the m previous points, x_k, x_{k-1}, ..., x_{k-m+1}, and the value of f at these points, to estimate x_{k+1}.

The general form for an m-step algorithm is

$$\begin{aligned} x_{k+1} = {} & a_0 x_k + a_1 x_{k-1} + \cdots + a_{m-1} x_{k-m+1} \\ & + h \{ b_{-1} f(x_{k+1}, t_{k+1}) + b_0 f(x_k, t_k) + b_1 f(x_{k-1}, t_{k-1}) \\ & \quad + \cdots + b_{m-1} f(x_{k-m+1}, t_{k-m+1}) \} \end{aligned} \tag{4.11}$$

where the t_{k-i} are evenly spaced with time-step h.

Different choices of the coefficients, a_i and b_i, lead to different integration algorithms. Note that a multi-step algorithm is explicit if and only if $b_{-1} = 0$; otherwise, it is implicit.

Equation (4.11) is very similar to standard interpolation routines. The a_i terms help to predict x_{k+1} from past values of the trajectory. The b_i terms use the additional information provided by f to increase the accuracy of the approximation.

Note that a multi-step algorithm requires only one new function evaluation—$f(x_k)$ for explicit algorithms, $f(x_{k+1})$ for implicit algorithms—per step and is, therefore, usually more efficient per step than a comparable single-step algorithm; however, a single-step algorithm (e.g., fourth-order Runge-Kutta) uses its extra function evaluations to approximate the vector field at intermediate points and thus, typically allows a larger step-size. In general, it is impossible to say whether a single- or multi-step algorithm is more efficient; it depends on the problem at hand.

There are many ways to choose the $2m + 1$ coefficients in (4.11). These algorithms typically come in families—each member of the family possessing a different order. *Kth-order* in this circumstance means that the integration algorithm is error-free for systems whose solution is a polynomial of order K or less. Most systems do not possess solutions that are polynomials, but since any analytic function may be approximated arbitrarily closely by a high-order polynomial, the requirement that an integration algorithm be exact for polynomials is a useful guideline. Besides, this criterion is simple enough to allow analytical results.

We present three different multi-step algorithms: Adams-Bashforth, Adams-Moulton, and Gear.

Adams-Bashforth The Kth-order Adams-Bashforth algorithm is generated by setting $m = K$, $b_{-1} = 0$, and $a_i = 0$ for $i = 1, \ldots, K$. The remaining coefficients are chosen such that the algorithm is exact

Order	
First	$x_{k+1} = x_k + hf(x_k, t_k)$
Second	$x_{k+1} = x_k + \frac{h}{2}\{3f(x_k, t_k) - f(x_{k-1}, t_{k-1})\}$
Third	$x_{k+1} = x_k + \frac{h}{12}\{23f(x_k, t_k) - 16f(x_{k-1}, t_{k-1}) + 5f(x_{k-2}, t_{k-2})\}$
Fourth	$x_{k+1} = x_k + \frac{h}{24}\{55f(x_k, t_k) - 59f(x_{k-1}, t_{k-1}) + 37f(x_{k-2}, t_{k-2}) - 9f(x_{k-3}, t_{k-3})\}$
Fifth	$x_{k+1} = x_k + \frac{h}{720}\{1901f(x_k, t_k) - 2774f(x_{k-1}, t_{k-1}) + 2616f(x_{k-2}, t_{k-2}) - 1274f(x_{k-3}, t_{k-3}) + 251f(x_{k-4}, t_{k-4})\}$
Sixth	$x_{k+1} = x_k + \frac{h}{1440}\{4277f(x_k, t_k) - 7923f(x_{k-1}, t_{k-1}) + 9982f(x_{k-2}, t_{k-2}) - 7298f(x_{k-3}, t_{k-3}) + 2877f(x_{k-4}, t_{k-4}) - 475f(x_{k-5}, t_{k-5})\}$

Table 4.1: Adams-Bashforth algorithms

for polynomials of order K. The Kth-order Adams-Bashforth algorithm is a K-step algorithm. Since $b_{-1} = 0$, the Adams-Bashforth algorithms are explicit in nature.

The formulae for orders one through six are presented in Table 4.1. Note that the first-order Adams-Bashforth algorithm is equivalent to the forward Euler algorithm.

Adams-Moulton To obtain the Kth-order Adams-Moulton algorithm, set $m = K - 1$, and $a_i = 0$ for $i = 1, \ldots, K - 1$. The remaining coefficients are chosen such that the algorithm is exact for polynomials of order K. The Kth-order Adams-Moulton algorithm is a $(K - 1)$-step algorithm. Since $b_{-1} \neq 0$, the Adams-Moulton algorithms are implicit in nature.

The formulae for orders one through six are presented in Table 4.2. Note that the first-order Adams-Moulton algorithm is equivalent to the backward Euler algorithm, and the second-order algorithm is the trapezoidal formula.

Gear's algorithm The Kth-order Gear algorithm is generated by setting $m = K$, and $b_i = 0$ for $i = 0, \ldots, K$. The remaining coef-

ORDER	
First	$x_{k+1} = x_k + hf(x_{k+1}, t_{k+1})$
Second	$x_{k+1} = x_k + \frac{h}{2}\{f(x_{k+1}, t_{k+1}) + f(x_k, t_k)\}$
Third	$x_{k+1} = x_k + \frac{h}{12}\{5f(x_{k+1}, t_{k+1}) + 8f(x_k, t_k) - f(x_{k-1}, t_{k-1})\}$
Fourth	$x_{k+1} = x_k + \frac{h}{24}\{9f(x_{k+1}, t_{k+1}) + 19f(x_k, t_k) - 5f(x_{k-1}, t_{k-1}) + f(x_{k-2}, t_{k-2})\}$
Fifth	$x_{k+1} = x_k + \frac{h}{720}\{251f(x_{k+1}, t_{k+1}) + 646f(x_k, t_k) - 264f(x_{k-1}, t_{k-1}) + 106f(x_{k-2}, t_{k-2}) - 19f(x_{k-3}, t_{k-3})\}$
Sixth	$x_{k+1} = x_k + \frac{h}{1440}\{475f(x_{k+1}, t_{k+1}) + 1427f(x_k, t_k) - 798f(x_{k-1}, t_{k-1}) + 482f(x_{k-2}, t_{k-2}) - 173f(x_{k-3}, t_{k-3}) + 27f(x_{k-4}, t_{k-4})\}$

Table 4.2: Adams-Moulton algorithms

ficients are chosen such that the algorithm is exact for polynomials of order K. The Kth-order Gear algorithm is a $(K-1)$-step algorithm. Since $b_{-1} \neq 0$, the Gear algorithms are implicit in nature. The formulae for orders one through six are presented in Table 4.3. Note that the first-order Gear algorithm is identical to the backward Euler algorithm.

ORDER	
First	$x_{k+1} = x_k + hf(x_{k+1}, t_{k+1})$
Second	$x_{k+1} = \frac{1}{3}\{4x_k - x_{k-1} + 2hf(x_{k+1}, t_{k+1})\}$
Third	$x_{k+1} = \frac{1}{11}\{18x_k - 9x_{k-1} + 2x_{k-2} + 6hf(x_{k+1}, t_{k+1})\}$
Fourth	$x_{k+1} = \frac{1}{25}\{48x_k - 36x_{k-1} + 16x_{k-2} - 3x_{k-3} + 12hf(x_{k+1}, t_{k+1})\}$
Fifth	$x_{k+1} = \frac{1}{137}\{300x_k - 300x_{k-1} + 200x_{k-2} - 75x_{k-3} + 12x_{k-4} + 60hf(x_{k+1}, t_{k+1})\}$
Sixth	$x_{k+1} = \frac{1}{147}\{360x_k - 450x_{k-1} + 400x_{k-2} - 225x_{k-3} + 72x_{k-4} - 10x_{k-5} + 60hf(x_{k+1}, t_{k+1})\}$

Table 4.3: Gear algorithms

4.2 Integration error

The output of a continuous-time system is a time waveform, and the output of an integration algorithm is a sequence of points. The important question is: Do the numbers that come out of the computer bear any resemblance to the trajectory?

There are two types of error. The *local error* is the error introduced by a single step of the integration algorithm. *Global error* is the overall error caused by repeated application of the integration formula.

What is of interest to users of an integration algorithm is the global error. Unfortunately, owing to the complexity of calculating the global error, most of the analytical work to date has focused on estimating the local error.

4.2.1 Local errors

The *local error* ϵ is defined as the error per algorithm step,

$$\epsilon := \|x_{k+1} - \phi_{t_k+h}(x_k, t_k)\|. \tag{4.12}$$

For m-step algorithms, it is assumed that the past m points, x_{k-i} for $i = 0, \ldots, m-1$, are exact, that is, $x_{k-i} = \phi_{t_k - ih}(x_k, t_k)$. Thus, local error does not account for the propagation of errors from one step to the next.

There are two types of local error: round-off error and truncation error.

Round-off error

Round-off error is the inevitable error that results from performing real arithmetic on a digital computer. The magnitude of the round-off error depends on the hardware used, and it is difficult to predict how different algorithms will react when run on different types of machines.

Note, however, that the round-off error depends on the number and type of arithmetic operations per step and is, therefore, independent of the integration step-size.

Once a computer has been chosen, the only way to reduce the local round-off error is to increase the precision of the floating-point representation. Typically, single-precision representations use 32 bits and are accurate to about seven decimal places; most double-precision representations use 64 bits and are accurate to about fifteen decimal places.

Truncation error

Truncation error is the local error that would result if the algorithm could be implemented on an infinite-precision computer. In other words, it is the local error that occurs assuming there is no round-off error. It is also important to remember that for m-step algorithms, the past m points, $x_k, \ldots, x_{k-m+1}$, are assumed exact.

Truncation error derives its name from the algorithms based on the Taylor series (e.g., Runge-Kutta). These algorithms would be exact if the entire infinite series could be used, but error is introduced when the series is truncated to a finite number of terms.

The truncation error depends only on the algorithm—it is independent of the computer used—and is, therefore, open to rigorous analysis.

For a Kth-order Runge-Kutta algorithm, under suitable conditions, the local truncation error is

$$\epsilon_t = \alpha_K h^K \tag{4.13}$$

where $\alpha_K \in \mathbb{R}$ depends on K, f, and x_k, but is independent of h.

For a Kth-order multi-step algorithm, the local truncation error is

$$\epsilon_t = \beta_K h^{K+1} \tag{4.14}$$

where $\beta_K \in \mathbb{R}$ depends on m, K, f, x_k, and which multi-step algorithm is used, but is independent of h.

It is important to realize what these results imply. During the following discussion, it may be useful to refer to Fig. 4.4 which is a plot of the local truncation error versus step-size for four different orders of the Adams-Moulton algorithm.

First, for a given algorithm of a given order, the local truncation error decreases with the step-size. This result is expected since a discrete-time system can more closely approximates a continuous-time system as the sampling interval is decreased.

Second, if h is small enough, for a given family of algorithms and a given step-size, the higher-order algorithms are more accurate than the lower-order algorithms (see Fig. 4.4(b)). Equivalently, to

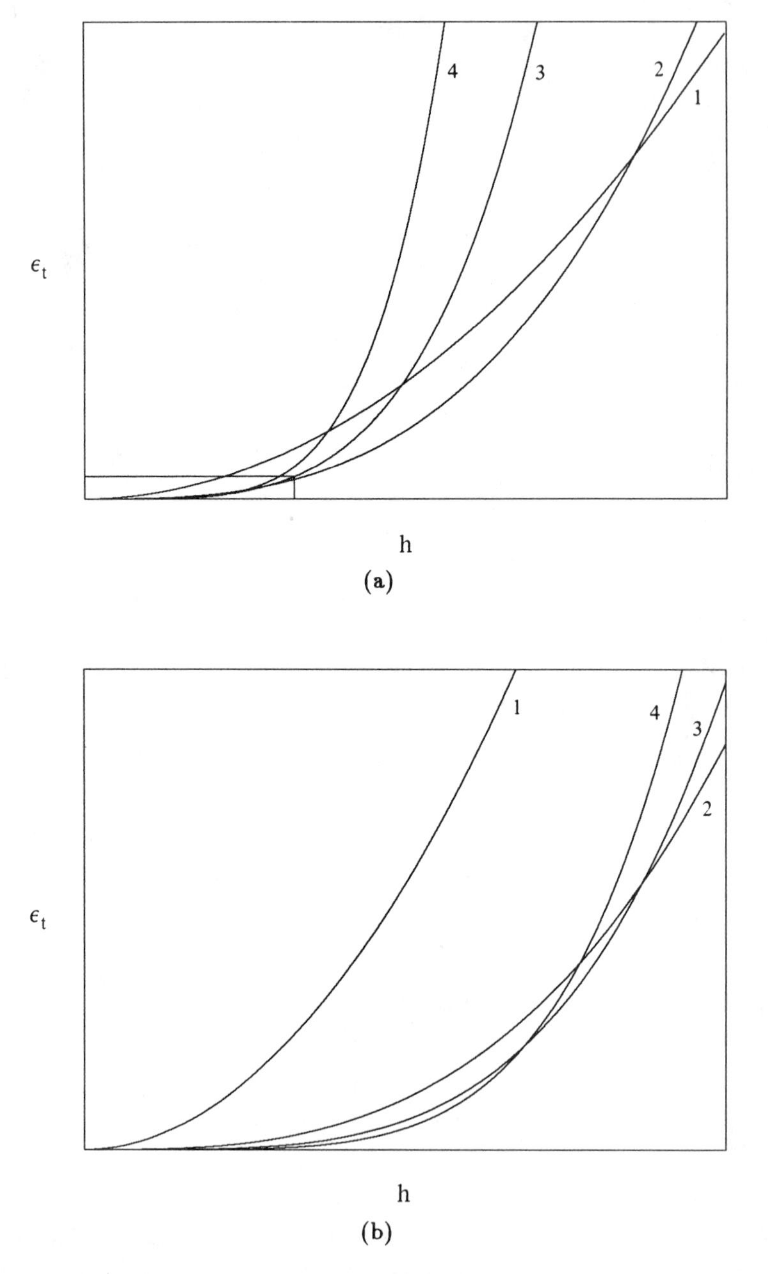

Figure 4.4: (a) Local truncation error ϵ_t versus step-size h for the Adams-Moulton algorithm; (b) is a magnification of the small box in (a).

achieve a given local truncation error, a larger step-size may be used with a higher-order algorithm than with a lower-order algorithm. This result, too, is as expected—a higher-order algorithm uses more information and should be more exact. Observe that this conclusion is valid only for small h. As is apparent from Fig. 4.4(a), if h is large enough, the first-order Adams-Moulton algorithm is more accurate than any other order.

4.2.2 Global errors

In this section we consider the global effects of the local errors. We assume the system is being integrated for 1 second from t_0 to (t_0+1) using a step-size $h \ll 1$. The total number of integration steps is $N = 1/h$.

Round-off error Global round-off error is simply the accumulation of the local round-off errors. If the local round-off error is ϵ_r, then the round-off error per unit time $\bar{\epsilon}_r$ is

$$\bar{\epsilon}_r = N\epsilon_r = \frac{\epsilon_r}{h}. \tag{4.15}$$

Thus the global round-off error is inversely proportional to the step-size. In other words, a larger step-size leads to a smaller error.

Truncation error Like the round-off error, the local truncation error accumulates with each step. For Kth-order single-step algorithms, the local truncation error is $\epsilon_t = \alpha_K h^K$. Neglecting the dependence of α_K on x_k, it follows that the truncation error per unit time $\bar{\epsilon}_t$ is

$$\bar{\epsilon}_t = N\epsilon_t = \alpha_K h^{K-1}. \tag{4.16}$$

For Kth-order multi-step algorithms, the truncation error per unit time is

$$\bar{\epsilon}_t = \beta_K h^K. \tag{4.17}$$

Fig. 4.5 shows a plot of $\bar{\epsilon}_t$ versus h for four different orders of the Adams-Moulton algorithm. Compare it with the plot of ϵ_t versus h in Fig. 4.4.

As Fig. 4.5 shows, one way to decrease $\bar{\epsilon}_t$ is to decrease the step-size. Of course, decreasing the step-size increases both $\bar{\epsilon}_r$ and the computation time of the integration.

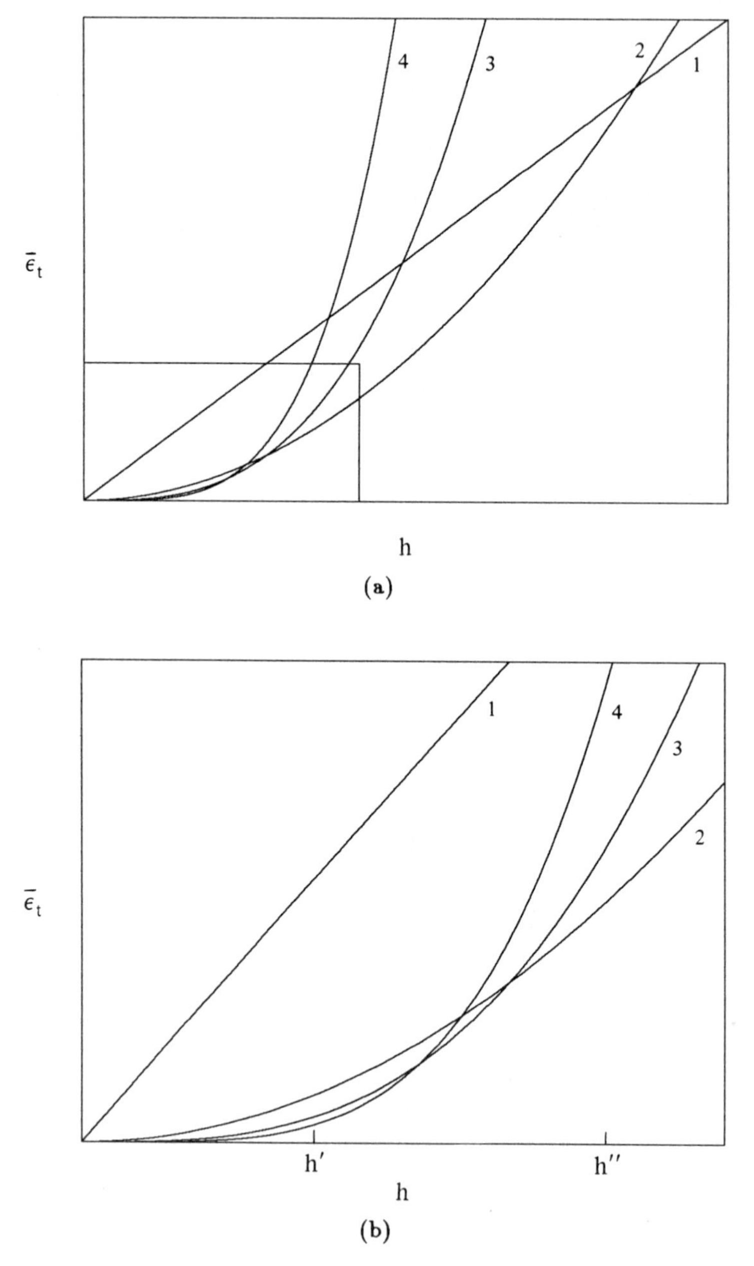

Figure 4.5: (a) Cumulative truncation error (per unit time) $\bar{\epsilon}_t$ versus step-size h for the Adams-Moulton algorithm. (b) is a magnification of the small box in (a).

Another approach is to keep the same step-size, but change the order of the integration algorithm. Whether the order should be increased or decreased depends on the circumstances. For example, suppose a third-order Adams-Moulton algorithm is used with a step-size of $h = h'$. According to Fig. 4.5(b), increasing the order to four decreases $\bar{\epsilon}_t$. On the other hand, if the step-size is $h = h''$, decreasing the order to two lessens the cumulative truncation error.

Most multi-step routines automatically adjust both their step-size and their order to achieve the desired accuracy with the largest possible step-size.

4.2.3 Numerical stability

An integration algorithm may have low round-off error and low truncation error, but be totally worthless because it is numerically unstable.

For an m-step integration algorithm, the calculation of local truncation error assumes that the previous m points, $x_k, \ldots, x_{k-m+1}$, lie on the trajectory $\phi_t(x_k)$. If, however, the previous points have been calculated using the integration algorithm, they will lie near but not exactly on the trajectory. Though this inaccuracy is initially caused by the truncation and round-off errors, the effect propagates through the rest of the integration, and the subsequent error is not accounted for by either of the local errors.

If the algorithm is numerically unstable, these initially small errors propagate through the integration and may become unbounded. The situation is analogous to an unstable equilibrium point. If the initial condition lies exactly on equilibrium point, the trajectory stays there forever. If, however, there is a slight perturbation, the trajectory diverges from the equilibrium point, never to return.

The standard method for testing numerical stability is to apply the integration algorithm to the first-order linear test equation

$$\dot{x} = \lambda x, \qquad x(0) = x_0 \tag{4.18}$$

with solution

$$\phi_t(x_0) = x_0 e^{\lambda t} \tag{4.19}$$

where x, x_0, and λ may be complex. Complex values are allowed since higher-order linear differential equations can have complex eigenvalues.

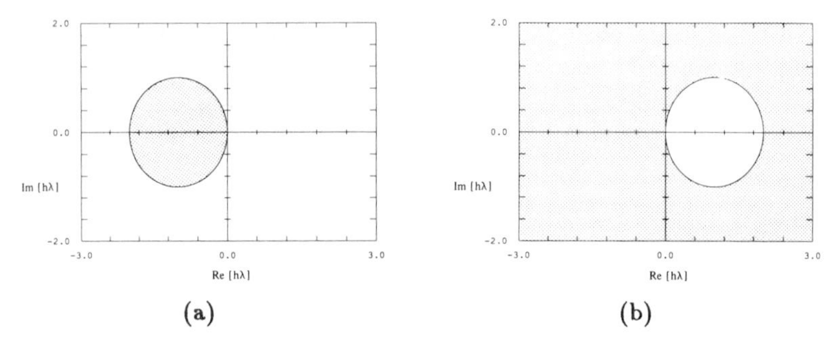

Figure 4.6: Regions of stability for the Euler algorithms. (a) Forward Euler; (b) backward Euler.

When an integration algorithm is applied to the system (4.18), the result is a linear, discrete-time system with a fixed point at the origin. The stability of the fixed point (see Section 3.2.1) determines the stability of the integration algorithm. If the fixed point is stable, the integration algorithm is *numerically stable*; if the fixed point is not stable, the algorithm is *numerically unstable*.

Example 1 Consider the forward Euler algorithm (4.3). Setting $f(x) = \lambda x$ we obtain

$$\begin{aligned} x_{k+1} &= x_k + h\lambda x_k \\ &= (1 + h\lambda) x_k. \end{aligned} \tag{4.20}$$

The characteristic multiplier of the fixed point is $m = 1 + h\lambda$, and the forward Euler algorithm is stable whenever $|1 + h\lambda| < 1$. This region, called the *region of stability*, is plotted in Fig. 4.6(a).

Example 2 Applying the backward Euler algorithm to (4.18) we obtain

$$\begin{aligned} x_{k+1} &= x_k + h\lambda x_{k+1} \\ &= \frac{1}{1 - h\lambda} x_k. \end{aligned} \tag{4.21}$$

The region of stability for the backward Euler algorithm is $|1 - h\lambda| > 1$ which is plotted in Fig. 4.6(b).

Remarks:

1. The region of stability is plotted in the complex plane. The x and y axes in Fig. 4.6 are $\text{Re}[h\lambda]$ and $\text{Im}[h\lambda]$.

2. There is an inverse relationship between the eigenvalue λ and the step-size h. If, for instance, λ is increased by a factor of 1000, then to stay at the same point in the stability region, the step-size should be reduced by 1000. This is reasonable because increasing λ speeds up the dynamics of the test system (4.18).
3. Compared with the backward Euler algorithm, the forward Euler algorithm has a limited region of stability. Typically an implicit algorithm has a larger region of stability than the corresponding explicit algorithm. Intuitively, this is due to the feedback present in an implicit algorithm.
4. This stability criterion guarantees that an algorithm is stable when integrating a linear system, but there is no guarantee that the algorithm will be stable for nonlinear systems. Fortunately, all the commonly used algorithms appear to be stable for a broad range of nonlinear systems.

For future reference, the regions of stability of the Adams-Bashforth, Adams-Moulton, and Gear algorithms are shown in Figs. 4.7–4.9.

4.3 Stiff equations

To maximize the computational efficiency of a given integration algorithm, it is desirable to minimize the number of time-steps by using the largest possible step-size. As we have seen there are two considerations when choosing the step-size: accuracy and stability.

Typically, the local error constraint specified by the user results in a step-size that lies well within the region of stability. For such systems, the numerical stability constraint is never active.

There are, however, systems for which the accuracy constraint permits a step-size that lies outside of the region of stability. Such a system is called *stiff*.[1] For stiff systems, the stability constraint is active and stability considerations limit the usable step-size.

Example Consider the uncoupled two-dimensional linear system[2]

$$\begin{bmatrix} \dot{x}_f \\ \dot{x}_s \end{bmatrix} = \begin{bmatrix} \lambda_f & 0 \\ 0 & \lambda_s \end{bmatrix} \begin{bmatrix} x_f \\ x_s \end{bmatrix} \tag{4.22}$$

[1]This definition is relative in that it depends on the algorithm used.

[2]f is for "fast" and s is for "slow."

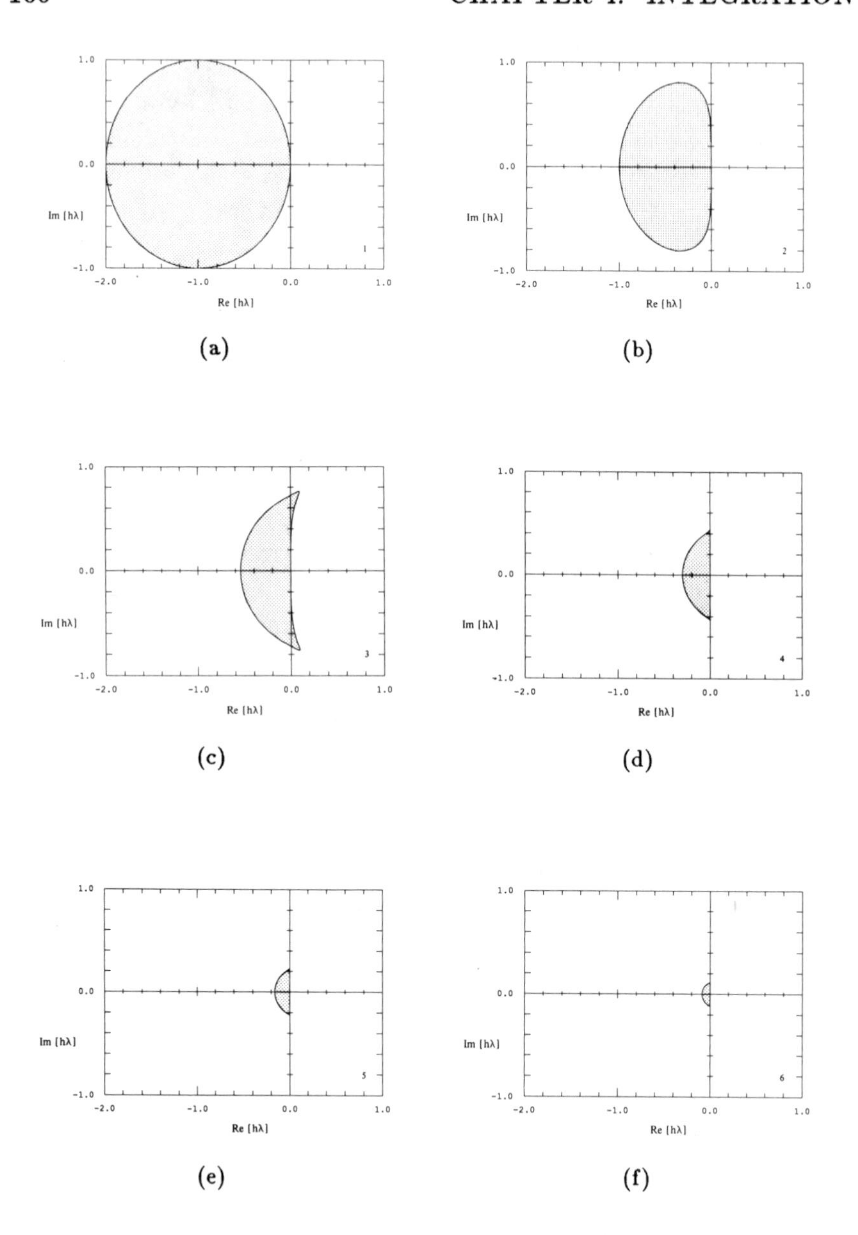

Figure 4.7: Regions of stability for the Adams-Bashforth algorithms for orders one through six.

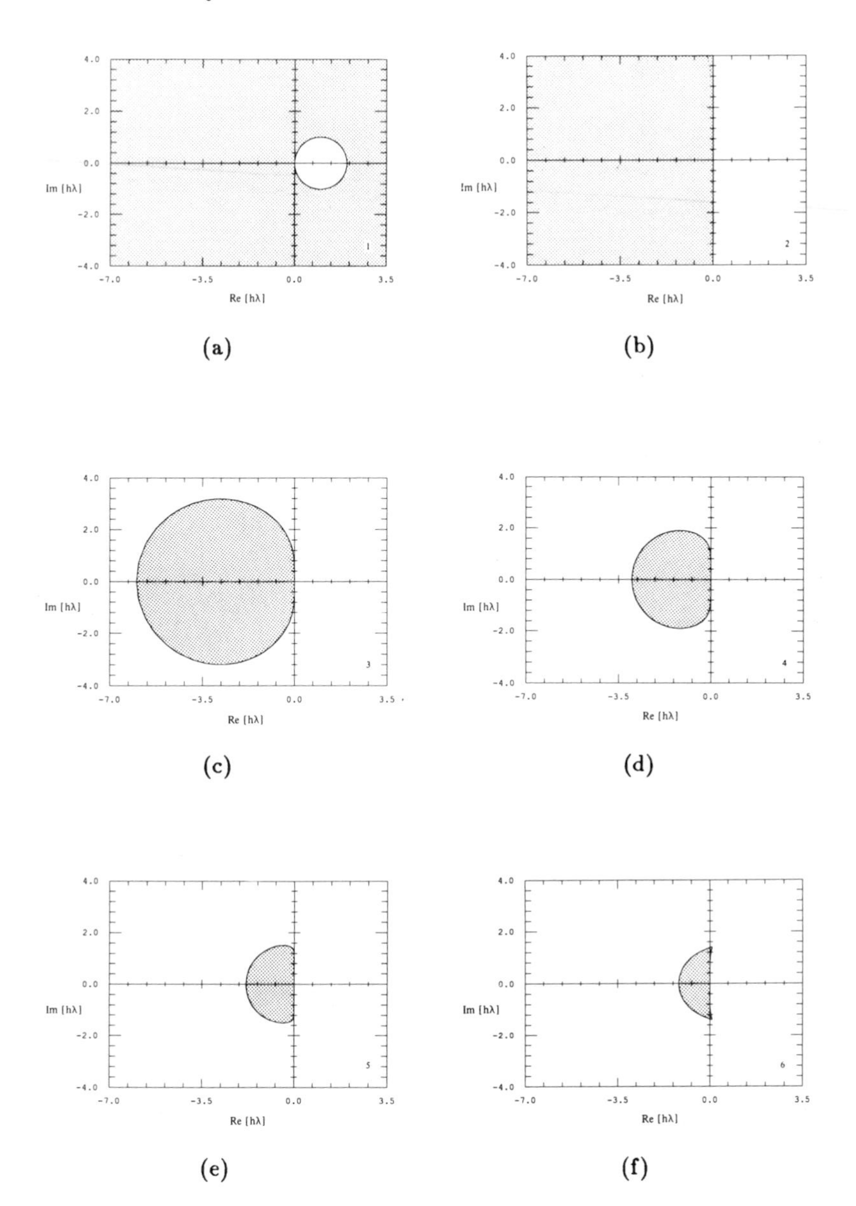

Figure 4.8: Regions of stability for the Adams-Moulton algorithms for orders one through six.

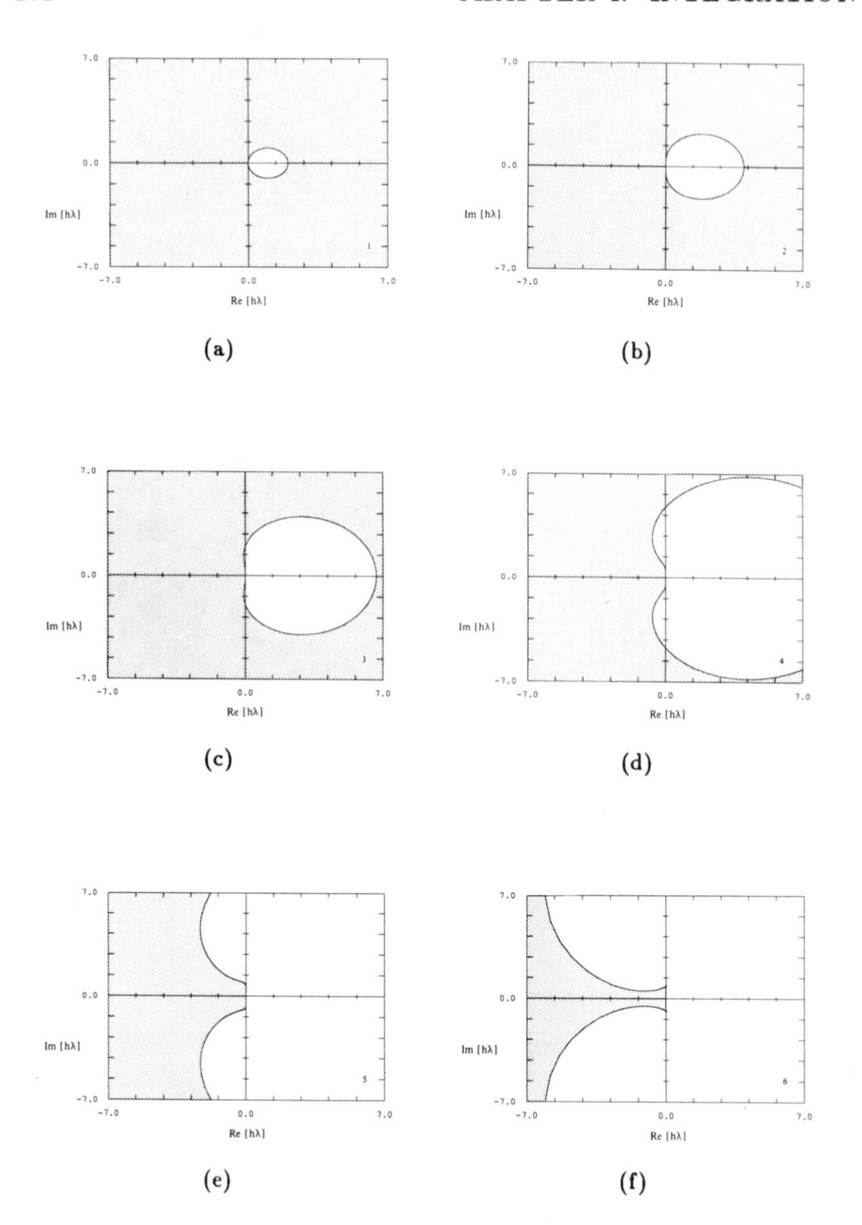

Figure 4.9: Regions of stability for the Gear algorithms for orders one through six.

with exact solution

$$\begin{bmatrix} x_f(t) \\ x_s(t) \end{bmatrix} = \begin{bmatrix} x_{f0}e^{\lambda_f t} \\ x_{s0}e^{\lambda_s t} \end{bmatrix}. \tag{4.23}$$

It is desired to find the solution of (4.22) by integration.

To make the discussion more specific, let $\lambda_f = -10^6\text{s}^{-1}$ and $\lambda_s = -1\text{s}^{-1}$. A conservative estimate of the length of the transient is ten time constants.[3] Therefore, $x_f(t) \approx 0$ for $t > t_f := 10^{-5}$s, and $x_s(t) \approx 0$ for $t > t_s := 10$s.

Suppose that when either x_f or x_s is integrated alone using the forward Euler algorithm, accuracy considerations require ten time-steps per time constant. Thus the time-step is $h_f := 10^{-7}$s for the fast component and $h_s := 0.1$s for the slow component.

These values suggest that when (4.22) is being integrated, a step-size of 10^{-7}s should be used over the initial interval, $0 < t < t_f$. Furthermore, since $x_f(t) \approx 0$ for $t > t_f$, the step-size may be increased to 0.1s during the interval $t_f < t < t_s$. Since the solution is approximately 0 for $t > t_s$, the integration stops at t_s. Using these step-sizes, the total number of steps is

$$\begin{aligned} N &= \frac{t_f}{h_f} + \frac{t_s - t_f}{h_s} \\ &= \frac{10^{-5}}{10^{-7}} + \frac{10 - 10^{-5}}{0.1} \\ &\approx 200. \end{aligned} \tag{4.24}$$

We can investigate the numerical stability of the forward Euler algorithm by calculating $h\lambda$ for the four combinations of eigenvalues and step-sizes. During $0 < t < t_f$, we have $h_f\lambda_f = -0.1$, and $h_f\lambda_s = -10^{-7}$. Both these values lie in the region of stability of the forward Euler routine. During the second time interval, the values are $h_s\lambda_s = -0.1$, and $h_s\lambda_f = -10^5$. The value -10^5 lies well outside the region of stability and it follows that the forward Euler algorithm is unstable using these step-sizes on (4.22).

The region of stability of the forward Euler algorithm limits the useful range of step-sizes to $-2 < h\lambda < 0$. Thus, in this example, the maximum step-size that can be used is $h_{max} = 2/\lambda_f = 2 \cdot 10^{-6}$s. The number of steps needed to integrate (4.22) using h_f during the

[3]For a first-order linear circuit with eigenvalue λ, the *time constant* is $-1/\lambda$.

first interval and h_{max} during the second is

$$\begin{aligned} N &= \frac{t_f}{h_f} + \frac{t_s - t_f}{h_{max}} \\ &= \frac{10^{-5}}{10^{-7}} + \frac{10 - 10^{-5}}{10^{-6}} \\ &\approx 10{,}000{,}000. \end{aligned} \tag{4.25}$$

Assuming that each integration step takes 10^{-3}s, the coupled integration of x_f and x_s takes 2.8 hours. Compare this with the execution time of 0.2s when x_s is integrated alone. Not only does the integration of the coupled system take an unreasonably long time, but the cumulative effect of the round-off error after 10^7 steps is quite large.

Now suppose that the backward Euler algorithm is used instead of the forward Euler algorithm. Since the region of stability of the backward Euler algorithm contains the entire left-half plane, there is no stability limit on the step-size (assuming $\lambda > 0$). Thus, the stiff system (4.22) may be integrated efficiently with the backward Euler algorithm.

Algorithms that can integrate stiff systems efficiently and accurately are called *stiffly stable.* Examination of the regions of stability presented in Figs. 4.7–4.9 shows that the only other stiffly stable algorithms presented in this book are the trapezoidal algorithm and the Gear algorithms. In fact, the integration of stiff systems was the motivation for, and the main purpose of, the Gear algorithms.

The system (4.22) is contrived, but there are many naturally occurring stiff systems, nonlinear as well as linear. All stiff systems share the feature of disparate time scales—time scales so different that stability considerations limit the step-size more than accuracy considerations.[4]

4.4 Practical considerations

In this section we distinguish between an integration algorithm and an integration routine. An *algorithm* is a mathematical entity; it

[4]With nonlinear systems, time constants are not well-defined, but an estimate of the time scales of a nonlinear system may be found by looking at the eigenvalues of the Jacobian $Df(x)$ evaluated at several typical points x.

is the recursion formula for x_{k+1} (e.g., equation (4.3)). A *routine* lives in the world of software; it is the code needed to implement an algorithm in a useful, accurate, and efficient manner.

In all but the simplest implementations, the integration algorithm itself is a small, albeit essential, part of the entire routine. The majority of the routine is devoted to error control, to the user interface, and for implicit algorithms, to solving the integration formula for x_{k+1}.

4.4.1 Variable step-size and order

Any useful integration routine automatically adjusts the step-size in an attempt to use the largest step-size that satisfies a user-supplied error tolerance. Many routines adjust the order of the integration algorithm as well.

There are several benefits to the automation of step-size and order selection:

1. Using the optimal step-size/order combination minimizes the number of integration steps, thereby reducing both the computation time and the accumulated round-off error.

2. Manually choosing the step-size is quite difficult. Recall that the step-size varies inversely with the time constant of the system. For a given integration error, it is common for two different systems to require drastically different step-sizes.

3. Since the user is interested in the integration error and not the step-size, specifying an error tolerance is more natural than specifying the step-size.

4. The error is controlled. Without automatic step-size selection, the user has no idea as to the accuracy of the result.

It appears from Fig. 4.5 that, given $\bar{e}_t$, there is one best step-size and one best order. Why, then, is there a need to alter the step-size at every step? This paradox is explained by recalling that the α_K in (4.13) and the β_K in (4.14) depend on the vector field near x_k (actually on the $(K+1)$th time derivative of the trajectory near x_k). It follows that there is a best step-size and a best order at each step, but the optimal values can change from step to step.

A typical step-size control algorithm works as follows. The user supplies an integration error tolerance E_{int}. The integration routine chooses a step-size, calculates x_{k+1}, and then estimates the error. If the error exceeds E_{int}, x_{k+1} is thrown away, the step-size is reduced, and the process repeats. If, on the other hand, the estimated error is less than E_{int}, x_{k+1} is accepted. If the error is much less than E_{int}, the step-size is increased for the next step.

In some cases, the integration error is estimated by a formula similar to (4.14). In the case of the Runge-Kutta-Fehlberg (4,5) routine (see Shampine and Watts [1977] and [1979]), the error is estimated by comparing the results of fourth- and fifth-order Runge-Kutta algorithms. In both cases, the error estimate is used to predict what the next step-size should be.

An m-step routine requires that $x_k, \ldots, x_{k-m+1}$ be evenly spaced in time. Whenever the step-size is changed, additional computation is required to calculate m new points with spacing equal to the new step-size. This overhead is acceptable, however, since an integration routine without automatic step-size control is extremely inefficient, if not completely worthless.

For variable-order routines, a similar approach is used to adjust the order. If, given the current step-size, a different order gives a more accurate result, the order is changed.

4.4.2 Output points

With variable step-size routines, there is the possibility of conflict between the times the routine selects and the times at which the user requests output.

Suppose the current time is t_k, and the current step-size is h_k. Left alone, the next point the integration routine calculates is x_{k+1} at time $t_{k+1} = t_k + h_k$. Suppose the user requests an output point at time $\hat{t}$ where $t_k < \hat{t} < t_{k+1}$.

There are two common solutions to this problem. The simplest is to set $t_{k+1} = \hat{t}$ thereby restricting the step-size to $\hat{t} - t_k$. The second is to calculate the solution at time t_{k+1} and then use an interpolation formula to calculate the solution at time $\hat{t}$. Routines of the first type are called *extrapolating routines* and those of the second type are called *interpolating routines*.

Multi-step algorithms are based on polynomial formulas that are similar to interpolation formulas, and most multi-step routines are interpolating routines. Single-step routines like Runge-Kutta are extrapolating routines.

To illustrate the difference between interpolating and extrapolating routines, assume that the integration routine selects $h_k = 0.1$s, and the user requests output points ten times more often, at intervals of 0.01s.

An interpolating routine takes an integration step with the natural step-size of 0.1s, and then uses interpolation nine times to calculate the points at the times the user requested. Since the output requests do not change the internal step-size of an interpolating routine, the integration process is completely unaffected by output requests. Furthermore, an interpolation is much quicker than an integration step, so interpolating routines are efficient even when many closely spaced output points are requested.

On the other hand, the accuracy and efficiency of an extrapolating routine suffer when closely spaced output points are requested. Instead of taking one step of 0.1s, an extrapolating routine takes ten steps of 0.01s each. This extra computation takes ten times as long. Though the smaller step-size results in higher accuracy, the larger step-size meets the user's requirements, so the additional accuracy is superfluous. Extrapolating routines are ill-suited for applications that require closely spaced output points.

It is sometimes the case that one wants to integrate up to t_f, but not past it. One application is locating the boundary crossing in a piecewise-defined system. Interpolating routines approximate the trajectory by a polynomial and they fail when the trajectory is not smooth. Since a trajectory is not smooth where the vector field is discontinuous, as it may be at a region boundary in a piecewise-defined system, such a system should be integrated up to but not past a boundary. To be useful in this situation, many interpolating routines have an option that turns the "integrate past and interpolate back" feature off.

To give the user more control over the step-size, many routines also allow the user to specify minimum and maximum values for the step-size.

4.4.3 Solving implicit equations

Implicit algorithms have stability and accuracy advantages over explicit routines, but they require additional code to solve for x_{k+1}. The standard form of the equation that must be solved is the fixed point equation

$$x_{k+1} = F(x_{k+1}). \tag{4.26}$$

There are two methods commonly used to calculate x_{k+1}: functional iteration and the Newton-Raphson algorithm.

Functional iteration It can be shown (see Chua and Lin [1975]) that if $f(x)$ satisfies a Lipschitz condition, if the step-size is small enough, and if the initial guess $x_{k+1}^{(0)}$ is close enough to x_{k+1}, then the sequence $\{x_{k+1}^{(i)}\}_{i=0}^{\infty}$ where

$$x_{k+1}^{(i)} = F(x_{k+1}^{(i-1)}), \qquad i = 1, 2, \ldots \tag{4.27}$$

converges to x_{k+1}. This method of calculating x_{k+1} is called *functional iteration.*

The better the initial guess, the more quickly the iterations converge. When an explicit algorithm like Adams-Bashforth is used to calculate $x_{k+1}^{(0)}$, one or two iterations of the implicit algorithm are sufficient to yield an accurate result. The combination of an explicit predictor and a functionally-iterated implicit algorithm is called a *predictor-corrector* pair.

This approach is easy to program, and each step is relatively efficient. Note, however, that the step-size must be reduced to ensure that the $x_{k+1}^{(i)}$ converge. The overall efficiency is lowered proportionately.

Newton-Raphson Rewrite (4.26) as

$$0 = H(x_{k+1}) := F(x_{k+1}) - x_{k+1}. \tag{4.28}$$

With this formulation, the Newton-Raphson algorithm may be used to calculate x_{k+1}. As with the functional iteration approach, an explicit algorithm is used to calculate $x_{k+1}^{(0)}$. The Newton-Raphson algorithm is iterated until it converges or until a maximum number of iterations is reached. If the iteration limit is reached, the integration step is rejected, and the step-size is reduced.

One advantage of the Newton-Raphson approach is that larger step-sizes may be selected. The Newton-Raphson algorithm also converges more quickly than the method of functional iterations.

One disadvantage is that the Newton-Raphson algorithm requires the Jacobian DH. The Jacobian may be obtained either analytically or by numerical approximation. Due to the overhead of finding the Jacobian and of solving an $n \times n$ system of linear equations, one integration step using the Newton-Raphson algorithm is typically slower than one step of a functionally iterated integration routine. It is hoped, however, that the efficiency gained through the larger step-size more than offsets the per-step decrease in speed.

4.4.4 Error considerations

In this section, we discuss the relationship between the global integration error and the integration tolerance specified by the user.

A standard way of judging the accuracy of the output of an integration routine is to re-integrate the system using a tighter error tolerance. The initial time interval over which the two results agree is then assumed to be accurate. This technique relies on the assumption that lowering the error tolerance lowers the global error.

This assumption is not always true. We have already encountered one example—an extrapolating routine with output requests so frequent that they limit the step-size. More specifically, assume that to satisfy a requested error tolerance E, the extrapolating routine requires a step-size $h = \sqrt{E}$. The user specifies an initial error tolerance of 10^{-2}, and requires output points every 10^{-3}s. To meet the error tolerance, the routine wants to select $h = 0.1$s, but due to the output requests, it actually uses $h = 10^{-3}$s. With this smaller step-size, the effective error tolerance is 10^{-6}. Clearly, if the system is re-integrated with any error tolerance greater than or equal to 10^{-6}, the values of the output points will not change. In other words, for $E \geq 10^{-6}$, the error of the integration routine is independent of E.

It is important to realize that in this situation, the integration routine is acting exactly as it should. It is supplying results that meet the error tolerance specified by the user.

To some readers, this example may seem contrived—if the user requests output points less often, the problem disappears. There are situations, however, where it is not the user that restricts the step-size. A good example is the system (B.8) used to calculate the solution of the variational equation. In this system, the original differential equation and the variational equation are integrated simultaneously. The step-size the integration routine selects to integrate the coupled system is usually smaller than the step-size that would be selected to integrate the original differential equation alone. Even if the same error tolerance is used, care must be taken when comparing values of $\phi_t(x)$ obtained from the integration of (B.8) and from the integration of (4.1) alone. Since the step-size reduction is not caused by output requests, this effect is not limited to extrapolating routines—interpolating routines can be affected, too.

4.4.5 Integrating chaotic systems

The integration of a chaotic system poses a special problem. Sensitive dependence on initial conditions implies that an arbitrarily small error eventually affects the macroscopic behavior of the system. Error is inherent in any integration algorithm. Does it follow that integration of a chaotic system is meaningless? Perhaps there are a few researchers who would answer yes, but the impressive contributions of numerical simulations in chaotic research show the answer is definitely no.

What *is* true is that simulations of chaotic systems require careful interpretation and should always be verified. If possible, two or more different routines should be used to integrate the same system. Since the routines introduce different errors and the errors propagate differently, this is a good check on the validity of the integration. Intuition and physical reasoning should also be used to verify the reasonableness of an integration result.

There is one task which an integration routine cannot perform. It cannot estimate with any accuracy the state of a chaotic system after a long period of integration. What is meant by "long" depends on the system. Recall that perturbations in a chaotic system grow initially as $e^{\lambda_1 t}$ where λ_1 is the largest Lyapunov exponent (see Section 3.3). If the per-step error is ϵ, it follows that at time t the error $E(t)$ due to the initial step is $E(t) \approx \epsilon e^{\lambda_1 t}$ which, when solved for t, becomes

$$t(E) = \frac{\ln(E/\epsilon)}{\lambda_1}. \tag{4.29}$$

If the length scale of the system is L, than integration past time $t(L)$ is completely useless for predicting the position of a point on the trajectory. As an example, let $\lambda_1 = 1\text{s}^{-1}$, $\epsilon = 10^{-4}$, and $L = 1$. Then integration cannot be used to predict $\phi_t(x)$ for $t > 2.3$s.

This limit to the interval of integration may seem constraining, but there are several applications not affected by it. There are two reasons.

The first reason is that when a system is being integrated over a relatively short period of time T, the integration tolerance can usually be tightened to force $t(E) \ll T$. An application of this type is

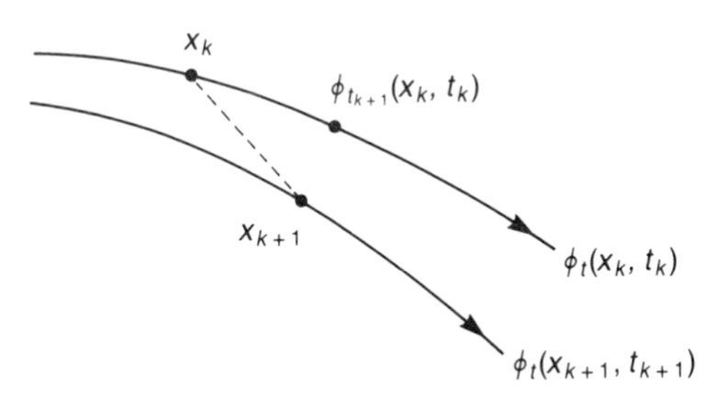

Figure 4.10: The effect of local error at the $(k+1)$th step is to hop from $\phi_t(x_k, t_k)$ to $\phi_t(x_{k+1}, t_{k+1})$.

the shooting method for locating periodic solutions (see Section 5.5). For this application, T is the period of the periodic solution.

The second reason is that a strange attractor is attracting. If the local integration error ϵ is reasonably small, then the output points x_k of the integration routine may not lie anywhere near $\phi_{t_k}(x_0)$, but they do approach the attractor. This is seen by the following argument. When stepping from x_k to x_{k+1}, the effect of the integration error is to switch from the trajectory $\phi_t(x_k, t_k)$ passing through x_k to the trajectory $\phi_t(x_{k+1}, t_{k+1})$ passing through x_{k+1} (see Fig. 4.10). If ϵ is reasonably small, the error introduced is small enough to keep x_{k+1} in the basin of attraction of the attractor, and the new trajectory $\phi_t(x_{k+1}, t_{k+1})$ approaches the attractor. Since it is an attractor, the contraction (toward the attractor) outweighs the expansion (away from the attractor), and it follows that the sequence of integration points approaches the attractor.

Therefore applications that do not require one specific trajectory to be followed but only that the output points lie on the attractor, are amenable to numerical integration. One example of this type of application is locating a strange attractor by repeated iterations of the Poincaré map. Another is the calculation of Lyapunov exponents using the algorithm of Section 3.4.3. From the definition of the Lyapunov exponents, it may appear that precisely one trajectory should be followed, but recall that the definition is an average over all time. By the principle of ergodicity, the definition may also be formulated as an average over all points on the attractor. Thus the trajectory-hopping caused by the local error does not affect the Lyapunov exponent algorithm.

4.4.6 Start-up costs

There is always extra computation involved at the beginning of an integration. This overhead, called the *start-up cost*, has two sources: initialization of a multi-step algorithm and initialization of the step-size selection procedure.

An m-step algorithm cannot be used until the m previous points, x_k, x_{k-1}, ..., x_{k-m+1}, are known. As an example of multi-step start-up, consider the Adams-Moulton family of algorithms. The second-order algorithm is a one-step algorithm, and may be used to calculate x_1. At the next step, the third-order algorithm is used to calculate x_2. Successively higher-step algorithms are used until $m-1$ points have been calculated, at which time, the m-step algorithm may be used.

Step-size selection also has to be initialized. The first step-size selected is usually far from optimal, and it may take the selection algorithm several iterations to find an efficient step-size.

To minimize the effect of start-up costs, an integration routine usually has two modes of operation. The *initialization mode* is used to start a new integration, either from a new initial condition or a new initial time. This mode incurs the start-up costs. The *continuation mode* resumes the integration from the last point calculated. It bypasses start-up, and is useful when the user requests intermediate points during the course of the integration.

4.5 Summary

- *Single-step algorithm*: A single-step integration algorithm requires only the current point x_k.

- *Multi-step algorithm*: An m-step integration algorithm requires the current point x_k and the $m-1$ previous points, x_{k-1}, ..., x_{k-m+1}.

- *Explicit algorithm*: An integration algorithm is explicit if its formula can be solved explicitly for x_{k+1}.

- *Implicit algorithm*: An integration algorithm is implicit if its formula cannot be solved explicitly for x_{k+1}. Implicit algorithms are generally more accurate and more stable than their explicit counterparts.

- *Local error*: The local error of an integration algorithm is the per-step error, $||x_{k+1} - \phi_{t_k+h}(x_k, t_k)||$, where it is assumed that the input to the integration algorithm is exact. Local error has two components: round-off error and truncation error.

- *Global error*: The global error of an integration program at time t_k is $||x_k - \phi_{t_k}(x_0, t_0)||$. Global error has two components: the accumulation of the local errors and the error caused by the propagation of the local errors through the rest of the integration.

- *Round-off error*: Round-off error is the error caused by performing real arithmetic on a digital computer. Local round-off error is independent of the step-size. The global round-off error is equal to the product of the number of steps taken and the local round-off error.

- *Truncation error*: The truncation error of an integration algorithm is the error that would be present if the algorithm could be implemented with infinite-precision arithmetic. The local truncation error depends on the step-size. The global truncation error is the accumulation of the local truncation errors.

- *Numerical stability*: An integration algorithm is numerically stable if the discrete-time system obtained by applying the algorithm to the test equation, $\dot{x} = \lambda x$, is stable. λ and x may be complex numbers.

- *Region of stability*: The region of stability of an integration algorithm comprises those points in the complex $h\lambda$-plane for which the algorithm is numerically stable.

- *Stiff system*: A continuous-time system is stiff if stability considerations limit the step-size more than accuracy considerations. Typically, a stiff system has disparate time scales.

- *Stiffly-stable algorithm*: An integration algorithm is stiffly stable if it can integrate a stiff system efficiently and accurately.

- *Extrapolating routines*: If an output request is made for time $\hat{t}$ where $t_k < \hat{t} < t_k + h_k$, an extrapolating integration routine decreases the step-size to $h_k = \hat{t} - t_k$, and uses an integration step to calculate the output at $\hat{t}$. Extrapolating routines are inefficient when frequent output is requested.

- *Interpolating routines*: If an output request is made for time $\hat{t}$ where $t_k < \hat{t} < t_{k+1} = t_k + h_k$, an interpolating integration routine integrates to t_{k+1}, and uses an interpolation formula to calculate the output at $\hat{t}$. Interpolating routines are efficient when frequent output is requested.

- *Start-up costs*: Start-up costs are the costs incurred when an integration routine is started from a new initial condition. Start-up costs arise from initializing the step-size selection algorithm and initializing multi-step algorithms.

Chapter 5

Locating Limit Sets

The first step in analyzing a dynamical system is to determine the location and stability type of its limit sets. In this chapter, we present algorithms that locate equilibrium points, fixed points, closed orbits, periodic solutions, and two-periodic solutions.

5.1 Introduction

5.1.1 Brute-force approach

The easiest way to locate a limit set is to let the system run until it reaches the steady state. We call this method the *brute-force approach.*

The brute-force approach has several advantages. First, it is easy to program. For continuous-time systems, all that is required is integration; for discrete-time systems, all that is needed are repeated iterations of the map. Second, brute force is quite general; it can locate any of the different kinds of limit sets (equilibrium point, fixed point, etc.). Third, as long as the initial condition is in the basin of attraction of an attractor, the brute force method always converges.

On the other hand, brute force has some serious drawbacks. First, convergence can be exceedingly slow for lightly damped systems. Second, brute force can locate stable limit sets only; it fails for non-stable limit sets and for unstable limit sets.[1] Third, it is difficult for a pro-

[1] Brute force can locate an unstable limit set of a system that can run in reverse time. Continuous-time systems can always be integrated in reverse time, but not all discrete-time systems are invertible.

gram (and often the user) to tell when the steady state has been achieved.

Despite these drawbacks, brute force finds use in several applications. The most common is the manual search for limit sets. When a researcher looks for new behavior in a system, she typically uses a simulation program that displays trajectories or orbits. She changes the parameters or the initial condition, and lets the system run until she is satisfied it has reached the steady state. When the approximate location of an interesting limit set has been found, she applies more sophisticated methods to calculate its precise position. A second application of brute force is the generation of bifurcation diagrams as is discussed in Chapter 8.

5.1.2 Newton-Raphson approach

More sophisticated algorithms transform the problem of locating a limit set into the task of calculating the zeros of a system of nonlinear equations. The zeros may be found by any zero-finding algorithm. We restrict our discussion to the Newton-Raphson algorithm.

The main advantages of this approach are the rapid rate of convergence and the ability to locate non-stable limit sets as well as stable and unstable ones.

This approach suffers from the usual drawbacks of the Newton-Raphson algorithm: the initial guess must be close to the actual value and the Jacobian must be known or approximated.

All of the following algorithms use the Newton-Raphson algorithm. Instead of presenting the Newton-Raphson iteration for each new algorithm, we present only the function H whose zeros are sought and its Jacobian DH. The reader is referred to Appendix A for details on implementing the Newton-Raphson algorithm.

5.2 Locating equilibrium points

The algorithm Equilibrium points are the zeros of the vector field f. The Newton-Raphson algorithm can be applied directly to $H(x) := f(x)$ assuming that Df can be calculated.

Once convergence is achieved, the stability of the equilibrium point can be found by calculating the eigenvalues of $Df(x_{eq})$ using the QR algorithm.

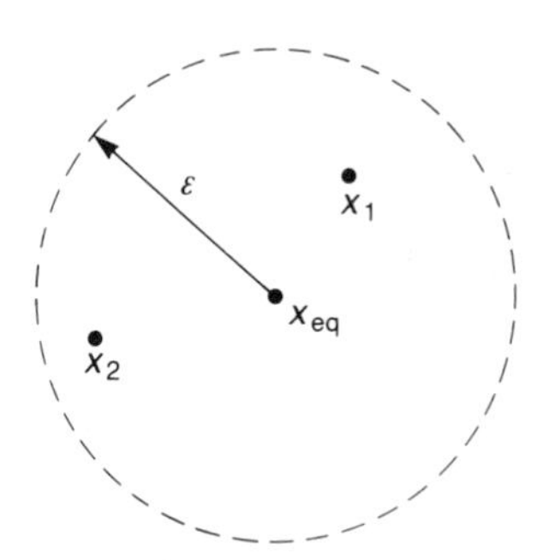

Figure 5.1: The two points, x_1 and x_2, lie within ϵ of x_{eq} and within 2ϵ of each other.

Duplicate equilibrium point test When a program locates more than one equilibrium point, it can check whether a newly found equilibrium point is a duplicate of any of the known equilibrium points by using a relative/absolute test (see equation (A.5)). If

$$\|x_{eq} - \hat{x}_{eq}\| < E_r \|x_{eq}\| + E_a, \tag{5.1}$$

then x_{eq} and $\hat{x}_{eq}$ correspond to the same equilibrium point.

As Fig. 5.1 shows, the values of the tolerances, E_r and E_a, should be larger than those used in the Newton-Raphson convergence test, but small enough to ensure that two distinct equilibrium points do indeed violate (5.1). Values ten times those used in the Newton-Raphson convergence test are typical.

5.3 Locating fixed points

The algorithm A fixed point x^* of a map $P: \mathbb{R}^p \to \mathbb{R}^p$ may be found by applying the Newton-Raphson algorithm to the function H defined by

$$H(x) := P(x) - x \tag{5.2}$$

with Jacobian

$$DH(x) = DP(x) - I. \tag{5.3}$$

The eigenvalues of $DP(x^*)$ are the characteristic multipliers of x^* and can be found using the QR algorithm.

Duplicate fixed point test A duplication test identical to the one presented in Section 5.2 for equilibrium points can be used to detect duplicate fixed points.

5.4 Locating closed orbits

The algorithm Each point x_k^* of a period-K closed orbit of a map P is a fixed point of the K-iterated map P^K, that is, $x_k^* = P^K(x_k^*)$ for $k = 1, \ldots, K$.

The closed orbit may be found by applying the Newton-Raphson algorithm to

$$H(x) := P^K(x) - x \tag{5.4}$$

with Jacobian

$$DH(x) = DP^K(x) - I. \tag{5.5}$$

$DP^K(x)$ is found in terms of DP as follows. Define $x_1 := x$ and $x_k := P(x_{k-1})$ for $k = 2, \ldots, K$. By the chain rule,

$$DP^K(x) = DP(x_K) \cdots DP(x_1). \tag{5.6}$$

The characteristic multipliers of the orbit are the eigenvalues of $DP^K(x_1^*)$ and can be found using the QR algorithm.

Duplicate orbit test Given two period-K orbits, a duplication test is performed by testing x_1^* of the first orbit against every point of the second using (5.1). If any point of the second orbit satisfies the test, the orbits are duplicates.

Minimum period test When the computer locates a period-K orbit, it is desirable to check whether K is the minimum period.

To find the minimum period of $\{x_1^*, \ldots, x_K^*\}$, compare x_1^* to each of $x_2^*, \ldots, x_{1+K/2}^*$ using (5.1).[2] If k is the smallest index such that x_1^* and x_k^* satisfy the test, then the minimum period of the orbit is $k-1$ and the orbit is $\{x_1^*, \ldots, x_{k-1}^*\}$. If none of the points of the orbit coincides with x_1^*, then K is the minimum period.

5.5 Locating periodic solutions

As usual, we consider the non-autonomous and autonomous cases separately.

[2]The index $1 + K/2$ is the worst case. If K has prime factors $p_1 < p_2 < \cdots < p_m$, then the highest index that needs to be tested is $1 + K/p_1$. The relative/absolute test is quick, however, and even when $p_1 > 2$, there is no practical advantage to using $1 + K/p_1$ instead of $1 + K/2$.

5.5.1 The non-autonomous case

The non-autonomous shooting method

The algorithm Let P_N be the Poincaré map associated with a non-autonomous system with minimum period T where

$$P_N(x) := \phi_{t_0+T}(x, t_0). \tag{5.7}$$

It was shown in Section 2.1.1 that a period-T solution of the non-autonomous system corresponds to a fixed point x^* of P_N. Thus a period-T solution can be located by finding a fixed point of P_N.

To find the fixed point using the Newton-Raphson algorithm, define

$$H(x) := P_N(x) - x \tag{5.8}$$

with Jacobian

$$DH(x) = DP_N(x) - I. \tag{5.9}$$

From (5.7)[3]

$$DP_N(x) = \Phi_{t_0+T}(x, t_0). \tag{5.10}$$

The Newton-Raphson algorithm applied to (5.8) is commonly called the *non-autonomous shooting method.* The name derives from the fact that at each iteration, the trajectory shoots forward T seconds.

The characteristic multipliers are the eigenvalues of $\Phi_{t_0+T}(x, t_0)$ and can be found using the QR algorithm.

Subharmonics The shooting method can be used to locate a Kth-order subharmonic by finding one of the corresponding K fixed points of the K-iterated Poincaré map

$$P_N^K(x) = \phi_{t_0+KT}(x, t_0). \tag{5.11}$$

In this case,

$$DH = DP_N^K(x) - I \tag{5.12}$$

and

$$DP_N(x) = \Phi_{t_0+KT}(x, t_0). \tag{5.13}$$

---

[3] Recall that Φ_t is the solution of the variational equation and is integrated simultaneously with ϕ_t (see Appendix B).

Minimum period test A test identical to the one given in Section 5.4 can be used to test whether the period of a subharmonic is minimal.

Duplicate periodic solution test A test identical to the one given in Section 5.3 (for fundamental solutions) or Section 5.4 (for subharmonics) should be used to detect the duplication of periodic solutions.

5.5.2 The autonomous case

We present two different algorithms. Both are generalizations of the non-autonomous shooting method.

The autonomous shooting method

This method, due to Mees [1981], is a variant of the autonomous shooting method first proposed by Aprille and Trick [1972]. Like the non-autonomous shooting method, it applies the Newton-Raphson algorithm to find zeros of

$$H(x, T) := \phi_T(x) - x. \tag{5.14}$$

The difference from the non-autonomous case is that there are now $n+1$ unknowns: the n components of x^* and the period T.

The algorithm Since (5.14) is a system of n equations in $n+1$ unknowns, the Newton-Raphson algorithm cannot be applied directly.

Instead, linearize $y = H(x, T)$ to obtain

$$\begin{aligned} \Delta y &\approx D_x H(x,T)\,\Delta x + D_T H(x,T)\,\Delta T \\ &= \{\Phi_T(x) - I\}\,\Delta x + f(\phi_T(x))\,\Delta T. \end{aligned} \tag{5.15}$$

In the hope of achieving $H(x + \Delta x, T + \Delta T) \approx 0$, choose Δx and ΔT such that $\Delta y = -H(x)$. Substitute this value of Δy in (5.15) to obtain

$$-H(x) = \{\Phi_T(x) - I\}\,\Delta x + f(\phi_T(x))\,\Delta T. \tag{5.16}$$

Since (5.16) is a system of n equations in the $n+1$ unknowns, Δx and ΔT, one more constraint must be found to give the system a unique solution. The constraint Mees chose restricts the state correction term Δx to be orthogonal to the trajectory, that is,

$$\langle f(x), \Delta x \rangle = 0. \tag{5.17}$$

This constraint, together with (5.16), results in a system of $n+1$ equations

$$\begin{bmatrix} \Phi_{T^{(i)}}(x^{(i)}) - I & f(\phi_{T^{(i)}}(x^{(i)})) \\ f(x^{(i)})^T & 0 \end{bmatrix} \begin{bmatrix} \Delta x^{(i)} \\ \Delta T^{(i)} \end{bmatrix} = \begin{bmatrix} x^{(i)} - \phi_{T^{(i)}}(x^{(i)}) \\ 0 \end{bmatrix} \tag{5.18}$$

where the superscripts have been added to indicate the iteration count. This relation is iterated in the typical Newton-Raphson fashion.

Due to the orthogonality constraint (5.17), this method is not the standard Newton-Raphson algorithm, that is, there is no function H such that DH is the matrix in (5.18). It is, however, similar enough that it inherits the convergence properties of the standard Newton-Raphson algorithm.

Minimum period test Let $\hat{x}$ and $\hat{T}$ be the values returned by the shooting method. To test whether $\hat{T}$ is the minimum period of the limit cycle, one would like to integrate from $\hat{x}$ and see whether the trajectory passes close to $\hat{x}$ in less than $\hat{T}$ seconds.

A practical way to do this is to use the cross-section Σ that is orthogonal to $f(\hat{x})$ and that passes through $\hat{x}$,

$$\Sigma := \{x : \langle f(\hat{x}), x - \hat{x} \rangle = 0\}. \tag{5.19}$$

Integrate from $\hat{x}$ until either time $\hat{T}$ is reached or the trajectory intersects Σ in the positive direction[4] at some point x_1 at some time $t_1 < \hat{T}$. If $\hat{T}$ is reached without a positive crossing, then $\hat{T}$ is the minimum period. If, however, the trajectory intersects Σ before $\hat{T}$, use (5.1) to check whether x_1 coincides with $\hat{x}$. If it does, then t_1 is the minimum period; if not, continue the integration from x_1 until $\hat{T}$ is reached or the trajectory again crosses Σ in the positive direction.

[4] Recall that, given a hyperplane, $\Sigma = \{x : \langle h, x - x_\Sigma \rangle = 0\}$, the direction of a hyperplane crossing at a point x is positive if $\langle h, f(x) \rangle \geq 0$.

Duplicate limit cycle test Let $\hat{x}_1$ and $\hat{T}_1$ be one set of values returned by the shooting method and $\hat{x}_2$ and $\hat{T}_2$ be another. Further assume that $\hat{T}_1$ and $\hat{T}_2$ are the minimum periods. If $\hat{T}_1$ and $\hat{T}_2$ satisfy a relative/absolute test, then it is possible that $\hat{x}_1$ and $\hat{x}_2$ lie on the same limit cycle.[5]

To test for duplicate limit cycles, let Σ be the hyperplane that passes through $\hat{x}_2$ and is orthogonal to $f(\hat{x}_2)$. Integrate the system from $\hat{x}_1$. If time $\hat{T}_1$ is reached before the trajectory intersects Σ in a positive direction (with respect to $f(\hat{x}_2)$), then the limit cycles are distinct. If, however, the trajectory intersects Σ in the positive direction at some point x_2' before time $\hat{T}_1$, perform a relative/absolute test to compare $\hat{x}_2$ with x_2'. If the test is satisfied, then the limit cycles are duplicates; if not, continue integrating from x_2' until $\hat{T}_1$ is reached or the trajectory again intersects Σ in the positive direction.

Poincaré map method

The Poincaré map method is a generalization of the non-autonomous shooting method in that it applies the Newton-Raphson algorithm to find fixed points of the one-sided Poincaré map P_+.

The algorithm Let x^* be a fixed point of P_+. Define

$$H(x) := P_+(x) - x. \tag{5.20}$$

Then x^* is a zero of H and may be found using the Newton-Raphson algorithm.

Let $y = P_+(x)$, and let T be the time it takes for the trajectory to travel from x to y. It is shown in Appendix D that

$$DP_+(x) = \left[I - \frac{1}{\langle h, f(y) \rangle} f(y) h^T \right] \Phi_T(x) \tag{5.21}$$

where h is a vector normal to the cross-section defining P_+.

To understand (5.21), recall that $DP_+(x)$ governs the behavior of small perturbations of x under one iteration of P_+. More precisely,

[5] It might seem that if the periods are the same, the limit cycles must be the same, for what are the odds of two limit cycles having the same period. The most common exception is when an odd-symmetric system possesses a non-symmetric limit cycle. In such a system, there exists a second limit cycle—the reflection through the origin of the first—with exactly the same period as the first.

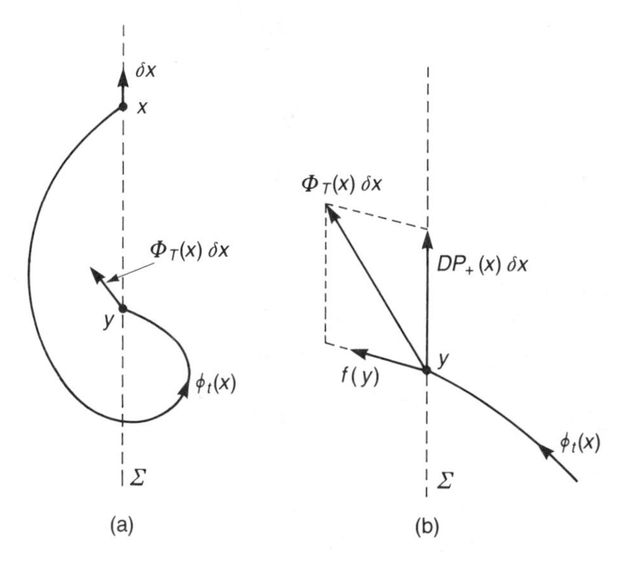

Figure 5.2: (a) A geometric interpretation of DP_+; (b) an enlargement of (a) near the point y.

given an infinitesimal perturbation δx of $x \in \Sigma$,

$$P_+(x + \delta x) = P_+(x) + DP_+(x)\,\delta x. \tag{5.22}$$

The range of P_+ is Σ, so $P_+(x + \delta x)$ must lie on Σ; equivalently, $\langle h, DP_+(x)\,\delta x\rangle$ must equal zero.

Multiply (5.21) by δx to obtain

$$\begin{aligned} DP_+(x)\,\delta x &= \Phi_T(x)\,\delta x - \frac{1}{\langle h, f(y)\rangle} f(y)\, h^T \Phi_T(x)\,\delta x \\ &= \delta y - \frac{\langle h, \delta y\rangle}{\langle h, f(y)\rangle} f(y) \end{aligned} \tag{5.23}$$

where $\delta y := \Phi_T(x)\,\delta x$.

The first term in (5.23) is the value of the perturbation δx at time T as it evolves in the flow.

To understand the significance of the second term in (5.23), remember that $P_+(x + \delta x)$ must lie on Σ. Since $P_+(x) + \delta y$ does not usually lie on Σ, it must be projected onto Σ (see Fig. 5.2). This is the task of the second term.

The natural direction of projection is $f(y)$, that is, along the trajectory that passes through y. To ease the notation, shift the origin to y. The projection of δy onto Σ is

$$\delta y - \alpha f(y) \tag{5.24}$$

where α is chosen such that $\langle h, \delta y - \alpha f(y) \rangle = 0$. Solve this expression for α to obtain

$$\alpha = \frac{\langle h, \delta y \rangle}{\langle h, f(y) \rangle} \tag{5.25}$$

which, after substitution into (5.24), yields (5.23).

Choosing Σ The cross-section Σ is specified by a vector h orthogonal to Σ and by a point $x_\Sigma \in \Sigma$. The initial guess $x^{(0)}$ must lie on Σ, so Σ and $x^{(0)}$ may not be specified independently.

If the user is not interested in a particular cross-section, he can specify only the initial guess $x^{(0)}$ and let the program automatically set $h = f(x^{(0)})$ and $x_\Sigma = x^{(0)}$.

If a specific cross-section is required, the user-supplied initial state, call it $\tilde{x}$, will most likely not lie on Σ, so the program needs to find $x^{(0)}$ itself. The simplest approach is to set $x^{(0)}$ to the point on Σ that is nearest to $\tilde{x}$. Specifically, $x^{(0)} = \tilde{x} - \alpha h$ where

$$\alpha = \frac{\langle h, \tilde{x} - x_\Sigma \rangle}{\|h\|^2}. \tag{5.26}$$

A limit cycle may intersect a given cross-section several times. Such a limit cycle generates a period-K orbit $\{x_1^*, \ldots, x_K^*\}$ of P_+. The x_k^* are fixed points of the K-iterated map P_+^K, and can be found by applying the Poincaré map method to P_+^K.

Minimum period test To test whether K is the minimum period of the closed orbit, the test presented in Section 5.4 can be used.

Duplicate limit cycle test To determine whether the limit cycle duplicates a previously found limit cycle, the test presented in Section 5.4 hould be used.

Discussion

The differences between the shooting method and the Poincaré map method are apparent in Fig. 5.3. In the shooting method, the orthogonality constraint specifies a cross-section. This cross-section, always orthogonal to $f(x^{(i)})$, changes position and orientation from iteration to iteration. On the other hand, for the Poincaré map method, the cross-section is fixed from the start. Also note that in each iteration of the shooting method, the trajectory is integrated for a known

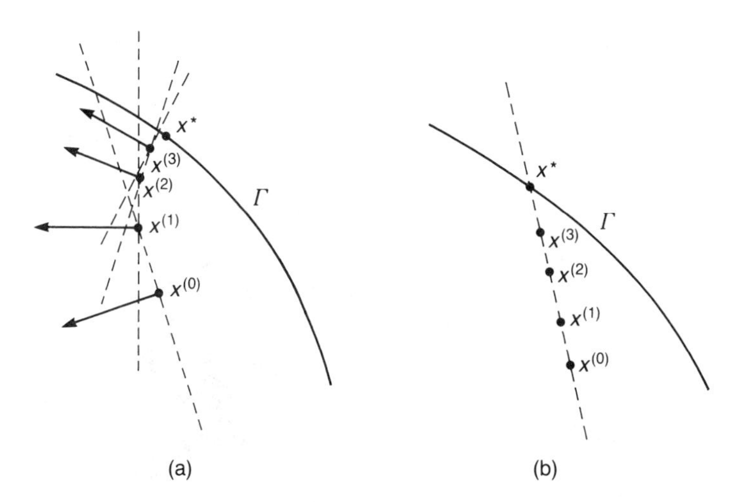

Figure 5.3: (a) The shooting method; (b) the Poincaré map method.

amount of time $T^{(i)}$. In the Poincaré map method, however, the trajectory is integrated until it intersects the cross-section in the proper direction—the integration time is not explicitly specified.

Convergence Both the shooting method and the Poincaré map method rely on the Newton-Raphson algorithm and are guaranteed to converge only when the initial guess is close to the exact solution. Furthermore, since almost every trajectory near a non-stable limit cycle diverges from the limit cycle, the convergence problem is compounded when searching for non-stable limit cycles. Both methods work best when the approximate position of the limit cycle is known. If the user has no idea where the limit cycle is, there is a chance of convergence, but it is slim.

An important advantage of the Poincaré map method is that the user does not have to specify an initial guess for the period. In the shooting method, even if $x^{(0)}$ lies on the limit cycle, a poor choice for $T^{(0)}$ may cause $\phi_{T^{(0)}}(x^{(0)})$ to lie far from $x^{(0)}$ and the shooting method could fail to converge.

Even though these methods require prior knowledge of the position of the limit cycle, there are two common situations where they are useful. The first is in an interactive environment where the investigator follows the iterations graphically and if the algorithm is not converging, can restart it with a new initial guess. The second application, presented in Chapter 8, is the generation of bifurcation diagrams. In a bifurcation diagram program, the steady-state behavior of the system is followed as a parameter of the system is varied. If, at each step, the parameter's variation is small, then, except at bifurcation points, the change in the steady state will also be small. Thus, the position of the limit cycle for the previous parameter value is an excellent estimate of the position of the limit cycle for the current parameter value.

The shooting method does not require the user to specify a cross-section Σ through which the limit cycle must pass. Initially, this may appear to be an advantage over the Poincaré map method but for our application, it is not. First, for either method, owing to convergence problems, $x^{(0)}$ must lie near the limit cycle, so some prior knowledge of the position of the limit cycle is necessary. Second, the shooting method was developed to locate a periodic solution of a flow. It does not allow one to study the behavior of the system on a specific cross-section. The Poincaré map method, however, does allow an investigator to study the limit sets of a Poincaré map directly.

Programming The shooting method is easier to program because the Poincaré map method requires P_+ which, in turn, requires the calculation of hyperplane crossings (see Section 2.4.2). Note, however, that the algorithms for detecting duplicate limit cycles and for testing the minimum period also require the location of hyperplane crossings, and if these algorithms are used, the coding advantage of the shooting method disappears.

Both methods calculate Φ_T, so the characteristic multipliers of the limit cycle may be easily determined with no additional integration.

Extraneous solutions There are two situations where the shooting method can converge to a solution that is not a limit cycle.

First, since equilibrium points satisfy (5.14), the shooting method can converge to an equilibrium point. Though this occurrence is easily detected—x^* is known to be an equilibrium point if $f(x^*) \approx 0$—it is a nuisance. The Poincaré map method will never converge to an equilibrium point except in the unlikely event that an equilibrium point lies on Σ.

Second, any point satisfies (5.14) with $T = 0$, and the shooting method can converge to this undesirable solution. Since the period is not implicitly used in the Poincaré map method, it is immune to this phenomenon.

Summary If a quick, easy method is required, use the shooting method. If a more robust routine is desired, or duplication and minimum period tests are required, the Poincaré map method is preferable.

Alternative methods

The shooting methods and the Poincaré map method are useful for finding periodic behavior in low-order systems, especially when the approximate location of the periodic solution is known.

In this section we outline two other techniques that are useful for finding periodic solutions. Like the shooting methods and the Poincaré map method, the first technique, called the finite-difference method, is a time-domain approach. The second technique, harmonic balance, performs the analysis in the frequency domain.

For simplicity, we discuss the non-autonomous case only. Both methods can be generalized to the autonomous case.

The finite-difference method Choose an integer $N > 0$. Define the time-step $h := T/N$ and the time points, $t_k := kh$ for $k = 1, \dots, N$. Let $\{x_1, \dots, x_N\}$ be a sequence of points. We wish to iterate this sequence until it approximates the periodic solution, that is, until $x_k \approx \phi_h(x_{k-1}, t_{k-1})$ for $k = 1, \dots, N$, where, to enforce periodicity, $x_0 := x_N$ and $t_0 := t_N$.

The iteration is performed by selecting an integration algorithm, using the integration formula to write an equation for each x_k in terms of f and the other x_j, and then applying the Newton-Raphson algorithm to the resulting nonlinear system of equations.

As an example, choose the trapezoidal algorithm (4.6),

$$x_{k+1} = x_k + \frac{h}{2}\{f(x_k, t_k) + f(x_{k+1}, t_{k+1})\}. \qquad (5.27)$$

Apply the trapezoidal rule to each x_k to obtain

$$\begin{aligned}
x_1 &= x_N + \frac{h}{2}\{f(x_N,t_N) + f(x_1,t_1)\} \\
x_2 &= x_1 + \frac{h}{2}\{f(x_1,t_1) + f(x_2,t_2)\} \\
&\vdots \\
x_k &= x_{k-1} + \frac{h}{2}\{f(x_{k-1},t_{k-1}) + f(x_k,t_k)\} \\
&\vdots \\
x_N &= x_{N-1} + \frac{h}{2}\{f(x_{N-1},t_{N-1}) + f(x_N,t_N)\}.
\end{aligned} \tag{5.28}$$

Subtract the left-hand side of (5.28) from both sides of (5.28). Define $H: \mathbb{R}^{nN} \to \mathbb{R}^{nN}$ as the right-hand side of the resulting system of equations. A solution of (5.28) is a zero of H, and can be found using the Newton-Raphson algorithm. The Jacobian of H is easily found to be

$$DH = \begin{bmatrix} a_{11} & 0 & \cdots & 0 & a_{1N} \\ a_{21} & a_{22} & \cdots & 0 & 0 \\ \vdots & \vdots & & \vdots & \vdots \\ 0 & 0 & \cdots & a_{N,N-1} & a_{NN} \end{bmatrix} \tag{5.29}$$

where each entry is an $n \times n$ matrix. The entries on the diagonal are

$$a_{kk} = -I + \frac{h}{2} Df(x_k, t_k) \tag{5.30}$$

and the non-zero off-diagonal entries are

$$a_{k,k-1} = I + \frac{h}{2} Df(x_{k-1}, t_{k-1}) \tag{5.31}$$

where $a_{1N} := a_{10}$.

Though DH is a large matrix, it is sparse, and the solution of $DH\Delta x = -H$ can be calculated efficiently using sparse matrix techniques (see Kundert [1986]).

In the shooting method, the trajectory is specified by the initial condition x_0, and the method iterates until the final state $\phi_T(x_0)$ matches the initial state. In the finite-difference approach, the trajectory is specified by the sequence $\{x_k\}$, and the method iterates until the sequence satisfies the integration formula. At each iteration, in other words, the shooting method satisfies the differential equation but does not satisfy periodicity; the finite-difference method satisfies periodicity but does not satisfy the differential equation.

The advantage of the finite-difference approach over the shooting method or Poincaré map method is that it is less sensitive to the initial guess and is, therefore, more suitable for finding non-stable periodic solutions. The drawback is that for large systems, it requires a large amount of memory. Also, it is not obvious how to choose the initial guess for the sequence $\{x_k\}$.

Harmonic balance The method of harmonic balance attempts to find a periodic solution by analyzing the system in the frequency domain. Let $x(t)$ be a time-periodic function with minimum period $T > 0$. $x(t)$ may be written in a Fourier series expansion as

$$x(t) = X^0 + \sum_{k=1}^{\infty} \{X^c(k)\cos(k\omega t) + X^s(k)\sin(k\omega t)\} \tag{5.32}$$

where $\omega := 2\pi/T$, and

$$\begin{aligned} X^0 &:= \frac{1}{T}\int_0^T x(t)\,dt \\ X^c(k) &:= \frac{2}{T}\int_0^T x(t)\cos(k\omega t)\,dt \\ X^s(k) &:= \frac{2}{T}\int_0^T x(t)\sin(k\omega t)\,dt. \end{aligned} \tag{5.33}$$

As an example of harmonic balance, consider the first-order system

$$\dot{x} = f(x) + u(t) \tag{5.34}$$

where $u(t)$ is a periodic forcing term with period T and Fourier coefficients $U(k)$. Let $X(k) := [X^0 \;\; X^c(k) \;\; X^s(k)]^T$ be the vector of Fourier coefficients for $x(t)$ and denote the Fourier transform operator by $\mathcal{F}$, that is, $X(k) = \mathcal{F}(x(t))$.

In the frequency domain, (5.34) becomes

$$\begin{aligned} 0 &= F^0(X(k)) + U^0 \\ \Omega X^s(k) &= F^c(X(k)) + U^c(k) \\ -\Omega X^c(k) &= F^s(X(k)) + U^s(k) \end{aligned} \tag{5.35}$$

where[6] $\Omega := \text{diag}(\omega, 2\omega, \ldots)$ and $F := \mathcal{F} \circ f \circ \mathcal{F}^{-1}$. As in (5.32), the superscript 0 denotes the constant term of the Fourier expansion, the superscript c denotes the cosine coefficients, and the superscript s denotes the sine coefficients.

Equation (5.35) is an infinite-dimensional system of equations. To solve this system on a computer, the Fourier series is truncated to K harmonics. The resulting system is

$$\begin{aligned} 0 &= F_K^0(X_K(k)) + U_K^0 \\ \Omega_K X_K^s(k) &= F_K^c(X_K(k)) + U_K^c(k) \\ -\Omega_K X_K^c(k) &= F_K^s(X_K(k)) + U_K^s(k) \end{aligned} \tag{5.36}$$

where $\Omega_K := \text{diag}(\omega, 2\omega, \ldots, K\omega)$, and $F_K := \mathcal{F}_K \circ f \circ \mathcal{F}_K^{-1}$. The subscript K indicates the truncated versions of X, U, and $\mathcal{F}$. Equation (5.36) is a system of $2K+1$ equations in $2K+1$ unknowns and, therefore, the Newton-Raphson algorithm can be used to solve for the $X(k)$. An expression for the Jacobian is derived easily. See Kundert *et al.* [1988] for details.

Since f is nonlinear, $F_K(X_K(k))$ cannot be evaluated directly in the frequency domain; instead, the calculation is performed in the time domain using the following three steps.

1. $x_K(t) = \mathcal{F}_K^{-1}(X_K(k))$.

2. $y_K(t) = f(x_K(t))$.

3. $F_K(X_K(k)) = \mathcal{F}_K(y_K(t))$.

$\mathcal{F}_K$ and $\mathcal{F}_K^{-1}$ are calculated efficiently using the fast Fourier transform and its inverse.

Harmonic balance can be viewed a frequency-domain version of the finite-difference method (see Kundert *et al.* [1987]), and it shares the same advantages over the shooting and Poincaré map methods. Furthermore, as will be seen in the next section, harmonic balance can be generalized to the case of quasi-periodic behavior.

The efficiency of harmonic balance depends on the number of harmonics that are needed to approximate the steady state with the desired accuracy. Harmonic balance is especially suited for systems

[6] $\text{diag}(a_1, \ldots, a_M)$ denotes the $M \times M$ matrix with zero entries everywhere except on the diagonal, where $a_{ii} = a_i$.

that are weakly nonlinear or whose steady state is nearly sinusoidal, since for these systems few harmonics are required.

Readers interested in a more complete discussion of harmonic balance are referred to the previously mentioned references as well as Ushida *et al.* [1988], Kundert and Sangiovanni-Vincentelli [1986], and Mees [1981].

5.6 Locating two-periodic solutions

We outline two methods for locating two-periodic behavior. The first, a generalization of finite differences, uses the Poincaré map. The second, the method of spectral balance, is a generalization of the method of harmonic balance.

Both methods are somewhat limited in generality, and the search for robust, efficient methods for locating two-periodic solutions of autonomous as well as non-autonomous systems is an active research area.

5.6.1 Finite differences

This method, due to Kevrekidis *et al.* [1985], attempts to locate an invariant closed curve of a map. Since, for a proper choice of cross-section, the limit set of a two-periodic trajectory is an invariant closed curve of the Poincaré map (see Section 2.2.3), this technique may be used to locate two-periodic behavior. The approach taken is a generalization of the finite-difference method for locating periodic solutions.

Consider a map $P: \mathbb{R}^+ \times S^1 \to \mathbb{R}^+ \times S^1$ of the plane to itself, where, for simplicity, we use polar coordinates. Let $r: S^1 \to \mathbb{R}^+$ be differentiable with a continuous derivative. Using polar coordinates, r defines a closed curve $\Gamma := \{(r(\theta), \theta) : 0 \leq \theta < 2\pi\}$ which encircles the origin (see Fig. 5.4).

Choose an integer $N \gg 1$, and define $\theta_k := 2(k-1)\pi/N$ and $r_k := r(\theta_k)$ for $k = 1, \ldots, N$. The Newton-Raphson algorithm will be used to iterate the sequence $\{r_k\}$ until it lies on an invariant closed curve.

Let Γ^* be an invariant closed curve of P, that is, $\Gamma^* = P(\Gamma^*)$. Define $(\hat{r}_k, \hat{\theta}_k) := P(r_k, \theta_k)$ for $k = 1, \ldots, N$. The sequence $\{r_k\}$ lies on Γ^* if and only if its image $\{\hat{r}_k\}$ lies on Γ^*.

To use the Newton-Raphson algorithm, the condition that $\{r_k\}$

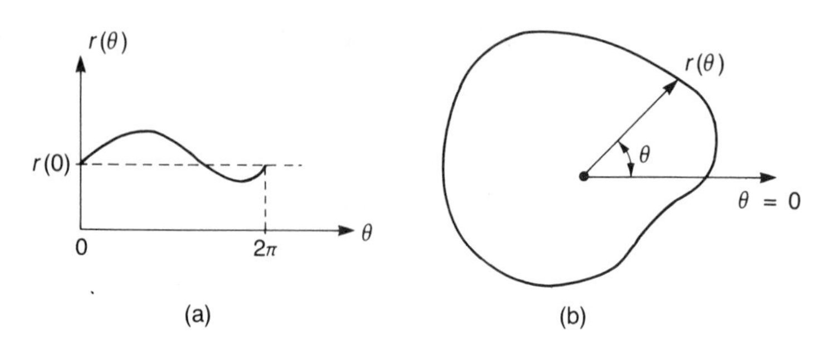

Figure 5.4: (a) The function $r: S^1 \to \mathbb{R}^+$; (b) using polar coordinates, r defines a closed curve in the plane.

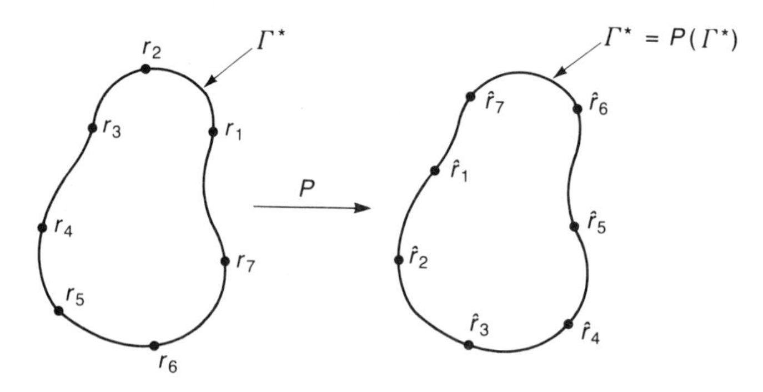

Figure 5.5: An invariant circle Γ^* of a map P. Due to the rotational effect of P, $P(x) \neq x$ for most $x \in \Gamma^*$.

and $\{\hat{r}_k\}$ lie on the same curve must be formulated in terms of the zeros of some function H. It would be nice if we could simply choose

$$H(r_1, \ldots, r_n) := \begin{bmatrix} r_1 \\ \vdots \\ r_N \end{bmatrix} - \begin{bmatrix} \hat{r}_1 \\ \vdots \\ \hat{r}_N \end{bmatrix} \tag{5.37}$$

but as Fig. 5.5 shows, P usually involves some rotation. Even if the r_k lie on Γ^*, θ_k does not equal $\hat{\theta}_k$, and, therefore, the r_k are not invariant under P.[7]

The problem is that we do not want to compare $\{r_k\}$ with $\{\hat{r}_k\}$, we want to compare the closed curves on which the sequences lie. Thus, some sort of interpolation must be used. Let $\hat{r}: S^1 \to \mathbb{R}^+$ be the radius function obtained by interpolation of $\{\hat{r}_k\}$ and $\{\hat{\theta}_k\}$, that

[7] If the r_k were invariant, they would correspond to fixed points, not an invariant closed curve.

is, $\hat{r}_k = \hat{r}(\hat{\theta}_k)$. Note that $\hat{r}$ differs from $P \circ r$ because $\hat{r}_k = P \circ r(\theta_k) = \hat{r}(\hat{\theta}_k)$.

The easiest way to check whether r and $\hat{r}$ agree is called *collocation.* Collocation considers r and $\hat{r}$ identical if they agree at the points θ_k for $k = 1, \ldots, N$. To apply Newton-Raphson, H is defined as

$$\begin{aligned} H(r_1, \ldots, r_N) &:= \begin{bmatrix} r(\theta_1) \\ \vdots \\ r(\theta_N) \end{bmatrix} - \begin{bmatrix} \hat{r}(\theta_1) \\ \vdots \\ \hat{r}(\theta_N) \end{bmatrix} \\ &= \begin{bmatrix} r_1 \\ \vdots \\ r_N \end{bmatrix} - \begin{bmatrix} \hat{r}(\theta_1) \\ \vdots \\ \hat{r}(\theta_N) \end{bmatrix}. \end{aligned} \tag{5.38}$$

The function $\hat{r}$ is based on interpolation of $\{\hat{r}_k\}$ and would be denoted more precisely by $\hat{r}(\theta; r_1, \ldots, r_N)$. H is not written as a function of θ because the θ_k are assumed fixed throughout.

Each element of the Jacobian *DH* has the form $h_{jk} = \delta_{jk} - \partial\hat{r}(\theta_j)/\partial r_k$ where δ_{jk} denotes the *Kronecker delta* defined to be 1 if $j = k$ and 0 otherwise. The exact form of the Jacobian depends on the type of interpolation used.

In most practical interpolation methods, the value for $\hat{r}(\theta_k)$ depends only on the nearby elements of $\{\hat{r}_k\}$. If, for example, linear interpolation is used, then

$$\hat{r}(\theta_j) = (1 - \alpha)\,\hat{r}_k + \alpha\,\hat{r}_{k+1} \tag{5.39}$$

where $\hat{\theta}_j < \theta_k < \hat{\theta}_{j+1}$, and $\alpha := (\theta_k - \hat{\theta}_j)/(\hat{\theta}_{j+1} - \hat{\theta}_j)$. Thus, ignoring the ones on the diagonal, *DH* is everywhere zero except for a narrow band of non-zero entries that winds through the matrix as shown in Fig. 5.6. *DH* is sparse and can be solved efficiently using sparse matrix techniques.

Upon convergence, the position of the non-zero band in *DH* is of special interest. Typically, the band lies away from the diagonal of the matrix due to the rotational effect of P. If, however, it does cross the diagonal, at row j, say, then P has a fixed point near (r_j, θ_j). This case corresponds to a saddle connection (Fig. 5.7) and does not indicate two-periodic behavior.

The approach outlined here is valid only when the function r exists. In cases where r is multi-valued or when maps of dimen-

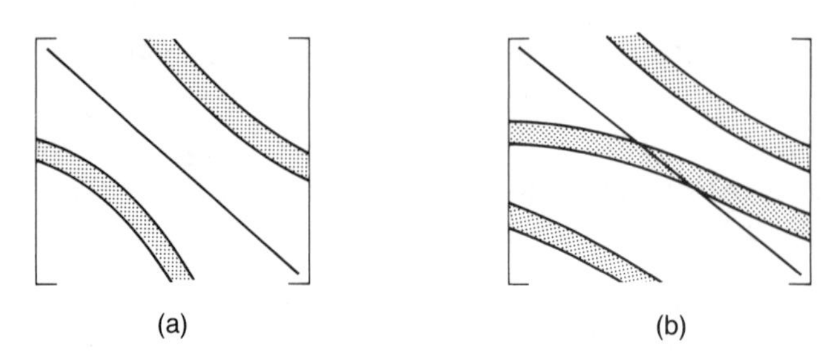

Figure 5.6: Structure of the Jacobian. (a) The Jacobian has a non-zero band that winds through the matrix; (b) if, after convergence, this band crosses the diagonal, there is a fixed point on the invariant curve.

Figure 5.7: An example of an invariant closed curve that contains a fixed point x^*.

sion greater than two are being studied, a parameterization using a variable other than θ must be used. Arc-length is a reasonable alternative.

5.6.2 Spectral balance

The method of spectral balance is a generalization of the method of harmonic balance. The version presented here can locate a two-periodic trajectory of a system that has two periodic forcing terms with incommensurate frequencies. It is easily generalized to the case of K incommensurate forcing terms.

For simplicity, consider the first-order system

$$\dot{x} = f(x) + u_1(t) + u_2(t) \tag{5.40}$$

where u_1 is periodic with period T_1, u_2 is periodic with period T_2, and T_1 and T_2 are incommensurate.

Suppose (5.40) has a two-periodic solution $x(t)$ with base frequencies $\hat{\omega}_1 = 2\pi/T_1$ and $\hat{\omega}_2 = 2\pi/T_2$. Let

$$\Lambda := \{\,|k_1\hat{\omega}_1 + k_2\hat{\omega}_2| : k_1, k_2 = 0, \pm 1, \ldots\} \tag{5.41}$$

be the countable set of frequencies corresponding to the various sums and differences of the two base frequencies. Since $x(t)$ is not periodic, it does not possess a Fourier series representation. It does, however, have a Fourier transform $\mathcal{F}(x(t))$ which is zero everywhere except at the frequencies in Λ.

Number the elements in Λ as ω_0 $(= 0)$, ω_1, ω_2, ... Define $X(k) := \mathcal{F}(x(T))$ and divide $X(k)$ into three parts: X^0 denotes the $\omega_0 = 0$ coefficient $X(0)$, $X^c(k)$ denotes the vector of cosine coefficients, and $X^s(k)$ denotes the vector of sine coefficients. With these definitions, $x(t)$ may be written as the Fourier expansion[8]

$$x(t) = X^0 + \sum_{k=1}^{\infty} \{X^c(k)\cos(\omega_k t) + X^s(k)\sin(\omega_k t)\}. \tag{5.42}$$

In terms of $X(k)$, (5.40) becomes

$$\begin{aligned} 0 &= F^0(X(k)) + U_1^0 + U_2^0 \\ \Omega X^s(k) &= F^c(X(k)) + U_1^c(k) + U_2^c(k) \\ -\Omega X^c(k) &= F^s(X(k)) + U_1^s(k) + U_2^s(k) \end{aligned} \tag{5.43}$$

where $\Omega := \text{diag}(\omega_1, \omega_2, \ldots)$, $F := \mathcal{F} \circ f \circ \mathcal{F}^{-1}$, $U_1 := \mathcal{F}(u_1)$, and $U_2 := \mathcal{F}(u_2)$. As in (5.42), the superscript 0 denotes the constant term of the Fourier expansion; the superscript c denotes the cosine coefficients; and the superscript s denotes the sine coefficients.

To solve (5.43) on a computer, the infinite series (5.42) is truncated to K frequencies. After truncation, (5.43) becomes

$$\begin{aligned} 0 &= F_K^0(X_K(k)) + U_{1K}^0 + U_{2K}^0 \\ \Omega_K X_K^s(k) &= F_K^c(X_K(k)) + U_{1K}^c(k) + U_{2K}^c(k) \\ -\Omega_K X_K^c(k) &= F_K^s(X_K(k)) + U_{1K}^s(k) + U_{2K}^s(k) \end{aligned} \tag{5.44}$$

where $\Omega_K := \text{diag}(\omega_1, \ldots, \omega_K)$, and $F_K := \mathcal{F}_K \circ f \circ \mathcal{F}_K^{-1}$. The subscript K indicates the truncated versions of X, U_1, U_2, and $\mathcal{F}$.

Equation (5.44) is a nonlinear system of $2K + 1$ equations in the $2K + 1$ unknown Fourier coefficients. If F_K can be efficiently evaluated, (5.44) may be solved using the Newton-Raphson algorithm.

[8] Equation (5.42) is not a Fourier series because the ω_k are not all harmonics of a single fundamental frequency. Thus the name *spectral* balance rather than *harmonic* balance.

As with harmonic balance, F_K is evaluated in the time domain in three steps:

1. $x_K(t) = \mathcal{F}_K^{-1} X_K(k)$.

2. $y_K(t) = f(x_K(t))$.

3. $F_K(X_K) = \mathcal{F}_K y_K(t)$.

The key to performing this procedure efficiently is the calculation of the Fourier and inverse Fourier transforms. Since $x(t)$ is not periodic, the discrete Fourier transform cannot be used.

Choose N distinct time points, $t_1, \ldots, t_N$. From (5.42),

$$\begin{bmatrix} 1 & a_{11}^c & a_{11}^s & \cdots & a_{1K}^c & a_{1K}^s \\ \vdots & \vdots & \vdots & & \vdots & \vdots \\ 1 & a_{N1}^c & a_{N1}^s & \cdots & a_{NK}^c & a_{NK}^s \end{bmatrix} \begin{bmatrix} X_K^0 \\ X_K^c(1) \\ X_K^s(1) \\ \vdots \\ X_K^c(K) \\ X_K^s(K) \end{bmatrix} = \begin{bmatrix} x(t_1) \\ \vdots \\ x(t_N) \end{bmatrix} \tag{5.45}$$

where $a_{jk}^c := \cos(\omega_k t_j)$ and $a_{jk}^s := \sin(\omega_k t_j)$.

Equation (5.45) is a system of N equations in $(2K+1)$ unknowns. Thus if $N = (2K+1)$ time-points are specified and if the matrix is non-singular, (5.45) can be solved for the Fourier coefficients. Inverting (5.45) gives $\mathcal{F}_K^{-1}$.

For the Fourier transform defined by (5.45) to be useful, the matrix must be well-conditioned. Typically, choosing evenly spaced time-points t_i results in an ill-conditioned matrix. An algorithm for choosing a set of $2K+1$ time points that gives near orthogonal columns is presented in Kundert *et al.* [1988]. A different tack is taken by Ushida and Chua [1984] who choose $N > 2K+1$ evenly-spaced time points. The resulting, over-specified system is solved by a least-squares approach.

5.7 Locating chaotic solutions

Other than brute force, there is currently no practical method for locating chaotic solutions.

5.8 Summary

- *Brute force*: Integrate or iterate the system until the steady state is achieved. Brute force is general and simple, but it cannot locate non-stable limit sets or unstable limit sets of non-invertible maps. Slow convergence and the inability to indicate convergence are additional disadvantages.

- *Locating equilibrium points*: Apply the Newton-Raphson algorithm directly to $H(x) := f(x)$.

- *Locating fixed points*: Apply the Newton-Raphson algorithm to $H(x) := P(x) - x$.

- *Locating closed orbits*: Apply the Newton-Raphson algorithm to $H(x) := P^K(x) - x$.

- *Locating periodic solutions*: Periodic solutions of non-autonomous systems can be located using the non-autonomous shooting method, finite differences, or harmonic balance. Limit cycles may be located using the autonomous shooting method, the Poincaré map method, finite differences, or harmonic balance.

- *The non-autonomous shooting method*: The non-autonomous shooting method locates a period-T solution by applying the Newton-Raphson algorithm to $H(x) := \phi_{t_0+T}(x, t_0) - x$.

- *The autonomous shooting method*: The autonomous shooting method locates a limit cycle by applying the Newton-Raphson algorithm to $H(x, T) = \phi_T(x) - x$.

- *The Poincaré map method*: The Poincaré map method locates a limit cycle by finding fixed points of the Poincaré map P_+ or its Kth iterate P_+^K.

- *Finite differences (periodic case)*: The method of finite-differences locates a periodic solution by iterating on a periodic sequence of points. The sequence represents a time-sampled version of the periodic solution. The iteration equations are found by applying the Newton-Raphson algorithm to the nonlinear system that results when an integration algorithm is applied to each point in the sequence.

- *Harmonic balance*: The method of harmonic balance represents a periodic solution by a truncated Fourier series of sines and cosines. Using this series, the differential equation is transformed into a nonlinear system of equations in the frequency domain. The nonlinear system is solved using the Newton-Raphson algorithm.

- *Locating two-periodic solutions*: Two-periodic solutions may be located using finite differences or the method of spectral balance.

- *Finite differences (two-periodic case)*: The method of finite differences locates two-periodic behavior by finding an invariant closed curve of the Poincaré map. The method iterates a periodic sequence of points until the sequence lies on an invariant closed curve. The iteration equations are obtained by applying the Newton-Raphson algorithm to a system of equations that results from interpolation of the image of the periodic sequence.

- *Spectral balance*: Spectral balance is a generalization of harmonic balance. A two-periodic solution is represented by a truncated Fourier expansion of sines and cosines. Using this expansion, the differential equation is transformed into a nonlinear system of equations in the frequency domain. The nonlinear system is solved using the Newton-Raphson algorithm.

- *Locating chaotic solutions*: There is currently no method except brute force for locating chaotic solutions.

Chapter 6

Stable and Unstable Manifolds

In this chapter, we discuss the stable and unstable manifolds of an equilibrium point and of a fixed point. Stable and unstable manifolds are a useful tool for both the theorist and the simulator of nonlinear systems. As we shall see, under the proper conditions, the structure of the manifolds indicates the presence of Smale horseshoes which, in turn, implies sensitive dependence on initial conditions.

This chapter relies on concepts from differential topology. Appendix C contains a brief review of this subject.

6.1 Definitions and theory

6.1.1 Continuous-time systems

Consider an nth-order autonomous system with flow ϕ_t.

Let L be a non-chaotic hyperbolic limit set of ϕ_t. The *stable manifold* of L, denoted $W^s(L)$, is the set of points x such that $\phi_t(x)$ approaches L as $t \to \infty$. The *unstable manifold* $W^u(L)$ is the set of points whose trajectory approaches L in reverse time.

In the terminology of limit sets, $W^s(L)$ is the set of points whose ω-limit set is L. $W^u(L)$ is the set of points whose α-limit set is L.

Remarks:

1. The stable and unstable manifolds are invariant under the flow ϕ_t, as is their intersection.

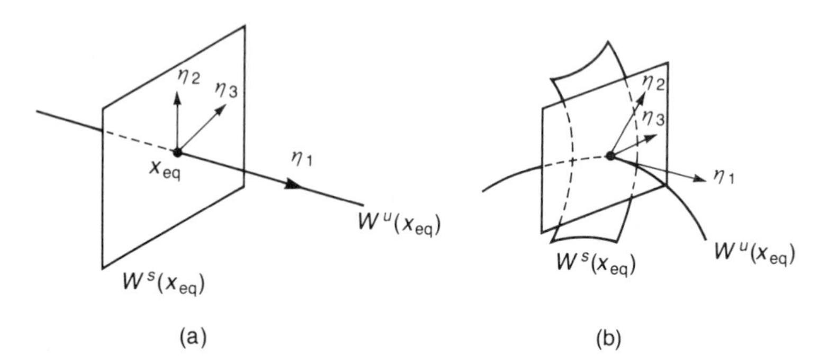

Figure 6.1: The stable and unstable manifolds of an equilibrium point in (a) a linear system; (b) a nonlinear system.

2. It will be shown in Section 6.1.2 that the stable and unstable manifolds of a limit set are not always manifolds. The terminology stable and unstable "manifolds" is not, however, totally misleading: $W^s(L)$ and $W^u(L)$ are always immersed manifolds.

As a simple example of stable and unstable manifolds of an equilibrium point, consider a third-order linear system with real eigenvalues, $\lambda_1 > 0 > \lambda_2 > \lambda_3$, and corresponding eigenvectors η_1, η_2, and η_3. The solution to this system is

$$\phi_t(x_0) = c_1 e^{\lambda_1 t}\, \eta_1 + c_2 e^{\lambda_2 t}\, \eta_2 + c_3 e^{\lambda_3 t}\, \eta_3 \tag{6.1}$$

where the $c_i \in \mathbb{R}$ are determined by the initial condition x_0. The only limit set of this system is the non-stable equilibrium point at the origin. The stable manifold of the equilibrium point is the plane spanned by η_2 and η_3 (see Fig. 6.1(a)). The unstable manifold is the line spanned by η_1. Observe that the two manifolds intersect transversally at the equilibrium point.

In the preceding example, the stable and unstable manifolds are linear subspaces of the state space. Typically, in a nonlinear system, the stable and unstable manifolds are not linear subspaces. In general, the most that can be said of their global topology is that they are immersed manifolds. Near a hyperbolic limit set, however, the geometry of the manifolds is completely determined by the linearized system.

Consider a hyperbolic equilibrium point x_{eq} of a nonlinear system. Let n_s be the number of stable eigenvalues (i.e., eigenvalues with a negative real part) and $n_u(= n - n_s)$ be the number of unstable eigenvalues of $Df(x_{eq})$. It can be proved that the stable manifold $W^s(x_{eq})$

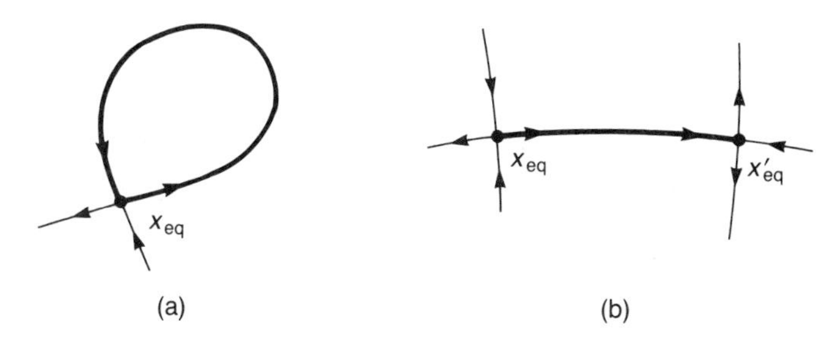

Figure 6.2: (a) A homoclinic trajectory. (b) a heteroclinic trajectory.

has dimension n_s and is tangent at x_{eq} to the linear space spanned by the real and imaginary parts of the stable eigenvectors (i.e., those eigenvectors corresponding to the stable eigenvalues). The unstable manifold has dimension n_u and is tangent at x_{eq} to the linear space spanned by the real and imaginary parts of the unstable eigenvectors. An example is shown in Fig. 6.1(b).

Homoclinic and heteroclinic trajectories The intersection of the stable and unstable manifolds of an equilibrium point is the union of all the trajectories that approach x_{eq} in both forward and reverse time. This intersection always includes one constant trajectory, namely, $\phi_t(x_{eq}) \equiv x_{eq}$. A non-constant trajectory that lies in $W^s(x_{eq}) \bigcap W^u(x_{eq})$ is called a *homoclinic trajectory.* A homoclinic trajectory of a second-order system is shown in Fig. 6.2(a).

Remark: The stable (unstable) manifold of an unstable (stable) equilibrium point is the equilibrium point itself. Thus, neither a stable nor an unstable equilibrium point can have a homoclinic trajectory.

Let x_1 and x_2 be distinct non-stable equilibrium points. A trajectory that lies in $W^s(x_1) \bigcap W^u(x_2)$ or in $W^s(x_2) \bigcap W^u(x_1)$ is called a *heteroclinic trajectory.* A heteroclinic trajectory approaches one equilibrium point as $t \to \infty$ and a different equilibrium point as $t \to -\infty$. Fig. 6.2(b) shows a heteroclinic trajectory of a second-order system.

Let $\hat{x} \neq x_{eq}$ be a point in $W^s(x_{eq}) \bigcap W^u(x_{eq})$. By invariance, the trajectory passing through x must lie in $W^s(x_{eq}) \bigcap W^u(x_{eq})$ for all t, from which it follows that $W^s(x_{eq}) \bigcap W^u(x_{eq})$ has dimension greater than or equal to one. Since

$$\dim T_{\hat{x}}(W^s(x_{eq})) + \dim T_{\hat{x}}(W^u(x_{eq})) = n \tag{6.2}$$

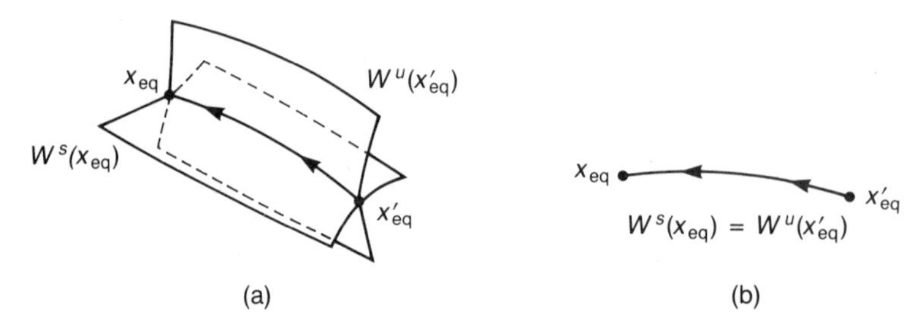

Figure 6.3: (a) A structurally stable heteroclinic trajectory; (b) A non-structurally stable heteroclinic trajectory.

and

$$\dim \left[T_{\hat{x}}(W^s(x_{eq})) \bigcap T_{\hat{x}}(W^u(x_{eq}))\right] \geq 1, \tag{6.3}$$

$T_{\hat{x}}(W^s(x_{eq}))$ and $T_{\hat{x}}(W^u(x_{eq}))$ cannot span $\mathbb{R}^n$ and, therefore, the stable and unstable manifolds of an equilibrium point are never transversal at any point other than the equilibrium point. The implication is that a homoclinic trajectory is never structurally stable.

On the other hand, consider $\hat{x} \in W^s(x_{eq}) \bigcap W^u(x'_{eq})$ where x_{eq} and x'_{eq} are distinct equilibrium points. In this case,

$$\dim T_{\hat{x}}(W^s(x_{eq})) + \dim T_{\hat{x}}(W^u(x'_{eq})) \tag{6.4}$$

can be larger than n. It follows that heteroclinic trajectories can be structurally stable though they do not have to be. Examples of structurally stable and non-structurally stable heteroclinic trajectories are shown in Fig. 6.3.

Homoclinic and heteroclinic trajectories are of interest because of Šilnikov's Theorem. In its basic form, Šilnikov's Theorem applies to a third-order system with a homoclinic trajectory at a non-stable equilibrium point x_{eq}. The equilibrium point has one real eigenvalue λ and two complex-conjugate eigenvalues, $\alpha \pm j\omega$. This situation is illustrated in Fig. 6.4.

Šilnikov's Theorem (Šilnikov [1965]) *Let ϕ_t be the flow of a third-order autonomous system with a homoclinic trajectory Γ. If $|\alpha| < |\lambda|$, then ϕ_t can be infinitesimally perturbed to ϕ'_t such that*

(i) ϕ'_t has a homoclinic trajectory Γ' near Γ;

(ii) the Poincaré map defined by a cross-section transversal to Γ' has a countable set of Smale horseshoes.

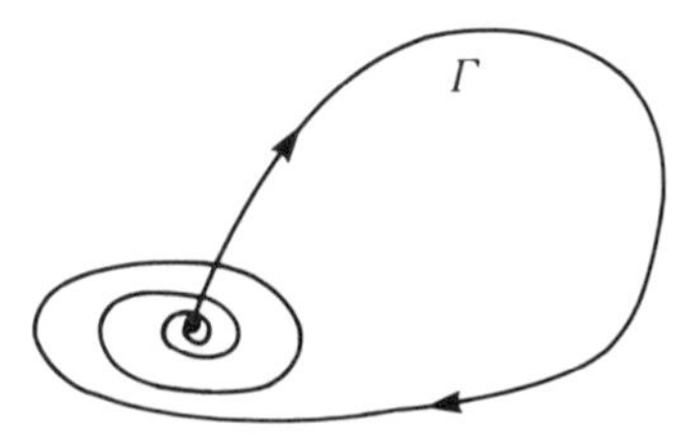

Figure 6.4: A Šilnikov-type homoclinic trajectory with $\lambda > 0$ and $\alpha < 0$.

At first glance, Šilnikov's Theorem may seem so weak as to be useless. Who cares that there is *one* perturbed flow that has horseshoes. To be useful, what is needed is something like "almost every perturbed flow has horseshoes." Fortunately, Šilnikov's Theorem is stronger than it first appears. The reason is that horseshoes are structurally stable. Thus, any sufficiently small perturbation of ϕ_t' creates a flow that also has horseshoes.

The limit set of a Smale horseshoe map is non-stable and, therefore, not attracting. It follows that the existence of a Smale horseshoe does not imply the existence of a chaotic attractor. The existence of Smale horseshoes does imply, however, that there is a region in state space that experiences sensitive dependence on initial conditions. Thus, even when there is no strange attractor in the flow, the dynamics of the system can appear to be chaotic until the steady state is achieved. Care must be taken in simulations so that a "chaotic" transient is not misinterpreted as a chaotic steady state.

We now present a brief explanation of why the Šilnikov structure implies the existence of horseshoes. Without loss of generality, assume that the equilibrium point is at the origin, and consider the case where the real eigenvalue is positive and the real part of the complex eigenvalues is negative. Change coordinates such that the linearized system has its real eigenvector along the z-axis, and the real and imaginary parts of the complex eigenvectors span the x-y plane.

Enclose the equilibrium point in a cylindrical "can" as shown in Fig. 6.5. Call the side of the can Σ_1 and the top of the can Σ_2. Let u_1 be the point of intersection of the homoclinic trajectory and Σ_1, and u_2 be the point of intersection of the homoclinic trajectory and Σ_2.

The homoclinic trajectory connects u_1 and u_2, so by Theorem D.1 there exists a diffeomorphism P_2 that maps an open neighborhood $U_2 \subset \Sigma_2$ of u_2 onto an open neighborhood $U_1 \subset \Sigma_1$ of u_1.

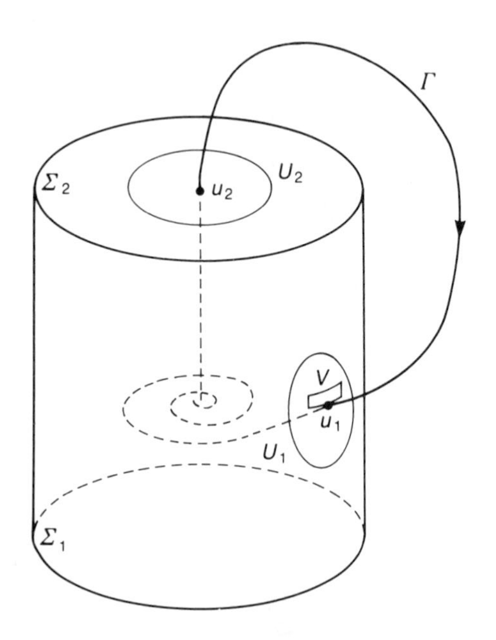

Figure 6.5: The poincare map $P = P_2 \circ P_1$.

Let $V \subset U_1$ be a small rectangular open region with u_1 on its lower boundary (see Fig. 6.5). We will argue that there exists a (Poincaré) map, $P: V \to U_1$, and that P contains a horseshoe structure.

P is composed of two maps: $P_1: V \to U_2$, and $P_2: U_2 \to U_1$. If the cylindrical can is chosen small enough, the behavior of P_1 is determined solely by the linearization of the system at the origin. The complex eigenvalues cause the x-y components of a trajectory to spiral in toward the z-axis, and the real eigenvalue causes the z-component to grow. Thus, trajectories emanating from V eventually intersect Σ_2. Furthermore, a point $u \in V$ that is closer to the x-y plane takes a longer time to reach Σ_2 and, therefore, $P_1(u) \to u_2$ as u approaches the x-y plane. It follows that $P_1(V)$ is a spiral-like region that wraps around u_2. If V is chosen small enough, $P_1(V)$ lies within U_2.

The map P_2 is a diffeomorphism and, therefore, $P_2 \circ P_1(V)$ retains its spiral nature. It follows that $P(V) \subset U_1$ is a spiral-like region with the spiral centered at u_1 as is shown in Fig. 6.6.

V undergoes stretching in the vertical direction (due to $\lambda > 0$) and contraction in the horizontal direction (due to $\sigma < 0$). Thus, it is not surprising that horseshoes are embedded in P. In fact, a

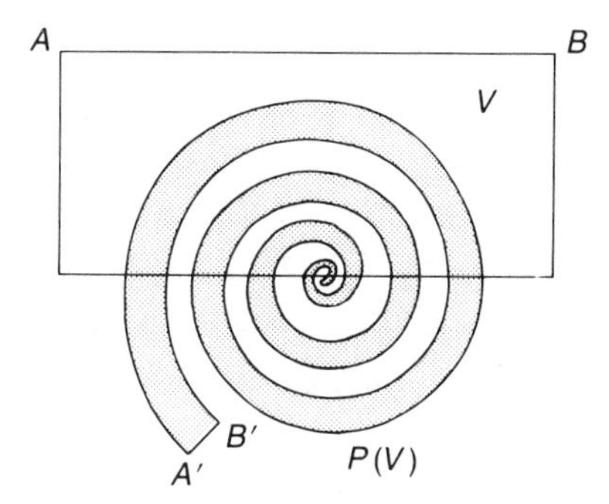

Figure 6.6: $P(V)$ shown superimposed on V. The points, A and B, map to A' and B'.

horizontal strip V' of V can be chosen such that V' and $P(V')$ have a horseshoe structure as in Fig. 6.7.

The full proof of Šilnikov's Theorem establishes that all the neighborhoods can be chosen as we have indicated and that the horseshoe structure of Fig. 6.7 is actually a Smale horseshoe.

Šilnikov's theorem can be generalized to other types of systems. For example, Chua *et al.* [1986] prove the existence of horseshoes in the double scroll equation by proving the existence of a *heteroclinic* trajectory. Mees and Sparrow [1987] present an overview of several generalizations of Šilnikov's Theorem.

6.1.2 Discrete-time systems

Consider a diffeomorphism $P: \mathbb{R}^p \rightarrow \mathbb{R}^p$.

Let L be a non-chaotic hyperbolic limit set of P. The *stable manifold of* L, denoted by $W^s(L)$, is defined as the set of all points x such

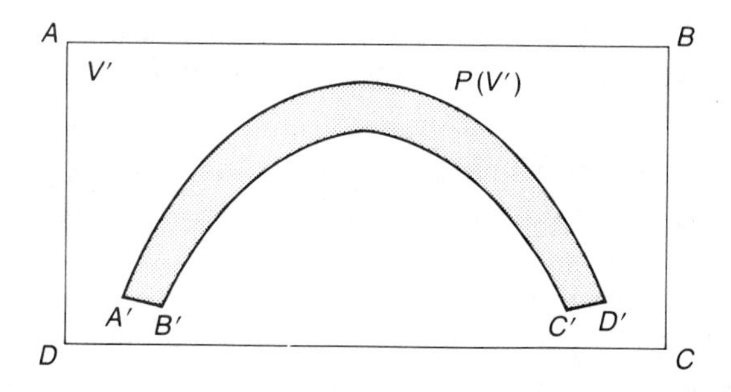

Figure 6.7: A horseshoe structure embedded in P.

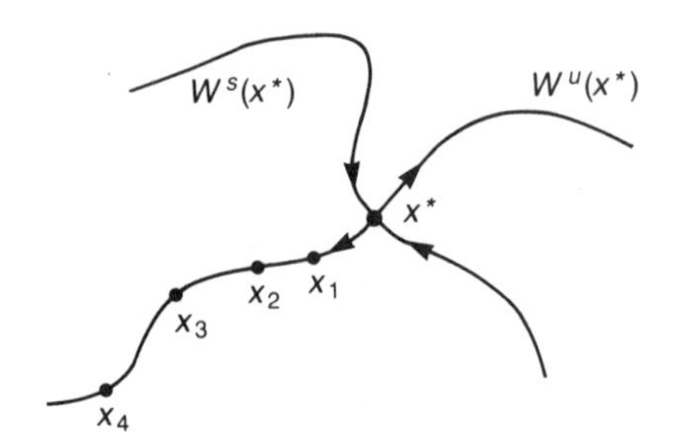

Figure 6.8: The stable and unstable manifolds of a fixed point. A typical orbit on the unstable manifold is also shown.

that $P^k(x)$ approaches L as $k \to \infty$. The *unstable manifold* $W^u(L)$ is defined as the set of points that approach L as $k \to -\infty$.

Remarks:

1. The stable and unstable manifolds are invariant under P, as is their intersection.
2. As in the continuous-time case, the stable and unstable manifolds are not guaranteed to be manifolds though they are always immersed manifolds.

Consider a hyperbolic fixed point x^* of P. Just as the local geometry of the stable and unstable manifolds of an equilibrium point is determined by $Df(x_{eq})$, the structure of the stable and unstable manifolds near x^* is determined by $DP(x^*)$. Let $DP(x^*)$ have n_s stable eigenvalues (i.e., eigenvalues with magnitude less than 1), and $n_u\,(= p - n_s)$ unstable eigenvalues. It can be shown that $W^s(x^*)$ has dimension n_s and is tangent to the span of the real and imaginary parts of the stable eigenvectors (i.e., those eigenvectors corresponding to the stable eigenvalues) and that $W^u(x^*)$ has dimension n_u and is tangent to the span of the real and imaginary parts of the unstable eigenvectors. The one-dimensional manifolds of a fixed point of a second-order system are shown in Fig. 6.8.

A word of caution is in order. The manifolds in Fig. 6.8 look like trajectories, and it is tempting to think of them as such. They are not. Suppose that P is a Poincaré map and that the non-stable fixed point x^* corresponds to a non-stable periodic solution Γ of the underlying flow. The stable and unstable manifolds of Γ are surfaces as shown in Fig. 6.9. $W^s(x^*)$ and $W^u(x^*)$ are cross-sections of the stable and unstable manifolds of Γ. Thus, the manifolds in Fig. 6.8

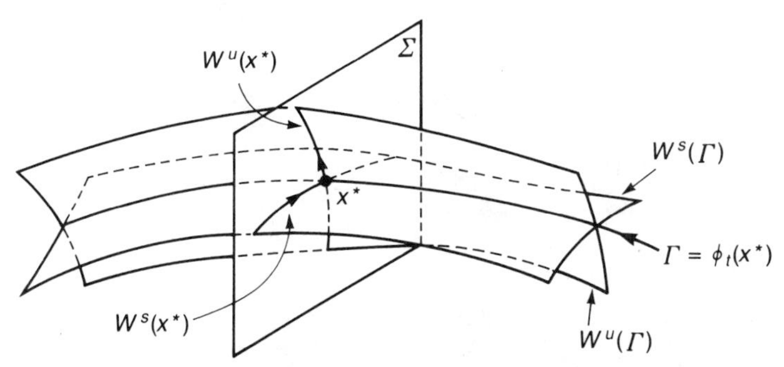

Figure 6.9: The stable and unstable manifolds of a fixed point of a Poincaré map correspond to cross-sections of the stable and unstable manifolds of the underlying periodic solutionΓ.

are not trajectories nor are they single orbits; they are the union of an infinite number of distinct orbits.

Homoclinic orbits Let x^* be a hyperbolic non-stable fixed point of a diffeomorphism $P: \mathbb{R}^p \to \mathbb{R}^p$. Suppose $W^s(x^*)$ and $W^u(x^*)$ intersect at a point $x_0 \neq x^*$. Let $\{x_k\}_{k=-\infty}^{\infty}$ be the orbit through x_0. $\{x_k\}$ is called a *homoclinic orbit* and each x_k is called a *homoclinic point.* Since x_0 lies in both the stable and unstable manifolds, so does the homoclinic orbit $\{x_k\}$. It follows that if the stable and unstable manifolds intersect at a point other than x^*, then they intersect at an infinite number of points.

Fig. 6.10(a) shows a two-dimensional map whose stable and unstable manifolds are identical. This structure is called a *homoclinic connection.* The manifolds do not intersect transversally and, there-

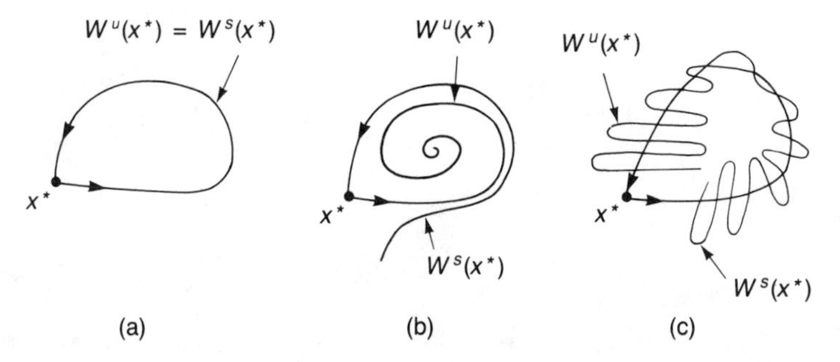

Figure 6.10: (a) A homoclinic connection; (b) the homoclinic connection is destroyed by almost any perturbation; (c) an intersection with transversal homoclinic points.

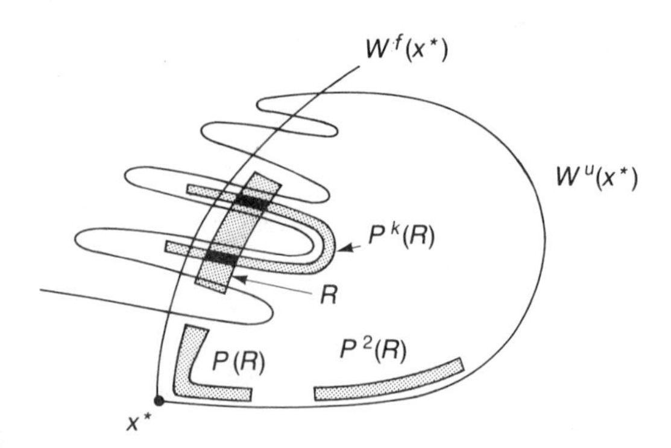

Figure 6.11: When transversal homoclinic points exist, Smale horseshoes are embedded in the map.

fore, almost any perturbation destroys the connection. Fig. 6.10(b) shows the stable and unstable manifolds resulting from a typical perturbation.

Much more complicated intersections can and do occur. Consider the transversal intersection of the manifolds shown in Fig. 6.10(c). When the manifolds intersect transversally, the homoclinic points are called *transversal homoclinic points* and the associated orbits are called *transversal homoclinic orbits.*

The fact that one transversal homoclinic point implies an infinity of homoclinic points leads to the extreme stretching and folding of the manifolds shown in Fig. 6.10(c). Stretching and folding are key mechanisms for chaos and, indeed, the existence of a transversal homoclinic point implies the existence of a horseshoe-like map embedded in P. This result is proved in the Smale-Birkhoff Homoclinic Theorem. After we present the Theorem, we give a short plausibility argument for the two-dimensional case.

The Smale-Birkhoff Homoclinic Theorem (Guckenheimer and Holmes [1983]) *Let $P\colon \mathbb{R}^p \to \mathbb{R}^p$ be a diffeomorphism with a hyperbolic fixed point x^*. If $W^s(x^*)$ and $W^u(x^*)$ intersect transversally at a point other than x^*, then P has an embedded horseshoe-like map.*

Plausibility argument: Consider a fixed point x^* of a two-dimensional map P with transversally intersecting manifolds as shown in Fig. 6.11. Choose a thin rectangular region R close to the stable manifold. We will follow R under repeated application of the map P. The first several applications of P move R toward the fixed point. As the region nears x^*, it contracts in the direction of the stable man-

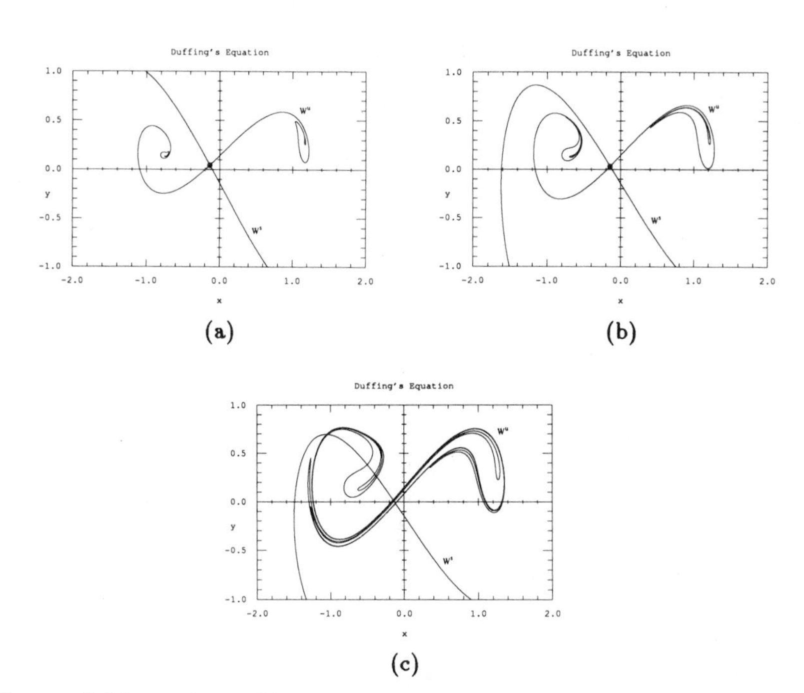

Figure 6.12: The stable and unstable manifolds for Duffing's equation with $\gamma = 0.3$, $\omega = 1.0$, and (a) $\epsilon = 0.65$; (b) $\epsilon = 0.45$; (c) $\epsilon = 0.25$.

ifold and expands in the direction of the unstable manifold. Under repeated applications of P, the region travels away from the fixed point. The iterates of R stay near the unstable manifold and for some integer K, $P^K(R)$ intersects R in a horseshoe-like manner.

In light of the Smale-Birkhoff Theorem, many researchers distinguish strange attractors from non-strange attractors by requiring that a strange attractor have a transversal homoclinic point.

Fig. 6.12 shows the stable and unstable manifolds of a fixed point of the Poincaré map associated with Duffing's equation (1.13) as a parameter is changed. In Figs. 6.12(a,b) there are no homoclinic points and the system is not chaotic for these parameter values. In Fig. 6.12(c), the stable and unstable manifolds contain transversal homoclinic points and the system exhibits chaotic steady-state behavior. Notice the similarity between the unstable manifold of Fig. 6.12(c) and the attracting set shown in Fig. 2.7(a). For this system, there is obviously an intimate relationship between the attractor and $W^u(x^*)$. It has been hypothesized that the strange attractor of this system is the closure of the unstable manifold.

We now show that the stable and unstable manifolds are not actually manifolds when they contain a transversal homoclinic point. Let $\{x_k\}$ be a homoclinic orbit. Since $x_k \to x^*$ as $k \to \infty$, given any neighborhood U of x^*, there exists a K such that $x_k \in U$ for $k > K$. At each x_k, the unstable manifold intersects the stable manifold transversally. It follows that there exists no neighborhood of x^* such that $W^u(x^*) \bigcap U$ is diffeomorphic to an open subset of Euclidean space. Thus, $W^u(x^*)$ is not a manifold. Analogous reasoning shows that $W^s(x^*)$ is not a manifold, either. $W^s(x^*)$ and $W^u(x^*)$ are, however, immersed manifolds.

6.2 Algorithms

In this section, we present algorithms for locating one-dimensional stable and unstable manifolds of continuous-time and discrete-time systems. Finding higher-dimensional manifolds is an open problem.

For a two-dimensional system, the stable and unstable manifolds of a saddle point are both one-dimensional and, therefore, the algorithms of this section are most commonly used to find both the stable and unstable manifolds of a second-order system. The algorithms, however, possess no inherent limit to the order of the system, and there are situations—the homoclinic trajectory in Šilnikov's Theorem is one—where one might want to apply the algorithms to higher-order systems.

As usual, we present the continuous-time case first, followed by the discrete-time case.

6.2.1 Continuous-time systems

Consider a hyperbolic equilibrium point x_{eq} of an nth-order continuous-time system $\dot{x} = f(x)$. Let $Df(x_{eq})$ have one positive eigenvalue λ_u, with corresponding eigenvalue η_u, and let the remaining $n-1$ eigenvalues have negative real parts.

For such a system, the unstable manifold $W^u(x_{eq})$ is a one-dimensional manifold. Remove x_{eq} from the unstable manifold to form two half-manifolds, $W^{u+}(x_{eq})$ and $W^{u-}(x_{eq})$. Each half-manifold is itself an invariant manifold consisting of a single trajectory: $W^{u+}(x_{eq})$ approaches x_{eq} along η_u, and $W^{u-}(x_{eq})$ approaches x_{eq} along $-\eta_u$ (see Fig. 6.13).

Let x_0 be a point on $W^{u+}(x_{eq})$. Integration from x_0 in both

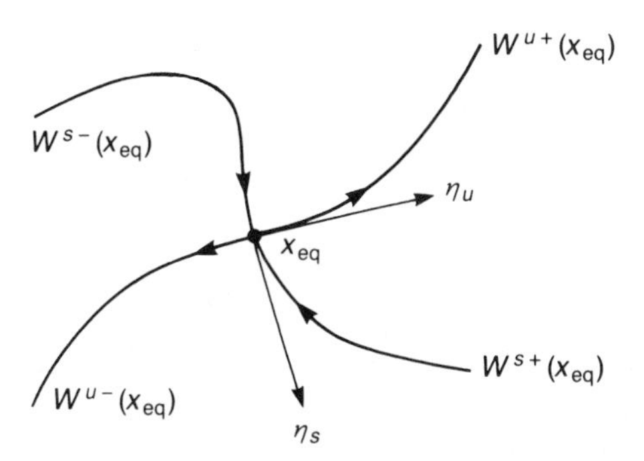

Figure 6.13: The half-manifolds at an equilibrium point.

forward and reverse time gives $W^{u+}(x_{eq})$. Thus, calculating a half-manifold can be broken into two tasks: locating a point on $W^{u+}(x_{eq})$ and integrating from this point.

Locating a point on $\mathbf{W^{u+}(x_{eq})}$ Near the equilibrium point, $W^{u+}(x_{eq})$ is given, to first order, by the unstable eigenvector η_u. Therefore, choose a point close to x_{eq} that lies on η_u. More specifically, for some small $\alpha > 0$, choose the point

$$x_\alpha := x_{eq} + \alpha\eta_u. \tag{6.5}$$

Care should be taken when choosing α. If α is too large, x_α might not lie close enough to $W^{u+}(x_{eq})$ in which case the trajectory through x_α is not a good approximation to $W^{u+}(x_{eq})$. If α is too small, two problems can occur. First, the trajectory may spend a large amount of time near x_{eq} causing the integration error to accumulate with little motion along the trajectory. Second, as Fig. 6.14 shows, a value of α that is of the same order of magnitude as the error in x_{eq} can cause x_α to lie relatively far from $W^{u+}(x_{eq})$, maybe even on $W^{u-}(x_{eq})$.

Automatic selection of $\boldsymbol{\alpha}$ Near x_{eq}, ϕ_t is approximated by the linearized flow

$$\phi_t^L(x) := x_{eq} + e^{Df(x_{eq})t}x. \tag{6.6}$$

The best value of α is the largest value such that (6.6) is accurate at x_α. More precisely, choose some $T > 0$, and define $P(x) := \phi_T(x)$ and $P_L(x) := \phi_T^L(x)$. α is chosen iteratively as follows. Set α to some initial value and keep halving α until $P(x_\alpha)$ and $P_L(x_\alpha)$ satisfy

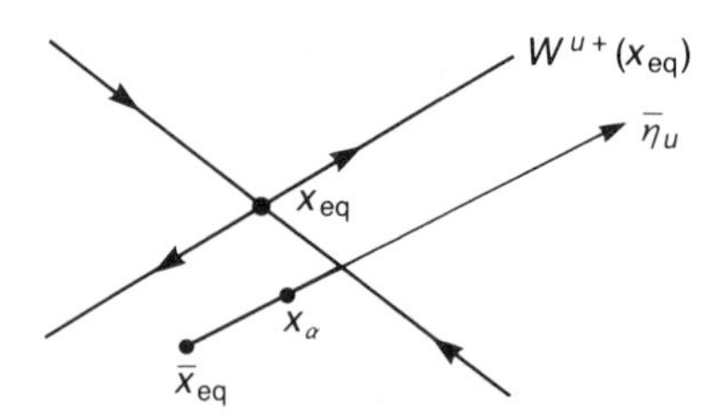

Figure 6.14: α may be a poor choice when it is on the same order of magnitude as the error in the position of x_{eq}. $\bar{x}_{eq}$ is the approximate position of the equilibrium point x_{eq}.

a relative/absolute test. Pseudo-code is presented in Fig. 6.15 for a routine **find_alpha** that implements the α-selection algorithm.

Remarks:

1. Since x_α lies in the unstable eigenspace of the linearized system,

$$P_L(x_\alpha) = x_{eq} + \alpha e^{\lambda_u T} \eta_u \tag{6.7}$$

2. T is chosen to be several time constants, that is, $T = M/\lambda_u$ for a small integer $M > 0$. With this value of T, (6.7) becomes

$$P_L(x_\alpha) = x_{eq} + \alpha e^{M} \eta_u. \tag{6.8}$$

3. Typically $\alpha_{max} = 1.0$. To avoid the situation of Fig. 6.14, choose α_{min} such that

$$||x_{eq} - x_{\alpha_{min}}|| \gg \hat{E}_r ||x_{eq}|| + \hat{E}_a \tag{6.9}$$

where $\hat{E}_r$ and $\hat{E}_a$ are the tolerances used in finding the equilibrium point x_{eq}. Solve (6.9) for α_{min} to obtain

$$\alpha_{min} \gg \hat{E}_r \frac{||x_{eq}||}{||\eta_u||} + \hat{E}_a \frac{1}{||\eta_u||}. \tag{6.10}$$

4. There is a machine accuracy constraint on α_{min}. Roughly speaking, if x_{eq} and η_u have comparable magnitudes, $-\log_{10} \alpha$ digits of accuracy are lost when forming $x_{eq} + \alpha\eta_u$. α_{min} should be chosen to ensure that machine accuracy limitations are not reached.

```
begin find_alpha(x_eq[], M, η_u[])
      choose E_r, E_a, α_min, α_max
      set α = 2.0 α_max
      repeat
            set α = α/2.0
            if (α ≤ α_min) then
               return α_min
            endif
            set x[] = x_eq[] + α η_u[]
            set px[] = P(x[])
            set p_L x[] = x_eq[] + α e^M η_u[]
      until (|| px[] − p_L x[] || < E_r ||px[]|| + E_a)
      return α
end find_alpha
```

Figure 6.15: Pseudo-code for find_alpha.

5. We have found from experience that for λ_u small, the unstable manifold possesses a fairly high degree of curvature, and, therefore, a relatively small value of α is selected. If, on the other hand, λ_u is large, then $W^u(x_{eq})$ is virtually straight near x_{eq} and find_alpha returns a larger value of α.

Integrating $\mathbf{W^{u+}(x_{eq})}$ Removal of x_α from $W^{u+}(x_{eq})$ splits the half-manifold into two segments. The larger segment, the one away from x_{eq}, is found by integrating from x_α in forward time. Owing to the definition of x_α the smaller segment, the one that approaches x_{eq}, is the line segment connecting x_{eq} and x_α. Pseudo-code for a routine find_half_manifold is presented in Fig. 6.16.

```
begin find_half_manifold(x_eq[], η_u[], h)
      choose M, N
      set α = find_alpha(x_eq[], M, η_u[])
      set x_α[] = x_eq[] + α η_u[]
      output x_eq[]
      output x_α[]
      for k from 1 to N
          output φ_kh(x_α[])
      endfor
end find_half_manifold
```

Figure 6.16: Pseudo-code for find_half_manifold.

Examples of manifolds found using this algorithm are presented in Chapter 10.

Remarks:

1. $W^{u+}(x_{eq})$ can be found by calling `find_half_manifold(`x_{eq}`[],` $-\eta_u$`[])`.
2. `find_half_manifold` can be used to find one-dimensional stable manifolds by running the system in reverse time.
3. For simplicity, `find_half_manifold` quits when a given number of points are found. In a real program, a more sophisticated control scheme should be used. For instance, `find_half_manifold` could stop when the manifold grows too long, when the manifold approaches a stable equilibrium point, or when some time-out criterion such as maximum number of steps is reached. The control scheme presented in Section 10.1.2 can also be used in this situation.

The question of how long the algorithm can run before its output becomes unreliable is a difficult one, and more work needs to be done in this area. A common technique for determining the reliability of the solution is to calculate the half-manifold using one set of tolerance values and then recalculate it using a tighter set of tolerances. The portion of the manifold where the two simulations disagree is known to be inaccurate (at least for the looser tolerances). Unfortunately, this technique does not guarantee that the portion of the manifold where the two simulations agree is accurate, so this approach should be used with care.

6.2.2 Discrete-time systems

The basic idea for calculating one-dimensional manifolds of a map is similar to that used for a flow; however, the fact that a one-dimensional manifold of a map does not correspond to a single trajectory makes the discrete-time case more complicated.

Consider a diffeomorphism $P: \mathbb{R}^p \to \mathbb{R}^p$ with a hyperbolic fixed point x^*. Let $DP(x^*)$ have one unstable eigenvalue m_u, $|m_u| > 1$, with corresponding eigenvector η_u, and let the remaining $p-1$ eigenvalues have magnitude less than 1. For such a system, $W^u(x^*)$ is an (immersed) one-dimensional manifold.

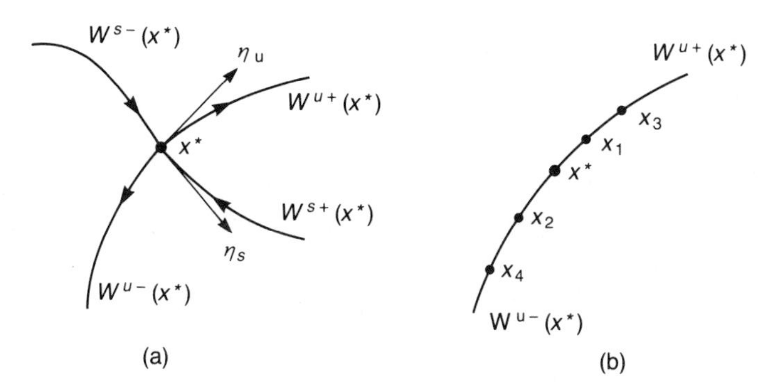

Figure 6.17: (a) Half-manifolds of a map. (b) a typical orbit on the unstable manifold when $m_u < -1$.

Remove x^* from $W^u(x^*)$ to form two half-manifolds as shown in Fig. 6.17(a). $W^{u+}(x^*)$ approaches x^* along η_u, and $W^{u-}(x^*)$ approaches x^* along $-\eta_u$. Observe that unlike the continuous-time case, the half-manifolds for discrete-time systems are not always invariant. In particular, when m_u is negative, a non-constant orbit on the unstable manifold bounces back and forth between $W^{u+}(x^*)$ and $W^{u-}(x^*)$ (see Fig. 6.17(b)).

In the following, we assume $m_u > 1$. Let $\bar{x} \in W^{u+}(x^*)$ be a point on the half-manifold, and let $\overline{W} \subset W^{u+}(x^*)$ be the set of points on the half-manifold that lie between $\bar{x}$ and $P(\bar{x})$. In other words, $\overline{W}$ is a subset of the half-manifold that is one iterate long. Then, under the assumption that $m_u > 1$, $W^{u+}(x^*)$ is the union

$$\bigcup_{i=-\infty}^{\infty} P^i(\overline{W}), \tag{6.11}$$

of all the forward and reverse iterates of the set $\overline{W}$ (see Fig. 6.18).

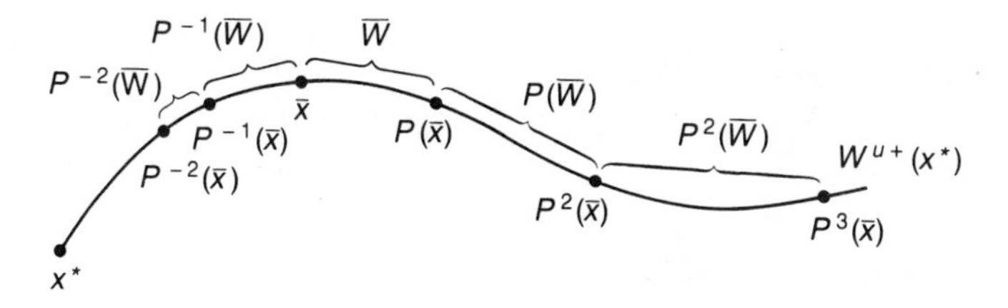

Figure 6.18: The forward and reverse iterates of $\overline{W}$ determine $W^{u+}(x^*)$.

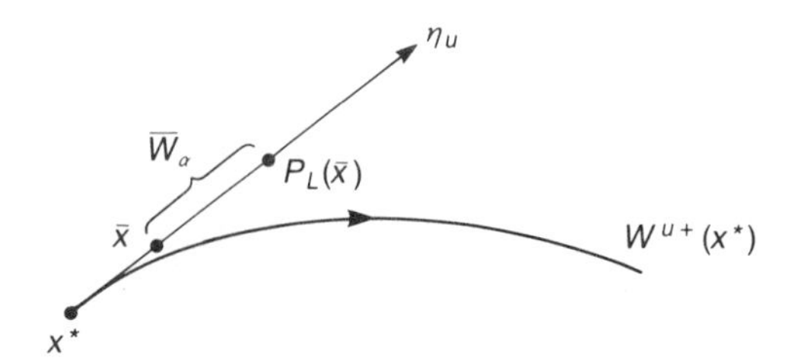

Figure 6.19: The definition of $\overline{W}_\alpha$.

Thus, calculating $W^{u+}(x^*)$ is reduced to two tasks: finding $\overline{W}$ and iterating it.

A suitable $\overline{W}$ can be found by choosing an $\bar{x}$ that is close to x^* and that lies on η_u

$$\bar{x} = x^* + \alpha\eta_u \tag{6.12}$$

for some small $\alpha > 0$. Near x^*, $P(x)$ is approximated by its linearization

$$P_L(x) := x^* + DP(x^*)(x - x^*). \tag{6.13}$$

It follows that $P(\bar{x}) \approx x^* + \alpha m_u \eta_u$, and $\overline{W}$ is approximated by the line segment $\overline{W}_\alpha$ connecting $\bar{x}$ and $P_L(\bar{x})$ (see Fig. 6.19).

With a few changes, the routine `find_alpha` which was presented in Fig. 6.15 for the continuous-time case can be modified to select a suitable value of α for the discrete-time case. First, change x_{eq} everywhere to x^*. Second, replace the factor e^M by the unstable eigenvalue m_u. And, of course, the map P now refers to the diffeomorphism under study.

We now present two algorithms for calculating $W^{u+}(x^*)$ assuming that α is known.

The standard algorithm Take N uniformly spaced points on $\overline{W}$,

$$\bar{x}_i := x^* + \left[1 + \frac{m_u - 1}{N}(i-1)\right]\alpha\eta_u, \qquad i = 1, \ldots, N, \tag{6.14}$$

where N is some large integer, typically around 100. Repeatedly apply P to the N points to obtain a set of points that lies approximately on $W^{u+}(x^*)$. If N is chosen large enough, then the $\bar{x}_i$ are closely spaced and give a good indication of the unstable half-manifold.

This method is simple to understand and to program, but possesses one major flaw. Even though $\bar{x}_i$ and $\bar{x}_{i+1}$ lie near one another, their iterates, $P^j(\bar{x}_i)$ and $P^j(\bar{x}_{i+1})$, can be quite far apart, resulting in a poor approximation to $W^{u+}(x^*)$. This stretching is especially common in chaotic systems.

Suppose that for an accurate estimation of $W^{u+}(x^*)$, it is required that none of the distances $\|P^j(\bar{x}_{i+1}) - P^j(\bar{x}_i)\|$ exceed $\epsilon > 0$. Further suppose that this constraint is satisfied for all $j < j'$ but not for $j = j'$. With the standard algorithm, the only way to achieve the desired tolerance for $j = j'$ is to increase N to some value $N' > N$. This increase, however, leads to increased computation; $N' - N$ additional points must be carried through the first $j' - 1$ iterations even though the extra points are needed only at the j'th iteration.

A better algorithm We now introduce an algorithm that overcomes this objection. Much like a variable step-size integration routine, this algorithm varies N from iteration to iteration. Pseudo-code for the routine `half_manifold` is presented in Fig. 6.20, and the supporting routines are presented in Figs. 6.21 and 6.22.

`half_manifold` produces a sequence of points $\{y_i\}_{i=1}^{n_y}$ that lies on $W^{u+}(x^*)$ such that y_{i+1} is close to y_i and farther along the half-manifold than y_i (in the sense of arc-length measured from x^*) for $1 \leq i < n_y$.

Remarks:

1. $W^{u-}(x^*)$ can be found by calling `half_manifold(`x^*`[]`, m_u, $-\eta_u$`[])`.
2. If $m_u < -1$, $W^u(x^*)$ can be found by applying `half_manifold` to the twice-iterated map $\hat{P} := P \circ P$. This method treats x^* as a period-two closed orbit of P with unstable eigenvalue $\hat{m}_u = m_u^2 > 1$. Using the technique presented on page 164, $W^{u+}(x^*)$ and $W^{u-}(x^*)$ can be calculated simultaneously.
3. One-dimensional stable manifolds can be calculated by running the system in reverse time. Of course, this only works if P^{-1} is available. For a Poincaré map, P^{-1} is obtained by integrating the system in reverse time. If P is not a Poincaré map, if it is represented, for instance, by discrete-time state equations or is obtained by interpolating experimentally obtained data, then P^{-1} may be impossible to calculate and might not even exist.

The main routine The algorithm uses two two-dimensional arrays, x[][] and px[][]. x[][] contains n_x n-vectors and defines a

```
begin half_manifold(x*[], m_u, η_u[])
      choose E_r, E_a, n_pts
      call initialize(x*[], m_u, η_u[])
      set i = 1
      while (i < n_pts) do
            if (n_px < 2) then
               call interpolate
            endif
            set l = ⟨px[1][], px[1][]⟩
            set d = ⟨px[2][] − px[1][], px[2][] − px[1][]⟩
            if (d < E_r l + E_a) then
               call accept
               set i = i + 1
            else
               call interpolate
            endif
      endwhile
end half_manifold

begin initialize(x*[], m_u, η_u[])
      set α = find_alpha(x*[], m_u, η_u[])
      set x̄[] = x*[] + αη_u[]
      call insert_x_entry(1, x̄[])
      call insert_px_entry(1)
      call insert_x_entry(2, px[1][])
      call insert_px_entry(2)
end initialize
```

Figure 6.20: Pseudo-code for `half_manifold` and `initialize`.

portion of $W^{u+}(x^*)$ that is one iteration long, that is, $x[n_x][\,] = P(x[1][\,])$. $x[\,][\,]$ can be thought of as a window that slides along $W^{u+}(x^*)$ as the algorithm progresses. $px[\,][\,]$ contains n_{px} n-vectors where $n_{px} \le n_x$. $px[\,][\,]$ is P of the first n_{px} entries of $x[\,][\,]$, that is, $px[i][\,] = P(x[i][\,])$ for $i = 1, \ldots, n_x$.

`half_manifold` initializes the arrays and then enters a loop. At the beginning of the loop, `half_manifold` checks whether $px[2][\,]$ has been calculated[1] and, if not, calls `insert_px_entry` to calculate $px[2][\,]$. `half_manifold` then uses a relative/absolute test to check whether $px[2][\,]$ is near $px[1][\,]$. If so, the step is accepted. If not, an interpolation routine is called to increase the resolution.

[1] $px[1][\,]$ is guaranteed to exist.

```
begin accept
      call reduce
      output px[1][]
      call delete_x_entry(1)
      call insert_x_entry(n_x + 1, px[1][])
      call delete_px_entry(1)
end accept

begin interpolate
      set x_new[] = (x[1][] + x[2][])/2.0
      call insert_x_entry(2, x_new[])
      call insert_px_entry(2)
end interpolate

begin reduce
      choose Ê_r and Ê_a
      if (n_x < 3) then
         return
      endif
      while (n_px < 3)
            call insert_px_entry(n_px + 1)
      endwhile
      set x_1[] = px[1][] - px[2][]
      set x_3[] = px[3][] - px[2][]
      set l_1 = ⟨px[1][], px[1][]⟩
      set d_1 = ⟨x_1[], x_1[]⟩
      set d_3 = ⟨x_3[], x_3[]⟩
      if (d_1 + d_3 < E_r l_1 + E_a) then
         set r = ⟨x_1[], x_3[]⟩
         if (r < 0 and r^2 + l_1^2 l_3^2 < Ê_r r^2 + Ê_a) then
            call delete_x_entry(2)
            call delete_px_entry(2)
            call reduce
         endif
      endif
end reduce
```

Figure 6.21: Pseudo-code for **accept**, **interpolate**, and **reduce**.

```
begin insert_x_entry(k, v[])
      set n_x = n_x + 1
      set i_max = n_x - k
      for i from 1 to i_max
          set x[n_x - i][] = x[n_x - i - 1][]
      endfor
      set x[k][] = v[]
end insert_x_entry

begin insert_px_entry(k)
      set n_px = n_px + 1
      set i_max = n_px - k
      for i from 1 to i_max
          set px[n_px - i + 1][] = px[n_px - i][]
      endfor
      set px[k][] = P(x[k][])
end insert_px_entry

begin delete_x_entry(k)
      set n_x = n_x - 1
      for i from k to n_x
          set x[i][] = x[i + 1][]
      endfor
end delete_x_entry

begin delete_px_entry(k)
      set n_px = n_px - 1
      for i from k to n_px
          set x[i][] = x[i + 1][]
      endfor
end delete_x_entry
```

Figure 6.22: Pseudo-code for `insert_x_entry`, `insert_px_entry`, `delete_x_entry`, and `delete_px_entry`.

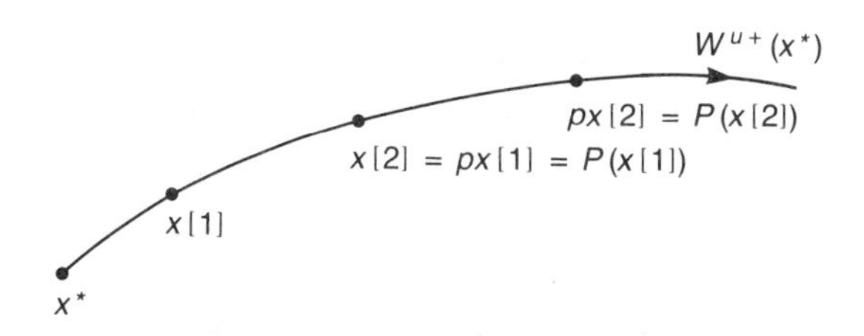

Figure 6.23: The initialization process.

Initialization `initialize` first calls `find_alpha` to find a suitable value of α and then initializes the arrays. Initially, each array contains two entries. $x[1][\,]$ is set to $\bar{x} := x^* + \alpha\eta_u$, $x[2][\,]$ and $px[1][\,]$ are both set to $P(\bar{x})$, and $px[2][\,]$ is set to $P(x[2][\,])$. The initialization process is illustrated in Fig. 6.23.

Acceptance Acceptance occurs when $px[1][\,]$ and $px[2][\,]$ lie near one another. `accept` first calls the routine `reduce` and then outputs the point $px[1][\,]$ as the next point on the half-manifold. Finally, the $x[\,][\,]$ and $px[\,][\,]$ arrays are advanced along the manifold by one step. More precisely, the $x[1][\,]$ entry is deleted, $px[1][\,]$ is appended to the $x[\,][\,]$ array, and then $px[1][\,]$ is deleted. This process is illustrated in Fig. 6.24. Note that `accept` leaves the number of entries in $x[\,][\,]$ unchanged but decreases the number of entries of $px[\,][\,]$ by one.

Interpolation Interpolation occurs when $px[1][\,]$ and $px[2][\,]$ are too far apart. An intermediate point on the manifold, one between $px[1][\,]$ and $px[2][\,]$, is calculated using linear interpolation on $x[1][\,]$ and $x[2][\,]$. The interpolation process is illustrated in Fig. 6.25. Note that both $x[\,][\,]$ and $px[\,][\,]$ gain an entry when interpolation occurs.

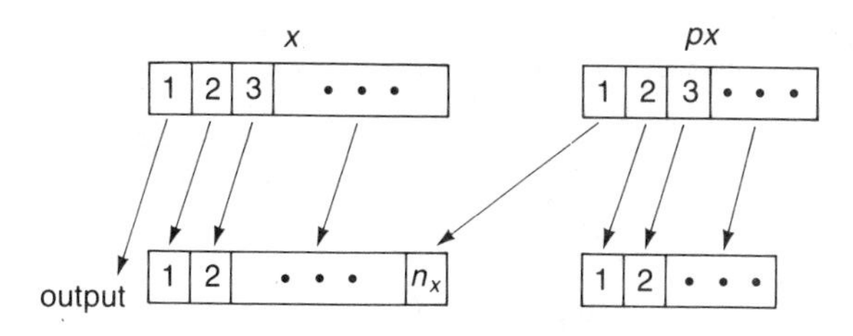

Figure 6.24: The acceptance process.

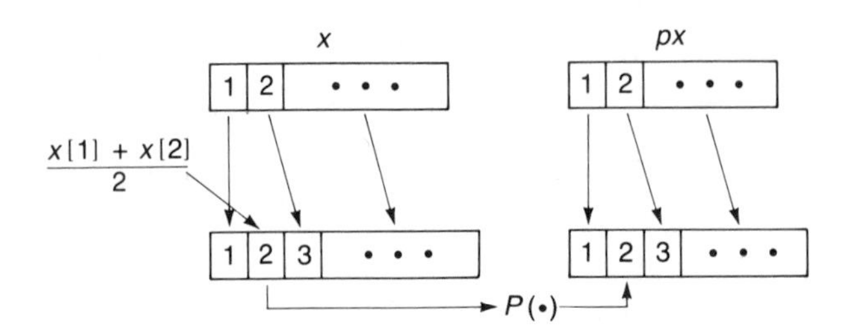

Figure 6.25: The interpolation process.

Reduction `interpolate` increases the accuracy of the result and is akin to decreasing the integration step-size. The `reduce` routine increases the efficiency of the algorithm and is analogous to increasing the integration step-size.

`reduce` reduces the number of entries in both $x[\,][\,]$ and $px[\,][\,]$, thereby reducing the number of output points, and since the entries of $px[\,][\,]$ eventually appear in $x[\,][\,]$, `reduce` also reduces the number of times P is calculated. If P is a Poincaré map, its evaluation requires integration and perhaps some hyperplane detection, so reducing the number of times P is evaluated is quite worthwhile.

`reduce` checks whether $px[1][\,]$, $px[2][\,]$, and $px[3][\,]$ are collinear. If so, $px[2][\,]$ is deleted because it can always be recalculated by linear interpolation. To keep the array indexing scheme intact, $x[2][\,]$ is deleted as well. The reduction process is illustrated in Fig. 6.26.

To check whether $px[1][\,]$, $px[2][\,]$, and $px[3][\,]$ are collinear, shift the origin to $px[2][\,]$. In the new coordinates, $px[1][\,]$ becomes $x_1 := px[1][\,] - px[2][\,]$, $x[2][\,]$ becomes the origin, and $px[3][\,]$ becomes $x_3 := px[3][\,] - px[2][\,]$. It is well known that, given two vectors, u and v,

$$\langle u, v \rangle = \|u\|\,\|v\| \cos\theta \tag{6.15}$$

where θ is the angle between the vectors and the norms are Euclidean norms. It follows that x_1, x_3, and the origin are collinear with the

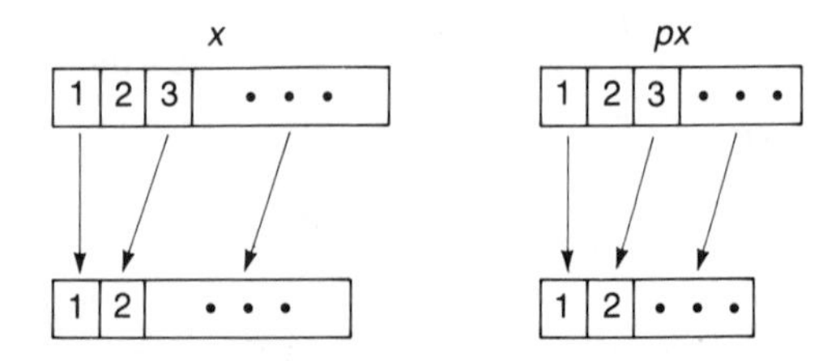

Figure 6.26: The reduction process.

origin between x_1 and x_3, if and only if

$$\langle x_1, x_3 \rangle = -\|x_1\| \, \|x_3\|. \tag{6.16}$$

The norms must be Euclidean so to avoid the square root, square both sides of (6.16). The sign information disappears in the squaring, and, therefore, `reduce` checks whether $\langle x_1, x_3 \rangle < 0$ and whether

$$\langle x_1, x_3 \rangle^2 + \|x_1\|^2 \|x_3\|^2 < \hat{E}_r \langle x_1, x_3 \rangle^2 + \hat{E}_a. \tag{6.17}$$

The tolerance values wear a hat to indicate that they are different from the tolerance values used in `half_manifold`. Note that some of the norms used by `reduce` have already been calculated in `half_manifold` and need not be recalculated.

To avoid unnecessary calculations, `reduce` first uses a relative-absolute test on $\|x_1\| + \|x_3\|$ to check whether x_1 is close to x_3. The tolerance values for this test are the ones used in `half_manifold`.

Array manipulation There are four array manipulation routines.

`insert_x_entry(`k`, `v`[])` calculates $v[\,]$ and inserts it as the kth entry of $x[\,][\,]$. To make room for the new entries, all entries with index greater than k are shifted one entry to the right.

`insert_px_entry(`k`)` inserts $P(x[k][\,])$ as the kth entry of $px[\,][\,]$. To make room for the new entries, all entries with index greater than k are shifted one entry to the right.

`delete_x_entry(`k`)` deletes the kth entry of $x[\,][\,]$. The remaining entries with index greater than k are shifted one entry to the left.

`delete_px_entry(`k`)` deletes the kth entry of $px[\,][\,]$. The remaining entries with index greater than k are shifted one entry to the left.

The $x[\,][\,]$ array can become quite large, as can the $px[\,][\,]$ array. When the dimension of the system is also large, the copying that occurs in `accept` and `interpolate` when $x[\,][\,]$ and $px[\,][\,]$ are shifted can be time consuming. There are several standard approaches to increase the efficiency in this situation. For example, two arrays $i_x[\,]$ and $i_{px}[\,]$ can be kept along with a big array $V[\,][\,]$ of real-valued vectors. In this scheme, $V[i_x[i]][\,]$ is equivalent to $x[i][\,]$, and $V[i_{px}[i]][\,]$ corresponds to $px[i][\,][\,]$. In the array manipulation routines, the entries in $V[\,][\,]$ are not copied in the `for` loops—the entries in the index arrays are. Copying integers is quicker than copying real-valued vectors, so this approach is more efficient. Another approach is to use linked lists in place of the arrays $x[\,][\,]$ and $px[\,][\,]$. Entries can be quickly added to or deleted from a linked list, and linked lists are especially suited to applications where the size of a list is not known beforehand.

Accuracy The accuracy of `half_manifold` depends on the accuracy of the interpolation. The linear interpolation used here is accurate only when the portion of $W^{u+}(x^*)$ that connects $x[1][\,]$ and $x[2][\,]$ is nearly straight. Since α is chosen such that $x[1][\,]$ and $x[2][\,]$ lie on η_u, the interpolation is accurate initially. The interpolation remains accurate as the algorithm proceeds if the error tolerances are small enough. What is meant by "small enough" depends on the system under study.

Closed orbits Let $\{x_1^*, \ldots, x_K^*\}$ be a hyperbolic period-K closed orbit of $P\colon \mathbb{R}^p \to \mathbb{R}^p$. Each of the x_k^* is a fixed point of the Kth-iterated map P^K and has a stable and an unstable manifold with respect to P^K defined in the usual manner. If $W^{u+}(x_1^*)$ is one-dimensional and if the corresponding eigenvalue of $DP(x_1^*)$ is positive, then $W^{u+}(x_1^*)$ can be calculated by applying the routine `half_manifold` to P^K. The $K-1$ other half-manifolds, $W^{u+}(x_k^*)$ for $k = 2, \ldots, K$, can be found simultaneously by making $px[\,][\,][\,]$ an $n_{px} \times K$ matrix of p-vectors. When $P^K(x[i][\,])$ is being calculated, each of the iterates $P^k(x[i][\,])$ is stored in $px[i][k+1][\,]$ for $k = 1, \ldots, K-1$, and the Kth iterate is stored in $px[i][1][\,]$. When a point is accepted, $px[1][k][\,]$ is output as the next point on $W^{u+}(x_k^*)$, for $k = 1, \ldots, K$. The intermediate iterates are not stored in $x[\,][\,]$—it is still a one-dimensional array of vectors—so `accept_K` inserts $px[1][1][\,]$ as the last entry in $x[\,][\,]$. Pseudo-code for a routine `accept_K` that performs the acceptance step for a period-K closed orbit is shown in Fig. 6.27. Fig. 6.27 also includes code for a routine `insert_px_entry_K`—a modified version of `insert_px_entry`—that illustrates the calculation of an entry in $px[\,][\,][\,]$. The modifications to `half_manifold`, `initialize`, `interpolate`, `reduce`, and the remaining array manipulation routines are straightforward.

Due to the increased size of $px[\,][\,][\,]$, one of the array management schemes discussed on page 163 is recommended.

```
begin accept_K
      for k from 1 to K do
          output px[1][k][]
      endfor
      call delete_x_entry(1)
      call insert_x_entry(n_x, px[1][1][])
end accept_K

begin insert_px_entry_K(k)
      set n_px = n_px + 1
      set i_max = n_px - k
      for i from 1 to i_max
          set px[n_px - i + 1][][] = px[n_px - i][][]
      set px[k][1][] = P(x[k][])
      for i from 2 to K do
          set px[k][k][] = P(px[k][k - 1][])
      endfor
end insert_px_entry_K
```

Figure 6.27: Pseudo-code for accept_K, the version of accept for period-K closed orbits.

6.3 Summary

- *Stable manifold (continuous-time)*: The stable manifold $W^s(L)$ of a limit set L of a flow ϕ_t is the set of all points x whose trajectory $\phi_t(x)$ approaches L as $t \to \infty$. The stable manifold is invariant under ϕ_t.

- *Unstable manifold (continuous-time)*: The unstable manifold $W^u(L)$ of a limit set L of a flow ϕ_t is the set of all points x whose trajectory $\phi_t(x)$ approaches L as $t \to -\infty$. The unstable manifold is invariant under ϕ_t.

- *Homoclinic trajectory*: A homoclinic trajectory is a non-constant trajectory that approaches the same equilibrium point as $t \to \pm\infty$. Homoclinic trajectories are always structurally unstable.

- *Heteroclinic trajectory*: A heteroclinic trajectory is a trajectory that approaches one non-stable equilibrium point as $t \to \infty$ and another as $t \to -\infty$.

- *Stable manifold (discrete-time)*: The stable manifold $W^s(L)$ of a limit set L of a diffeomorphism P is the set of all points x

whose orbit approaches L as $k \to \infty$. The stable manifold is invariant under P.

- *Unstable manifold (discrete-time)*: The unstable manifold $W^u(L)$ of a limit set L of a diffeomorphism P is the set of all points x whose orbit approaches L as $k \to -\infty$. The unstable manifold is invariant under P.

- *Homoclinic point*: Let x^* be a fixed point of a diffeomorphism P. A homoclinic point is a point $x \neq x^*$ that lies in $W^s(x^*) \bigcap W^u(x^*)$. The existence of one homoclinic point implies the existence of an infinity of homoclinic points.

- *Homoclinic orbit*: An orbit composed of homoclinic points.

- *Transversal homoclinic point*: A homoclinic point at which the manifolds intersect transversally. A map that possesses a transversal homoclinic point has horseshoe-like maps embedded in it and, therefore, exhibits sensitive dependence on initial conditions.

Chapter 7

Dimension

This chapter addresses the question of the dimension of a limit set, in particular, the dimension of a strange attractor. We will see that a strange attractor possesses non-integer dimension while the dimension of a non-chaotic attractor is always an integer.

After we discuss dimension, we present a remarkable result that permits an attractor to be reconstructed from a sampled time waveform of just one component of the state.

7.1 Dimension

There are several different types of dimension. The dimension of Euclidean space is familiar to everyone—it is the minimum number of coordinates needed to specify a point uniquely. The dimension of a dynamical system is the number of state variables that are used to describe the dynamics of the system. In differential topology, the dimension of a manifold is the dimension of the Euclidean space that the manifold resembles locally. None of these dimensions allows non-integer values and none of them can be used to describe strange attractors. The generic term for a dimension that allows non-integer values is *fractal dimension.* A set that has non-integer dimension is called a *fractal.* Almost all strange attractors are fractals.

We present five different types of fractal dimension. The most well-known is capacity. The four others are information dimension, correlation dimension, kth nearest-neighbor dimension, and Lyapunov dimension.

Though these five are the most commonly used dimensions, there are several other definitions. Some enlightening discussions on these other dimensions and their relationship to the ones presented in this chapter can be found in Young [1983], Farmer *et al.* [1983], Badii and Politi [1985], and Mayer-Kress [1986].

7.1.1 Definitions

Capacity

The simplest type of dimension is *capacity*. Cover an attractor A with volume elements (spheres, cubes, etc.) each with diameter ϵ. Let $N(\epsilon)$ be the minimum number of volume elements needed to cover A. If A is a D-dimensional manifold—D is necessarily an integer—then the number of volume elements needed to cover A is inversely proportional to ϵ^D, that is,

$$N(\epsilon) = k\,\epsilon^{-D} \tag{7.1}$$

for some constant k.[1] The capacity, denoted by D_{cap}, is obtained by solving (7.1) for D and taking the limit as ϵ approaches zero,

$$D_{cap} := \lim_{\epsilon \to 0} \frac{\ln N(\epsilon)}{\ln(1/\epsilon)}. \tag{7.2}$$

If the limit does not exist, then D_{cap} is undefined. Since a d-dimensional manifold locally resembles $\mathbb{R}^d$, D_{cap} of a manifold equals the topological dimension, which is an integer. For objects that are not manifolds, D_{cap} can take on non-integer values.

Example 7.1 The unit interval:
As volume elements, choose intervals of length $\epsilon = 1/3^k$. It takes $N(\epsilon) = 3^k$ of these volume elements to cover the unit interval $[0, 1]$ (see Fig. 7.1(a)). To refine the covering, let $k \to \infty$ to obtain

$$D_{cap} = \lim_{k \to \infty} \frac{\ln 3^k}{\ln 3^k} = 1. \tag{7.3}$$

As expected, the unit interval has dimension 1.

Example 7.2 The middle-third Cantor set:
This example will show that the middle-third Cantor set has non-integer dimension. Since strange attractors have a Cantor-set-like

[1] k depends on the geometry of the attractor and on the type of volume element used.

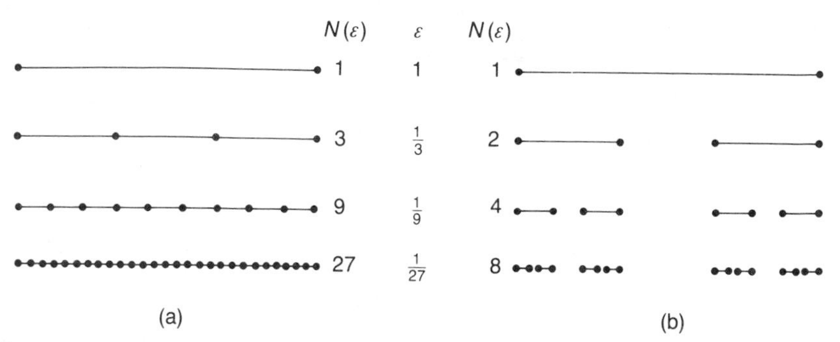

Figure 7.1: Two simple examples of capacity. (a) The unit interval; (b) the middle-third Cantor set.

structure, the example also explains why a strange attractor has a non-integer dimension.

The middle-third Cantor set is constructed iteratively by removing the middle third of the unit interval and then the middle third of the remaining two intervals, etc. (see Appendix F). To calculate the capacity of the middle-third Cantor set, cover it, at the kth step in the construction, with intervals of length $\epsilon = 1/3^k$ as shown in Fig. 7.1(b). At the kth step, the number of intervals required to cover the set is $N(\epsilon) = 2^k$, and

$$D_{cap} = \lim_{k\to\infty} \frac{\ln 2^k}{\ln 3^k} = \frac{\ln 2}{\ln 3} = 0.6309\ldots. \tag{7.4}$$

Hence, the Cantor set is something more than a point (dimension 0) but something less than an interval (dimension 1).

Remark: When the limit in (7.2) exists, the question arises whether another covering (e.g., spheres instead of cubes or, perhaps, a mixture of spheres of different sizes) can result in a different value of D_{cap}. The answer, unfortunately, is yes. To resolve this dilemma, we simply comment that capacity is closely related to Hausdorff dimension and that the definition of Hausdorff dimension implies that if different coverings result in different values of D_{cap}, then the minimum value over all coverings should be used (see Young [1983]).

Information dimension

Capacity is a purely metric concept. It utilizes no information about the time behavior of the dynamical system. Information dimension, on the other hand, is defined in terms of the relative frequency of visitation of a typical trajectory. The setting is the same as for capacity—a covering of $N(\epsilon)$ volume elements each with diameter ϵ. The *information dimension* D_I is defined by

$$D_I := \lim_{\epsilon \to 0} \frac{H(\epsilon)}{\ln(1/\epsilon)} \tag{7.5}$$

where

$$H(\epsilon) := -\sum_{i=1}^{N(\epsilon)} P_i \ln P_i. \tag{7.6}$$

P_i is the relative frequency with which a typical trajectory enters the ith volume element of the covering.

Readers familiar with information theory will recognize $H(\epsilon)$ as entropy—the amount of information needed to specify the state of the system to an accuracy of ϵ if the state is known to be on the attractor.

For sufficiently small ϵ, (7.5) can be rewritten as

$$H(\epsilon) = k\epsilon^{-D_I} \tag{7.7}$$

for some constant of proportionality k. In words, the amount of information needed to specify the state increases inversely with the D_Ith power of ϵ. Compare this equation with (7.1).

Example 7.3 Unit interval:
Assume the attractor is the unit interval and that the probability density is uniform. As before, choose intervals of length $\epsilon = 1/3^k$ as volume elements. It follows that $N(\epsilon) = 3^k$ and that $P_i = 1/3^k$. With these values, the entropy is

$$\begin{aligned} H(\epsilon) &= -\sum_{i=1}^{3^k} \frac{1}{3^k} \ln \frac{1}{3^k} \\ &= \ln 3^k \end{aligned} \tag{7.8}$$

and the information dimension is

$$D_I = \lim_{k \to \infty} \frac{\ln 3^k}{\ln 3^k} = 1. \tag{7.9}$$

which agrees with D_{cap}. In fact, it is easy to show that for uniform densities, D_{cap} and D_I always agree.

Example 7.4 Interval and point:
Let the set under study be the subset of $\mathbb{R}$ that consists of the unit interval $[0, 1]$ and the isolated point 2. Assume that the probability of finding a point at 2 is 1/2, and of finding a point on the interval is also 1/2. Further assume that the density is uniform over the interval.

It is easy to show that the capacity of this set is 1. This is the same result as for the unit interval itself and, therefore, capacity ignores the isolated point even though the probability of being at the point is the same as being on the interval.[2]

To find the information dimension of this set, choose intervals of length $\epsilon = 1/3^k$ as volume elements. It takes 3^k such elements to cover the unit interval but only one to cover the isolated point, so $N(\epsilon) = 3^k + 1$. $P_i = 1/2$ for the volume element covering the isolated point and $P_i = 1/(2 \cdot 3^k)$ for the remaining volume elements. Thus,

$$\begin{aligned} H(\epsilon) &= -\frac{1}{2}\ln\left(\frac{1}{2}\right) - \frac{1}{2}\sum_{i=1}^{3^k}\frac{1}{3^k}\ln\left(\frac{1}{2\cdot 3^k}\right) \\ &= \frac{1}{2}\ln 2 + \frac{1}{2}\ln(2\cdot 3^k) \\ &= \ln 2 + \frac{1}{2}\ln 3^k \end{aligned} \tag{7.10}$$

and

$$\begin{aligned} D_I &= \lim_{k\to\infty}\left(\frac{\ln 2}{\ln 3^k} + \frac{\ln 3^k}{2\ln 3^k}\right) \\ &= \frac{1}{2}. \end{aligned} \tag{7.11}$$

Thus, for this example, D_I is the average of the dimensions of the point and of the interval.

Though the set in the last example is not the limit set of a dynamical system, it does show how information dimension differs from capacity. Capacity tends to ignore lower dimensional subsets of the attractor. Information dimension, on the other hand, weights the lower-dimensional subsets according to the frequency of visitation of a typical trajectory.

Example 7.5 The middle-third Cantor set:
To find the information dimension of the middle-third Cantor set,

[2] It can be shown using (7.1) and (7.2) that $D_{cap}(S) = \max_i D_{cap}(S_i)$ for $S = S_1 \bigcup \cdots \bigcup S_k$ for some finite k.

it is necessary to define a probability density on the set. In each step of the construction of the Cantor set, the middle of an interval is removed leaving two smaller sub-intervals, one on the left and one on the right. Any point in the Cantor set is, therefore, identified uniquely by its left-right history, that is, whether at the kth step of the construction it was in the right or left sub-interval. For example, $\ell\ell\ell\ell\ldots$ is the left-most point of the Cantor set (i.e., 0), and $\ell rr\ldots$ is the right-most point of the first left sub-interval (i.e., 1/3).

The left-right history can be used to define a self-similar probability density. Let $0 \leq p_\ell \leq 1$ be the probability of being in the left sub-interval and $p_r = 1 - p_\ell$ be the probability of being in the right sub-interval. For example, the probability that a point lies in the segment $[2/27, 3/27]$ corresponding to $\ell\ell r$ is $p_\ell p_\ell p_r$. It is shown in Appendix F that the information dimension of the middle-third Cantor set with this probability density is

$$D_I = -\frac{p_\ell \ln p_\ell + p_r \ln p_r}{\ln 3}. \tag{7.12}$$

Observe that when $p_\ell = p_r = 1/2$, the information dimension agrees with the capacity. When $p_\ell \neq p_r$, however, the information capacity is always less than the capacity.

Remark: It can be shown that for any attractor, $D_I \leq D_{cap}$.

Correlation dimension

Another probabilistic type of dimension is the *correlation dimension* D_C. It, too, depends upon refining a covering of $N(\epsilon)$ volume elements of diameter ϵ, and is defined by

$$D_C := \lim_{\epsilon \to 0} \frac{\ln \sum_{i=1}^{N(\epsilon)} P_i^2}{\ln \epsilon} \tag{7.13}$$

where, as before, P_i is the relative frequency with which a typical trajectory enters the ith volume element.

To help interpret the numerator of (7.13), suppose N points of a trajectory have been collected, either through simulation or from measurements. Define the correlation as

$$C(\epsilon) := \lim_{N \to \infty} \frac{1}{N^2}\{\textit{ the number of pairs of points } (x_i, x_j) \textit{ such that } \|x_i - x_j\| < \epsilon\}. \tag{7.14}$$

Then

$$D_C = \lim_{\epsilon \to 0} \frac{\ln C(\epsilon)}{\ln \epsilon}. \tag{7.15}$$

To show the plausibility of (7.15), let n_i be the number of points lying in the ith volume element. Then

$$P_i = \lim_{N \to \infty} \frac{n_i}{N}. \tag{7.16}$$

Since the volume element has diameter ϵ, all the n_i points lie within ϵ of each other and form $n_i^2 - n_i$ pairs of points.[3] It follows that

$$\begin{aligned} C(\epsilon) &= \lim_{N \to \infty} \frac{1}{N^2} \sum_{i=1}^{N(\epsilon)} (n_i^2 - n_i) \\ &= \sum_{i=1}^{N(\epsilon)} \lim_{N \to \infty} \frac{n_i^2}{N^2} - \lim_{N \to \infty} \frac{n_i}{N^2} \\ &= \sum_{i=1}^{N(\epsilon)} P_i^2 - \lim_{N \to \infty} \frac{P_i}{N} \\ &= \sum_{i=1}^{N(\epsilon)} P_i^2 \end{aligned} \tag{7.17}$$

from which (7.15) follows.

An objection to the preceding derivation can be made because $n_i^2 - n_i$ is the number of points in a *single* volume element. It does not include pairs of points that are within ϵ of each other but that lie in different volume elements. In response to this objection, assume that there are actually μn_i^2 pairs of points within ϵ of one another where $\mu > 1$ is a correction factor. With this correction term, (7.15) becomes

$$\begin{aligned} D_C &= \lim_{\epsilon \to 0} \frac{\ln (\mu C(\epsilon))}{\ln \epsilon} \\ &= \lim_{\epsilon \to 0} \frac{\ln \mu}{\ln \epsilon} + \lim_{\epsilon \to 0} \frac{\ln C(\epsilon)}{\ln \epsilon} \\ &= \lim_{\epsilon \to 0} \frac{\ln C(\epsilon)}{\ln \epsilon} \end{aligned} \tag{7.18}$$

which agrees with (7.15).

[3] (x_i, x_i) is not counted as a pair and for $i \neq j$, (x_i, x_j) is treated as a pair different from (x_j, x_i).

Example 7.6 Interval and point:
Consider the set from Example 7.4. Use intervals of length $\epsilon = 1/3^k$ as volume elements. Then $N(\epsilon) = 3^k + 1$, $P_i = 1/(2 \cdot 3^k)$ for the volume elements covering the interval, and $P_i = 1/2$ for the single volume element covering the isolated point. Substitute these values into the numerator of (7.13) to obtain

$$\begin{aligned} \ln \sum_{i=1}^{N(\epsilon)} P_i^2 &= \ln \left[\frac{1}{4} + \sum_{i=1}^{3^k} \frac{1}{4 \cdot 3^{2k}} \right] \\ &= \ln \left[\frac{1}{4} \left(1 + \frac{1}{3^k} \right) \right]. \end{aligned} \tag{7.19}$$

which yields

$$\begin{aligned} D_C &= \lim_{k \to \infty} \frac{\ln \left[\frac{1}{4} \left(1 + \frac{1}{3^k} \right) \right]}{\ln(1/3^k)} \\ &= \lim_{k \to \infty} \frac{\ln(1/4)}{\ln(1/3^k)} \\ &= 0. \end{aligned} \tag{7.20}$$

Thus, unlike capacity and information dimension, the correlation dimension ignores the unit interval entirely. The reader may find it interesting to calculate D_C for this example using (7.15).

Example 7.7 The middle-third Cantor set:
Consider the middle-third Cantor set with probabilities as in Example 7.5. It is shown in Appendix F that the correlation dimension of the middle-third Cantor set is

$$D_C = -\frac{\ln(p_\ell^2 + p_r^2)}{\ln 3}. \tag{7.21}$$

Observe that when $p_\ell = p_r = 1/2$, the correlation dimension agrees with the information dimension and with the capacity. When, however, $p_\ell \neq p_r$, the correlation dimension is always less than the information dimension.

Remark: It can be shown that for any attractor, $D_C \leq D_I \leq D_{cap}$.

kth nearest-neighbor dimension

The kth nearest-neighbor dimension was formulated by Pettis *et al.* [1979]. The appeal of this dimension is that its definition is based firmly on probabilistic concepts.

Consider an attractor A embedded in $\mathbb{R}^n$. Let $x_1, \ldots, x_N$ be N randomly chosen data points lying on A. Let $r(k,x)$ be the distance between x and its kth nearest neighbor in $\{x_i\}$. Define $\bar{r}(k)$ as the mean of $r(k,x)$ taken over $\{x_i\}$, that is,

$$\bar{r}(k) := \frac{1}{N}\sum_{i=1}^{N} r(k,x_i). \tag{7.22}$$

Pettis *et al.* show that under reasonable assumptions and for large N, there exist functions g and c such that the *kth nearest-neighbor dimension* D_{nn} is well-defined by the equation

$$D_{nn} := \frac{\ln k + c(x_1,\ldots,x_N)}{g(k,D_{nn}) + \ln \bar{r}(k)}. \tag{7.23}$$

Pettis *et al.* show that $g(k,D_{nn})$ is small for all k and D_{nn}, so ignoring g, (7.23) can be rewritten as

$$\bar{r}(k) \approx e^{c(x_1,\ldots,x_N)} k^{1/D_{nn}} \tag{7.24}$$

which, given $\{x_1, \ldots, x_N\}$, implies that the average distance to the kth nearest neighbor is proportional to $k^{1/D_{nn}}$. This proportionality is intuitively satisfying, at least for manifolds. For example, let $\{x_1, \ldots, x_N\}$, be chosen from the interior of the unit circle using a uniform probability density. Let S be a smaller circle with radius r that is randomly positioned in the interior of the unit circle. The number $n(r)$ of points in the intersection of S and $\{x_i\}$ is, on the average, proportional to the area of S, that is,

$$n(r) \propto r^2. \tag{7.25}$$

Since a circle of radius $\bar{r}(k)$ contains, on the average, k points, it follows from (7.25) that $k \propto \bar{n}(r)^2$ which agrees with (7.24).

Owing to the complexity of calculating $\bar{r}(k)$, simple yet meaningful analytical examples are difficult to find. Pettis *et al.* give an example using a uniform density over the unit interval with $N = 3$.

Lyapunov dimension

Let $\lambda_1 \geq \cdots \geq \lambda_n$ be the Lyapunov exponents of an attractor of a continuous-time dynamical system. Let j be the largest integer such that $\lambda_1 + \cdots + \lambda_j \geq 0$. The *Lyapunov dimension* as defined by Kaplan and Yorke [1979] is

$$D_L := j + \frac{\lambda_1 + \cdots + \lambda_j}{|\lambda_{j+1}|}. \tag{7.26}$$

If no such j exists, as is the case for a stable hyperbolic equilibrium point, D_L is defined to be 0.

For an attractor, $\sum_{i=1}^{n} \lambda_i < 0$, so j is guaranteed to be less than n. For an attracting limit cycle, the generic situation is $\lambda_1 = 0 > \lambda_2 > \cdots > \lambda_n$. Thus, the Lyapunov dimension of a generic attracting limit cycle is 1. Similarly, the Lyapunov dimension of generic attracting K-periodic behavior is K.

If the attractor is chaotic, D_L is almost always a non-integer.[4] In a three-dimensional chaotic system with Lyapunov exponents $\lambda_+ > 0 > \lambda_-$,

$$D_L = 2 + \frac{\lambda_+}{|\lambda_-|}. \tag{7.27}$$

For an attractor, $\lambda_+ + \lambda_- < 0$ from which it follows that $2 < D_L < 3$.

Plausibility argument: The derivation of (7.26) by Kaplan and Yorke is not rigorous, and we present a plausibility argument. Consider an n-dimensional hyper-cube C evolving in a flow ϕ_t. Let the length of the sides of the hyper-cube be ϵ. With the proper change of coordinates and for ϵ small, the ith side of C evolves under the flow, on the average, as $\epsilon\, e^{\lambda_i t}$. To find the capacity of C cover C at time 0 with hyper-cubes of side ϵ. To refine this covering, let each of the sides of each of the volume elements contract at the constant rate $e^{\lambda_{k+1} t}$ where k is chosen such that $\lambda_{k+1} < 0$ (the λ_i are in decreasing order as before). The number of these contracting cubes it takes to cover C at time t is

$$\begin{aligned} N(t) &= \frac{\epsilon\, e^{\lambda_1 t}}{\epsilon\, e^{\lambda_{k+1} t}} \cdots \frac{\epsilon\, e^{\lambda_k t}}{\epsilon\, e^{\lambda_{k+1} t}} \\ &= e^{(\lambda_1 + \cdots + \lambda_k - k\lambda_{k+1})t}. \end{aligned} \tag{7.28}$$

[4] In the non-generic case of $\lambda_1 > 0$, $\lambda_2 = 0$, $\lambda_3 = -\lambda_1$, and $\lambda_4 < \lambda_3$, the attractor has Lyapunov dimension 3.

The sides of C that grow with rates $\lambda_{k+2}, \ldots, \lambda_n$, do not influence $N(t)$ because they are shrinking with respect to $e^{\lambda_{k+1}t}$.

The capacity is approximated by

$$\begin{aligned} D_{cap}(k) &= -\lim_{t\to\infty} \frac{\ln N(t)}{\ln(\epsilon\, e^{\lambda_{k+1}t})} \\ &= k - \frac{\lambda_1 + \cdots + \lambda_k}{\lambda_{k+1}}. \end{aligned} \tag{7.29}$$

It is shown by Frederickson *et al.* [1983] that $k = j$ yields the lowest value[5] of $D_{cap}(k)$ so define

$$\begin{aligned} D_L := D_{cap}(j) &= j - \frac{\lambda_1 + \cdots + \lambda_j}{\lambda_{j+1}} \\ &= j + \frac{\lambda_1 + \cdots + \lambda_j}{|\lambda_{j+1}|} \end{aligned} \tag{7.30}$$

to complete the plausibility argument.

Discussion

Given the different definitions of dimension, it is natural to ask what relationship they bear to one another. Are they equivalent? Is one more useful than another?

First, we warn the reader that dimension is an active research area and the relationships and meanings of the different dimensions are unclear, especially in experimental settings or when applied to simulations. Part of the problem arises from not having an exact definition of a strange attractor. Other problems are due to the difficulty of analyzing the statistical properties that are required for D_I, D_C, D_{nn}, and D_L. Until these issues are settled, it is difficult to make any rigorous statements regarding the dimension of a strange attractor. Thus, this discussion is necessarily speculative in nature.

One task for which dimension appears to be impractical is to describe the geometric structure of an attractor. It seems ridiculous that a single number can fully describe the complex structure of a strange attractor. It is shown in Appendix F that D_{cap}, D_I, and D_C are three members of a countably infinite family of dimensions called the Renyi dimensions. Moreover, Badii and Politi [1985] introduced

[5] Recall that if two different coverings result in different capacities, the minimum value should be used.

an entire continuum of dimensions in which, at least for self-similar sets, the Renyi dimensions are embedded. A different continuum of dimensions, the *multifractal spectrum*, has been used to measure apparently universal properties of attractors at the onset of chaos through period-adding bifurcations (see Glazier and Libchaber [1988] for a review of this work). The current state of knowledge is unclear as to how much information these different values of dimension carry. For example, to distinguish between two attractors on the basis of dimension alone, is it sufficient to know the values of all these different dimensions, of just a few, or is additional information needed?

Dimension can be used to classify strange versus non-strange attractors. Non-strange attractors have integer dimension, and almost all chaotic attractors have non-integer dimension. This classification scheme, though nice in theory, is fairly useless in practice. First, the dimension algorithms have low precision, making it difficult if not impossible to judge whether the dimension is an integer. Second, there are better ways to judge whether an attractor is chaotic (e.g., looking at it, or one of the shooting methods).

The main use of dimension is to quantify the complexity of an attractor. The dimension of an attractor gives a lower bound on the number of state variables needed to describe the dynamics on the attractor. In words familiar from physics, the dimension is a lower bound on the number of degrees of freedom of the attractor. For example, motion on a limit cycle (dimension 1) can be described by a first-order differential equation where the variable is arc-length along the circle or perhaps the angle of rotation. In the chaotic case, motion on an attractor with dimension 2.4 cannot be described by two state variables, but can be modeled, at least theoretically, by a third-order system.

If the main reason for finding the dimension of an attractor is to estimate the minimum number of variables needed to describe the steady-state dynamics, there seems to be no theoretical reason for choosing one type of dimension over another—D_{cap}, D_I, D_C, and D_{nn} give values close to one another. The main reasons for choosing one type of dimension over another are the ease and accuracy of its computation. Since new algorithms may be found, there is no cut and dried answer as to which dimension-finding algorithm is best. There exist algorithms of comparable efficiency for finding each of D_{cap}, D_C, and D_{nn}. These algorithms are presented in the next section.

As is obvious from the definition, Lyapunov dimension is somewhat different from the other four. It was originally conjectured that $D_L = D_{cap}$ as the plausibility argument is designed to show. Since Lyapunov exponents are probabilistic in nature—they are defined in terms of an average over time—and D_{cap} is not, the conjecture was changed to $D_L = D_I$. Numerical simulations do not appear to contradict this conjecture, but analytical examples exist where $D_L \neq D_I$ (see Grassberger and Procaccia [1983]). These examples, however, are not structurally stable and if perturbed, the relationship $D_L = D_I$ does hold. The exact relationship between D_L and D_I (and the other dimensions as well) is an active research topic.

7.1.2 Algorithms

For each type of dimension, there are several different algorithms for estimating it. We cannot cover all the different techniques and are satisfied to present one technique for each of D_{cap}, D_C, and D_{nn}. This is an active field of research and the algorithms presented here should not be taken as the best possible algorithms—most likely they will be superseded by more reliable algorithms in the near future. The algorithms presented do demonstrate, however, the wide variety of approaches that are available to tackle the problem of estimating dimension and they highlight the creativity and energy researchers have devoted to this area.

The input to the following dimension-finding algorithms is a finite sequence $\{x_1, \ldots, x_N\}$ of points on an attractor. Typically, the points are evenly spaced time-samples of one or more trajectories that have achieved the steady state. It is important to keep in mind that the dimension-finding algorithms are not given full information about an attractor; they are given only partial information about a finite number of trajectories on the attractor.

Capacity

The first attempts to calculate capacity were based directly on the definition. From (7.1),

$$\ln N(\epsilon) = D_{cap} \ln \frac{1}{\epsilon} + \ln k. \tag{7.31}$$

Hence, D_{cap} is the slope of a log-log plot of $N(\epsilon)$ versus $1/\epsilon$.

To calculate $N(\epsilon)$ for several values of ϵ simultaneously, choose some minimum value ϵ_0 of ϵ. Cover the n-dimensional state space with a grid of boxes (hyper-cubes) of side ϵ_0. In the computer, maintain a boolean array—all entries initialized to FALSE—where each entry corresponds to one of the boxes. For each data point x_i, calculate the indices of the box in which x_i lies and set the corresponding entry of the array to TRUE. $N(\epsilon_0)$ is the number of array entries that are TRUE. For $m = 2, \ldots, M$, the value $N(2^m\epsilon_0)$ can be calculated from the entries of the boolean array by partitioning the ϵ_0 grid into larger cubes of side $\epsilon_0 2^m$. $N(\epsilon_0 2^m)$ is the number of these larger cubes that contain at least one ϵ_0 cube with a TRUE boolean array entry. Note that the choice of $\epsilon = \epsilon_0 2^m$ yields evenly spaced points on the x-axis of the log-log plot.

This box-counting technique works, but has two drawbacks.

1. For systems of dimension greater than three, the memory requirements are excessive. The total number of entries in the boolean array is proportional to $(1/\epsilon_0)^n$. For $\epsilon_0 = 0.01$ and $n = 4$, there are on the order of 10^8 entries in the array.

2. A large amount of data is required to ensure that nearly every ϵ_0 cube that contains a point on the attractor is visited by the trajectory under study; otherwise, the estimate of D_{cap} will be low. This point is directly related to the fact that capacity does not distinguish between those parts of the attractor that are visited frequently and those parts that are rarely touched by a typical trajectory.

A more efficient algorithm for estimating D_{cap} was presented by Hunt and Sullivan [1986]. Given an attractor A embedded in $\mathbb{R}^n$, define

$$\bar{A}(\epsilon) := \text{volume}\,\{y : \text{dist}\,(y, A) < \epsilon\} \tag{7.32}$$

where

$$\text{dist}\,(y, A) := \inf_{x \in A} \|x - y\| \tag{7.33}$$

is the distance from y to A.

Let D_{cap} be the capacity of A and define

$$D_A := \lim_{\epsilon \to 0} \frac{\bar{A}(\epsilon)}{\ln \epsilon}. \tag{7.34}$$

We now show that

$$D_{cap} = n - D_A \tag{7.35}$$

and, therefore, that D_A can be used to calculate D_{cap}.

Cover A with $N(\epsilon)$ hyper-cubes of side ϵ. Every point y that is within ϵ of A lies in one of these $N(\epsilon)$ cubes or in a cube that neighbors one of these cubes. Each cube has $3^n - 1$ neighbors so

$$\bar{A}(\epsilon) \leq 3^n \epsilon^n N(\epsilon). \tag{7.36}$$

Take the log of both sides and divide by $\ln \epsilon$ (which is negative for ϵ small) to obtain

$$\frac{\ln \bar{A}(\epsilon)}{\ln \epsilon} \geq \frac{n \ln 3}{\ln \epsilon} + \frac{n \ln \epsilon}{\ln \epsilon} + \frac{\ln N(\epsilon)}{\ln \epsilon}. \tag{7.37}$$

Take the limit as $\epsilon \to 0$ to obtain

$$D_A \geq n - D_{cap}. \tag{7.38}$$

Now cover A with $N(\epsilon/\sqrt{n})$ hyper-cubes of side $\epsilon/\sqrt{n}$. Any point in these cubes is within ϵ of A so

$$\bar{A}(\epsilon) \geq \left(\frac{\epsilon}{\sqrt{n}}\right)^n N(\epsilon/\sqrt{n}). \tag{7.39}$$

Take the log of both sides and divide by $\ln(\epsilon/\sqrt{n})$ to obtain

$$\frac{\ln \bar{A}(\epsilon)}{\ln \epsilon - \ln \sqrt{n}} \leq \frac{n \ln(\epsilon/\sqrt{n})}{\ln(\epsilon/\sqrt{n})} + \frac{\ln N(\epsilon/\sqrt{n})}{\ln(\epsilon/\sqrt{n})}. \tag{7.40}$$

Take the limit as $\epsilon \to 0$ to obtain

$$D_A \leq n - D_{cap}. \tag{7.41}$$

Equation (7.35) follows from (7.38) and (7.41).

The advantage of D_A over D_{cap} is that D_A can be calculated more efficiently than D_{cap}. D_A is calculated by finding the slope of a log-log plot of $\bar{A}(\epsilon)$ versus ϵ. $\bar{A}(\epsilon)$ can be calculated efficiently using Monte Carlo techniques. It is not possible to present a full explanation of Monte Carlo techniques here, so we present only a brief description. Normalize the coordinates such that A sits in an n-dimensional unit hyper-cube in $\mathbb{R}^n$. Generate K random points $x_1, \ldots, x_K$ in this hyper-cube. Count the number N_A of these points that lie within ϵ of A. Then $\bar{A}(\epsilon) \approx N_A/M$. The most costly step in this technique is calculating whether each x_i lies within ϵ of A. Hunt and Sullivan [1986] present a tree-like data structure that implements this Monte Carlo technique efficiently on vector floating-point processors.

Correlation dimension

The correlation dimension can be found directly from (7.13) using a box-counting scheme to estimate P_i, but as explained earlier, this approach is inefficient. A faster algorithm, due to Grassberger and Procaccia [1983], is obtained using the correlation $C(\epsilon)$ defined in (7.14). From (7.15), the correlation dimension is the slope of a log-log plot of $C(\epsilon)$ versus ϵ. For a given ϵ, estimating $C(\epsilon)$ entails computing all the inter-point distances, $r_{ij} := \|x_i - x_j\|$, and counting the number $N_r(\epsilon)$ of $r_{ij} < \epsilon$ for $i, j = 1, \ldots, N$. Then, $C(\epsilon) = N_r(\epsilon)/N^2$.

The x-axis of the log-log plot is $\ln \epsilon$. To obtain a set of points that are evenly spaced in the x direction, a simple binning algorithm is used. Calculate $N(\epsilon)$ for values of ϵ that are geometrically spaced, that is, for ϵ_0, ϵ_0^2, ..., ϵ_0^K, for some $\epsilon_0 > 0$ and some integer $K > 0$. This task is best accomplished by maintaining a K-dimensional array $N_\epsilon[\,]$ of integers. $N_\epsilon[k]$ is the count of the inter-point distances that satisfy $\epsilon_0^{k-1} < r_{ij} < \epsilon_0^k$. Then

$$N(\epsilon_0^k) = \sum_{i=1}^{k} N_\epsilon[i], \qquad k = 1, \ldots, K. \tag{7.42}$$

The $N_\epsilon[\,]$ array is initialized to all zeros. The most straightforward way to fill $N_\epsilon[\,]$ is to choose ϵ_0 equal to some small integer b. For each of the inter-point distances r_{ij}, set k to the integer part of $\log_b r_{ij}$ and then increment $N_\epsilon[k]$ by one.[6] The drawback of this approach is that log() is an expensive floating-point operation.

The $N_\epsilon[\,]$ array can be filled more efficiently by taking advantage of the floating-point representation used in computers. A floating-point number r_{ij} is represented by a sequence of bits using a mantissa/exponent format

$$r_{ij} = \pm m\, b^e \tag{7.43}$$

where the exponent e is an integer, the base b is also an integer, typically some power of two, and the mantissa m is a fraction normalized such that $1/b \le m < 1$. For example, in the 32-bit IEEE standard, $b = 2$, the first bit stores the sign information, the next eight bits hold the exponent e, and the remaining twenty-three bits represent the mantissa. Using shifting and masking functions, the exponent can be retrieved without any time consuming floating-point operations.

[6] The value k is often negative so a constant offset must be added to ensure the index to $N_\epsilon[\,]$ is positive. More on this point shortly.

```
begin find_cor_dim(x[][], N)
      set K = e_max - e_min + 1
      set offset = 1 - e_min
      set k_min = K
      set k_max = 1
      for k from 1 to K
          set N_ε[k] = 0
      endfor
      for i from 1 to N - 1
          for j from i + 1 to N
              set r = || x[i][] - x[j][] ||
              set e to the exponent of r
              set k = e + offset
              set N_ε[k] = N_ε[k] + 1
              if (k < k_max) then
                 set k_max = k
              endif
              if (k > k_min) then
                 set k_min = k
              endif
          endfor
      endfor
      set sum = 0
      for k from k_min to k_max do
          set sum = sum + N_ε[k]
          plot x = k versus y = ln(2 sum/N^2)/ ln 2
      endfor
end find_cor_dim
```

Figure 7.2: Pseudo-code for find_cor_dim.

To take advantage of this floating-point representation, choose $\epsilon_0 = 2^{e_{min}}$ where e_{min} is the smallest exponent possible in the floating-point representation, and choose

$$K = e_{max} - e_{min} + 1 \tag{7.44}$$

where e_{max} is the largest possible exponent. For each inter-point distance r_{ij}, increment $N_\epsilon[e - e_{min} + 1]$ by one where e is the exponent of the floating-point representation of r_{ij}. This exponent is offset by $e_{min} - 1$ to ensure that the index is positive.

Pseudo-code for the algorithm is presented in Fig. 7.2. and the output of the algorithm for Henon's map is shown in Fig. 7.3.

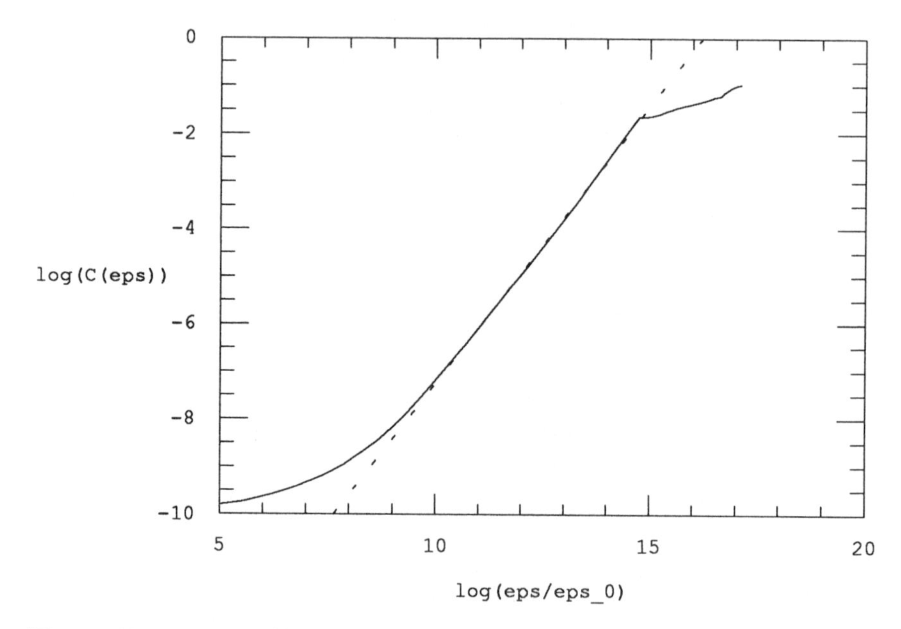

Figure 7.3: $\log_2 C(\epsilon)$ versus $\log_2 \epsilon$ for Henon's map: $x_{k+1} = y_k + 1 - 1.4x_k^2$, $y_{k+1} = 0.3x_k$. ϵ_0 is a constant that accounts for the biasing of the exponents—it does not affect the slope. The slope and, therefore, D_C is approximately 1.2. $N = 2000$ points were used in the calculation.

Remarks:

1. In the IEEE floating-point standard, the exponent is stored with a constant value, called a *bias*, added to it. The bias is chosen such that the exponent is never negative. Thus, $e_{min} = 0$ and *offset* $= 1$. When plotting, the slope is all that is sought and, therefore, the exponents (i.e., the k's) do not have to be debiased before plotting.

2. Owing to duplications, not all the inter-point distances are computed. Only the $(N^2 - N)/2$ values r_{ij} for $i = 1, \ldots, N-1$ and $j = i+1, \ldots, N$ are calculated. It follows that the entries in $N_\epsilon[\,]$ are half the actual count. This explains why *sum* is multiplied by two before plotting.

3. There will be some k_{min} and k_{max} such that $N_\epsilon[k] = 0$ for any $k > k_{max}$ and for any $k < k_{min}$. Only values of k from k_{min} to k_{max} are plotted.

4. If more points on the plot are desired, raise r to the mth power for some integer $m > 0$, and use $x = k/m$ when plotting. This

increases the number of points plotted by the factor m. A value of $m = 2$ is reasonable. It doubles the number of points plotted, and the cost is just one multiplication per r_{ij}.

kth nearest-neighbor dimension

If $g(k, D_{nn})$ were independent of k and D_{nn}, then the slope of a log-log plot of $\bar{r}(k)$ versus k would yield D_{nn}. $g(k, D_{nn})$, however, depends on k and D_{nn} and, therefore, a log-log plot of $\bar{r}(k)$ versus k is not a straight line. Fortunately, $g(k, D_{nn})$ can be approximated by (Pettis *et al.* [1979])

$$\begin{aligned} g(k, D_{nn}) \;\approx\; & \frac{D_{nn}-1}{2kD_{nn}^2} + \frac{(D_{nn}-1)(D_{nn}-2)}{12k^2D_{nn}^3} - \frac{(D_{nn}-1)^2}{12k^3D_{nn}^4} \\ & - \frac{(D_{nn}-1)(D_{nn}-2)(D_{nn}^2+3D_{nn}-3)}{120k^4D_{nn}^5}. \end{aligned} \tag{7.45}$$

This expression leads to an iterative method for calculating D_{nn}. Pseudo-code for this algorithm is presented in Figs. 7.4 and 7.5.

Remarks:

1. The indexing scheme in `find_ln_r` is designed so that no inter-point distance is calculated twice.
2. The routine `calculate_d` (not shown) calculates the current iterate of d using the reciprocal of the standard least-squares formula

$$d = \frac{k_{max}\sum_{k=1}^{k_{max}}(\ln k)^2 - \left(\sum_{k=1}^{k_{max}}\ln k\right)^2}{k_{max}\sum_{k=1}^{k_{max}}\ln(k)\, g\_ln\_r[k] - \sum_{k=1}^{k_{max}}\ln k \sum_{k=1}^{k_{max}} g\_ln\_r[k]} \tag{7.46}$$

3. Equation (7.23) is valid only when the nearest-neighbor distances are small. It follows that the algorithm is accurate only for small values of k_{max}. What is meant by "small" depends on the number of data points. If N is increased, k_{max} can be increased too because a larger N implies that the

```
begin find_dnn(x[][], N)
      choose k_max, i_max, E_r, and E_a
      set ln_r[] = find_ln_r(k_max, x[][], N)
      set i = 0
      set d = 0
      repeat
            set i = i + 1
            if (k = k_max) then
               exit--no convergence
            endif
            set d_old = d
            for k from 1 to k_max do
                set g_ln_r[k] = ln_r[k] + g(k, d)
            endfor
            set d = calculate_d(k_max, g_ln_r[])
      until (|d - d_old| < d E_r + E_a)
      return (d)
end find_dnn

begin find_ln_r(k_max, x[][], N)
      for k from 1 to k_max do
          for i from 1 to N do
              set nn[i][k] = 0
          endfor
      endfor
      for i from 1 to N - 1 do
          for j from i + 1 to N do
              set r = ||x[i][] - x[j][]||
              call sort(r, k_max, nn[i][])
              call sort(r, k_max, nn[j][])
          endfor
      endfor
      for k from 1 to k_max do
          set ln_r[k] = 0
          for i from 1 to N do
              set ln_r[k] = ln_r[k] + nn[i][k]
          endfor
          set ln_r[k] = ln(ln_r[k]/N)
      endfor
end find_ln_r
```

Figure 7.4: Pseudo-code for find_dnn and find_ln_r.

```
begin sort(r, k_max, nni[])
      set k = k_max
      while (k > 0 and nni[k] = 0.0) do
            set k = k - 1
      endwhile
      while (k > 0 and r < nni[k]) do
            if (k < k_max) then
               set nni[k + 1] = nni[k]
            endif
            set k = k - 1
      endwhile
      if (k < k_max) then
            set nni[k + 1] = r
      endif
end sort
```

Figure 7.5: Pseudo-code for a bubble-sort routine sort. Called by find_ln_r.

nearest neighbors are closer. Somorjai [1986] suggests choosing $k_{max} = \alpha\sqrt{N}$ with $\alpha \approx 0.5$. This rule-of-thumb yields $k_{max} = 16$ for $N = 1000$.

Lyapunov dimension

Calculation of the Lyapunov dimension requires calculation of the Lyapunov exponents. See Section 3.4.3 for details.

Discussion

Calculation of distances All three of the dimension algorithms presented above require the calculation of inter-point distances. A single distance calculation is not costly, but given N points, there are roughly $N^2/2$ distances that need to be computed. Since $N = 10000$ is not uncommon, the distance calculations are the most time consuming part of the algorithms.

There are a few techniques that can be used to decrease the number of distances that are calculated. One approach is to choose a small set of N_{ref} random reference points $\{x_{i_{ref}}\} \subset \{x_i\}$. The distances from these reference points to the other x_i are used by the algorithm. This reduces the number of distance calculations to approximately $N_{ref}(N - N_{ref}/2)$ thereby decreasing the number of distance calculations by a factor of $2N_{ref}/N$.

```
begin ell_one_norm(n, r_max, x[])
      set r = 0.0
      for i from 1 to n
          set r = r + |x[i]|
          if (r > r_max) then
             return r_max
          endif
      endfor
      return r
end ell_one_norm
```

Figure 7.6: Pseudo-code for ell_one_norm.

A second technique that improves the efficiency of the algorithms is to use the ℓ_1 norm to calculate r,

$$r = \|u\|_1 := |u_1| + \cdots + |u_n| \tag{7.47}$$

where $u = [u_1 \ \cdots \ u_n]^T$. The ℓ_1 norm avoids the multiplications and square root operation that are required to evaluate the Euclidean norm and reduces the number of floating-point operations by a factor of two.

A third technique is to calculate fully only those distances that are less than some limit r_{max}. The norm is usually calculated in a loop as is shown in Fig. 7.6. The distance calculation is aborted if d exceeds r_{max}. This approach can reduce the elapsed time for estimating the dimension by fifty percent. For D_{cap} and D_C, r_{max} is set to the largest value of ϵ that will be plotted. For D_{nn}, r_{max} is not constant; when r_{ij} is calculated, r_{max} is set to the maximum of the k_{max}th nearest neighbor of x_i and of x_j.

Calculation of the slope Typically, the slope of the log-log plot is determined using a least-squares fit,

$$m = \frac{K\sum_{i=1}^{K} x_i y_i - \sum_{i=1}^{K} x_i \sum_{i=1}^{K} y_i}{K\sum_{i=1}^{K} x_i^2 - \left(\sum_{i=1}^{K} x_i\right)^2} \tag{7.48}$$

where the points on the x-axis are $\{x_1, \ldots, x_K\}$ and the points on the y-axis are $\{y_1, \ldots, y_K\}$.

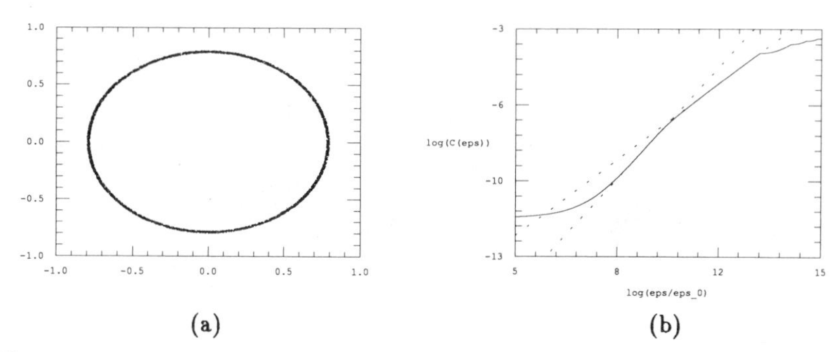

(a) (b)

Figure 7.7: A set with different dimensions at different length scales. (a) From a distance the set appears to be a circle; upon magnification, the set is seen to be an annulus; (b) a log-log plot of $C(\epsilon)$ versus ϵ.

Fig. 7.3 is typical of the log-log plots encountered when estimating dimension. It is clear from the figure that there is a limited range where the slope is approximately constant, and data only from this range should be used to estimate the slope. Unfortunately, there is currently no robust technique available to detect the useful portion of the plot automatically.

It can happen that the log-log plot has two or more regions with different slopes. Consider the set in Fig. 7.7 that, from a distance, appears to be a circle. Upon magnification, the set is seen to be a thin annulus, that is, it has non-zero width. At large length scales (i.e., for ϵ greater than the width of the annulus), the dimension of the set is 1. At smaller length scales (i.e., for ϵ less than the width of the annulus), the dimension is 2.

The dependence of the dimension on the length scale can be caused not only by the structure of the attractor, but by noisy data. Unless there is reason to believe otherwise, additive noise favors no particular direction in state space. Thus, additive noise tends to "fatten" the attractor. More precisely, if the attractor is embedded in $\mathbb{R}^n$, for length scales below the noise level, the dimension of the noisy attractor is n. As a simple example, let $x^* \in \mathbb{R}^2$ be a point on an attracting limit cycle. For some $\tau > 0$, let $\{x_i := \phi_{i\tau}(x^*)\}_{i=1}^{N}$ be the set of input points to a dimension-finding algorithm. Without noise, the x_i lie on a diffeomorphic copy of a circle, and the log-log plot has a single slope equal to 1. When noise is added to $\{x_i\}$, the noisy data lie in a thin annulus (as in Fig. 7.7(b)) and the log-log plot has two slopes. For ϵ greater than the noise level, the slope is 1. For ϵ less than the noise level, the slope is 2. See Ben-Mizrachi *et* *al.* [1984] and Ott *et al.* [1985] for a further discussion of the effects of noise on the estimation of dimension.

As a final note of caution, we observe that the statistically rigorous derivation of kth nearest-neighbor dimension shows that the log-log plot of $\bar{r}(k)$ versus k is not a straight line—there is the $g(k, D_{nn})$ correction term. This result makes one wonder whether it is unrealistic to expect the log-log plots for the other types of dimension to be straight lines. Until this question is answered, numerical estimates of dimension should be interpreted carefully and any unexpected or unusual results should be corroborated by other methods.

Accuracy The statistics and accuracy of these algorithms are not well-understood. How many data points are needed for an accurate estimate of the dimension? What is the mean and variance of the estimate? How do the statistics depend on the dimension of the embedding space?

Interested readers are referred to Caswell and Yorke [1986], Holzfuss and Mayer-Kress [1986], and Somorjai [1986] for more discussion on this topic.

The curse of dimensionality The algorithms presented in this section are reasonably accurate for low-dimensional attractors embedded in low-dimensional spaces (i.e., $\mathbb{R}^n$ for $1 \leq n \leq 5$). When the embedding dimension n increases, the accuracy and efficiency of the algorithms decrease. This is due to the so-called "curse of dimensionality." As the embedding dimension is increased, more and more of the embedding space is empty. Thus, to achieve statistically reliable estimates, the number of data points must increase. Furthermore, in higher dimensions, distances tend to be distributed over a more narrow range. For instance, when d is large, points distributed uniformly in the interior of a d-dimensional hypersphere tend to lie near the surface of the hypersphere[7] and as $d \to \infty$, the standard deviation of the inter-point distances approaches 0. Thus, any dimension algorithm that relies on statistics over a range of inter-point distances encounters difficulty with high-dimensional spaces.

One approach that bypasses the curse of dimensionality is the method of projection pursuit. This technique uses low-dimensional

[7] If r is a random variable uniformly distributed over the unit interval, then the random variable for the distance from the origin of points uniformly distributed in the interior of a d-dimensional unit hypersphere is $r^{1/d}$.

projections to form an estimate of the probability density which is then used to calculate any of the probabilistic dimensions. Interested readers are referred to Friedman *et al.* [1984], Huber [1985], and Somorjai [1986].

7.2 Reconstruction of attractors

In this section, we present a remarkable result, first proved by Takens [1980], that allows a strange attractor to be reconstructed from a sampled time waveform of just one component of the state. This is a useful technique in experimental settings. If data are gathered from measurements of a physical system, only one state variable needs to be measured, thereby cutting instrumentation and data storage costs. Moreover, for an infinite-order system or for a system where one or more of the state variables cannot be measured directly, reconstruction may be the only way to observe the attractor.

Let an attractor A of an nth-order system with flow ϕ_t be contained in an N-dimensional compact manifold $M \subset \mathbb{R}^n$. Define the reconstruction function $F: M \to \mathbb{R}^{2N+1}$ as

$$F(x) := [\,\phi_0^{(j)}(x) \;\; \phi_\tau^{(j)}(x) \;\; \cdots \;\; \phi_{2N\tau}^{(j)}(x)\,]^T \tag{7.49}$$

where $\phi_t^{(j)}(x)$ is the jth component of $\phi_t(x)$, j is arbitrary, and $\tau > 0$ is the sampling period, also arbitrary.

Generically, F is an embedding, that is, F diffeomorphically maps M onto some compact N-dimensional manifold $M' \subset \mathbb{R}^{2N+1}$.

This fact implies that given a sequence $\{y_k\} := \{\phi_{k\tau}^{(j)}(x)\}_{k=1}^{K}$ that corresponds to a uniformly time-sampled component of a trajectory that lies on an attractor A, the sequence of points

$$\begin{array}{c}
[\,y_0 \;\; y_1 \;\; \cdots \;\; y_{2N}\,]^T \\
[\,y_1 \;\; y_2 \;\; \cdots \;\; y_{2N+1}\,]^T \\
\vdots \\
[\,y_i \;\; y_{i+1} \;\; \cdots \;\; y_{i+2N}\,]^T \\
\vdots \\
[\,y_{K-2N} \;\; y_{K-2N+1} \;\; \cdots \;\; y_K\,]^T
\end{array} \tag{7.50}$$

lies on a diffeomorphic copy of A.

Several examples of reconstructed attractors are presented in Fig. 7.8. The reconstructed versions are definitely different from

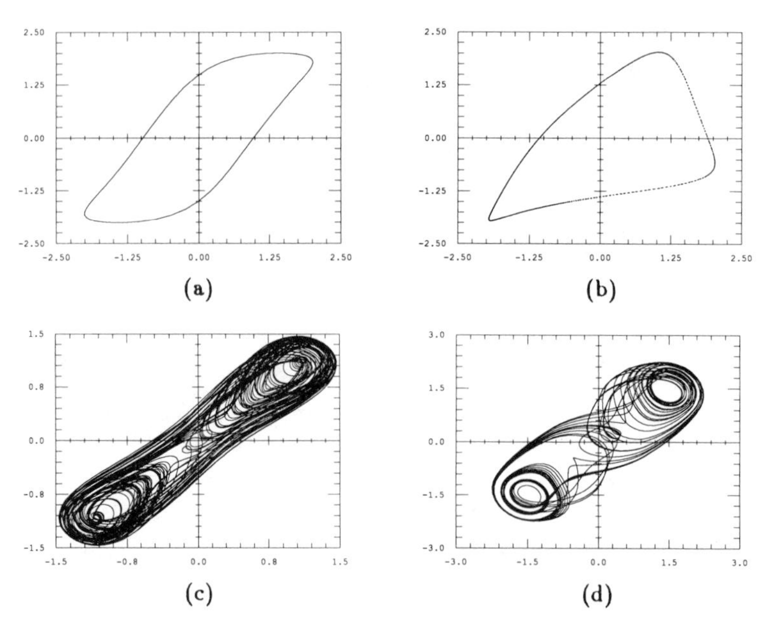

Figure 7.8: Several reconstructed attractors. (a) van der Pol limit cycle, $\tau = 0.6$s (c.f. Fig. 1.2); (b) van der Pol two-periodic solution (Poincaré map) (c.f. Fig. 2.5); (c) chaos in Duffing's equation, $\tau = 0.8$s (c.f. Fig. 1.8); (d) the double scroll, $\tau = 0.8$s (c.f. Fig. 1.9).

the originals, but can be seen to share the same qualitative features. More importantly, the dimension of the reconstruction is equal to the dimension of the original.

At first glance, it seems preposterous that one component of a trajectory carries enough information to reconstruct the entire attractor and, even if it does, that the simple function F accomplishes the feat. For F to be an embedding, it must be differentiable and possess a differentiable inverse. Differentiability of the flow with respect to the initial conditions guarantees that F is differentiable. What is surprising is that F is one-to-one, that is, that $F^{-1}: M' \to M$ exists. To show that invertibility is actually quite reasonable, we present two plausibility arguments.

Plausibility argument 1: This argument suggests that F is one-to-one by demonstrating the invertibility of a similar function.

Define $\hat{F}: \mathbb{R}^n \to \mathbb{R}^n$ by

$$\hat{F}(x) := [\,\phi_0^{(j)}(x)\ \phi_\tau^{(j)}(x) \quad \cdots \quad \phi_{(n-1)\tau}^{(j)}(x)\,]^T \tag{7.51}$$

for some $\tau > 0$. We will argue that $\hat{F}(x)$ identifies the trajectory $\phi_t(x)$ uniquely. Since there is a one-to-one correspondence between trajectories and initial conditions, $\hat{F}$ is one-to-one.

To identify uniquely a trajectory of an nth-order system, n independent pieces of information are required. This information can be specified in several ways. For example, one may:

1. Give the initial condition: $x = \phi_0(x)$.

2. Give the first $n-1$ derivatives of the jth component of the state: $[\phi_0^{(j)}(x) \;\; D_\tau\phi_0^{(j)}(x) \;\; \cdots \;\; D_\tau^{n-1}\phi_0^{(j)}(x)]^T$.

3. Give n samples of the jth component of the state: $[\phi_0^{(j)}(x) \;\; \phi_\tau^{(j)}(x) \;\; \cdots \;\; \phi_{(n-1)\tau}^{(j)}(x)]^T$.

Method 1 is the most familiar. It always works and is the standard way of specifying a trajectory.

Method 2 may be familiar to some readers. It is the standard way of specifying the initial condition of an nth-order scalar differential equation. Thus, method 2 works whenever the state equation can be transformed into a single nth-order scalar differential equation; it fails when the state equation is not sufficiently coupled. For example, consider the second-order linear system

$$\begin{bmatrix} \dot{x}_1 \\ \dot{x}_2 \end{bmatrix} = \begin{bmatrix} \alpha & \epsilon \\ \epsilon & \beta \end{bmatrix} \begin{bmatrix} x_1 \\ x_2 \end{bmatrix}. \tag{7.52}$$

When $\epsilon = 0$, x_1 and x_2 are completely uncoupled and, therefore, method 2 fails. In fact, when $\epsilon = 0$, $\phi_0^{(1)}(x)$ and $D_\tau\phi_0^{(1)}(x)$ cannot be specified independently. For the generic case, however, when $\epsilon \neq 0$, method 2 does specify a unique trajectory.

Like method 2, method 3 involves only one component of the state, so like method 2, it fails when the system is insufficiently coupled. In addition, method 3 involves time-sampling with a sampling period τ. If the trajectory is periodic with period τ, then each of the samples $\phi_{k\tau}^{(j)}(x)$ is identical, and the n samples do not specify independent information. Observe, however, that with any slight perturbation of τ, the n samples are independent and, generically, method 3 specifies a unique trajectory. It follows that $\hat{F}$ is generically one-to-one.

If $2N+1 = n$, the reconstruction function F is identical to $\hat{F}$, so for this value of N, F is generically one-to-one. When $2N+1 > n$, then the $2N+1-n$ additional components, $\phi_{n\tau}^{(j)}(x)$, ... $\phi_{2N\tau}^{(j)}(x)$,

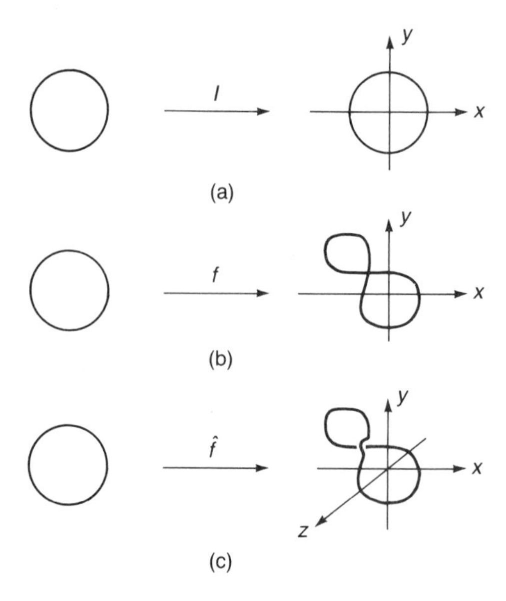

Figure 7.9: Maps of the circle into Euclidean space. (a) An embedding in $\mathbb{R}^2$; (b) a differentiable map $h: S^1 \to \mathbb{R}^2$ that is not an embedding; (c) when the range of h is changed to $\mathbb{R}^3$, almost any slight perturbation yields an embedding $\hat{h}: S^1 \to \mathbb{R}^3$.

of F are uniquely determined by the the first n components, $\phi_0^{(j)}(x)$, $\phi_\tau^{(j)}(x)$, ... $\phi_{(n-1)\tau}^{(j)}(x)$, and again F is generically one-to-one.

Plausibility argument 2: The preceding argument shows that F is generically one-to-one if $2N+1 \geq n$. To show that F is one-to-one for any n, we resort to some ideas from differential topology.

A classic result of differential topology is Whitney's Theorem: If M is an N-dimensional manifold, then there exists an embedding $f: M \to \mathbb{R}^{2N}$. For example, the circle, which is a one-dimensional manifold, can be embedded in $\mathbb{R}^2$ (see Fig. 7.9(a)), but not in $\mathbb{R}$.

Whitney's Theorem guarantees that M can be embedded in $\mathbb{R}^{2N}$, but it is of little direct use in showing that F is an embedding in $\mathbb{R}^{2N+1}$.

Modified version of Whitney's Theorem (Takens [1980]): If M is compact and $F: M \to \mathbb{R}^{2N+1}$ is twice continuously differentiable, then, generically, F is an embedding.

As an example, let $h: S^1 \to \mathbb{R}^2$ be a continuously differentiable map that transforms the circle into a figure eight (Fig. 7.9(b)). h is not an embedding because it is not one-to-one. Furthermore, no small perturbation of h can make it one-to-one. But if the dimension of the range is increased to three, almost every perturbation produces a one-to-one map $\hat{h}: S^1 \to \mathbb{R}^3$ (Fig. 7.9(c)) and the property of being one-to-one is generic.

Since the dimensions of the domain and range of the reconstruction map F satisfy the modified theorem and F is differentiable due to the differentiability of a trajectory with respect to the initial condition,[8] F is generically a diffeomorphism.

This argument ensures that $\mathbb{R}^{2N+1}$ is always a sufficient range in which to reconstruct an attractor. It is quite possible that a reconstruction space with dimension less than $2N + 1$ can be used. For example, as the first plausibility argument shows, if $n \leq 2N + 1$, then $\mathbb{R}^n$ is sufficient. For example, the reconstruction theorem guarantees that the van der Pol limit cycle can be reconstructed in $\mathbb{R}^3$, but a few quick sketches should convince the reader that, generically, F reconstructs the limit cycle in $\mathbb{R}^2$.

There are two questions remaining. How is τ picked, and what happens if N is not known?

Choosing τ Almost every τ works, but there are three practical limitations. First, if τ is too small, then $\phi^{(j)}_{k\tau}(x) \approx \phi^{(j)}_{(k+1)\tau}(x)$ and the reconstructed attractor is restricted to the diagonal of the reconstruction space. Second, if τ is too large and the system is chaotic, then $\phi^{(j)}_{k\tau}(x)$ and $\phi^{(j)}_{(k+1)\tau}(x)$ are uncorrelated (to working precision) and the structure of the reconstructed attractor disappears. Finally, if τ is close to some periodicity in the system, the component at that period will be under-represented in the reconstruction. This effect is similar to the stroboscopic action of the Poincaré map which freezes a particular periodic component. Fig. 7.10 shows a reconstructed attractor for various values of τ.

The algorithms for estimating the dimension of an attractor, require a large number of data points as input. The number of data points can be lessened if the set of data points is chosen to maximize the amount of information it contains. This idea was proposed by Holzfuss and Mayer-Kress [1986] and leads to an algorithm that

---

[8]That $\phi_t(x)$ is *twice* differentiable with respect to the initial condition is more difficult to show. In practice, the reconstruction seems to work well for some systems—piecewise-linear systems are a good example—that are not twice differentiable.

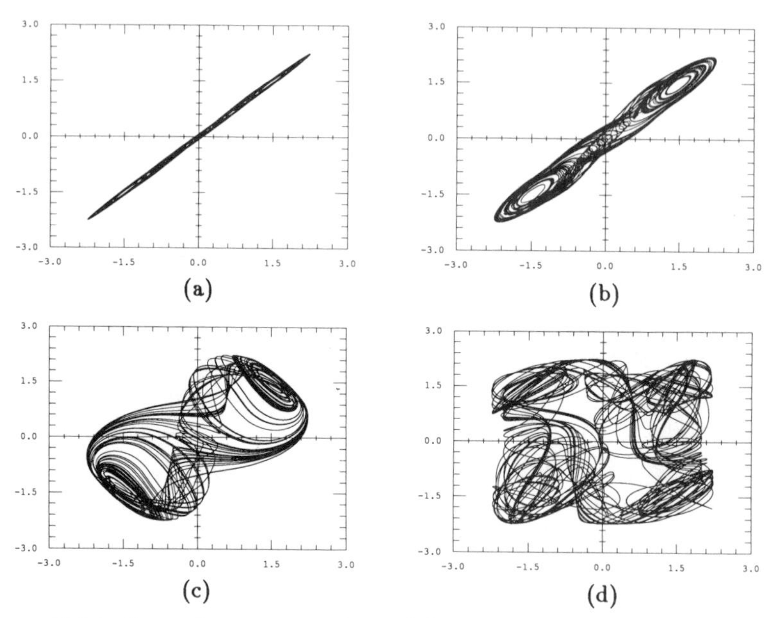

Figure 7.10: Several reconstructions of the double scroll attractor. (a) $\tau = 0.05$; (b) $\tau = 0.25$; (c) $\tau = 1.25$; (d) $\tau = 6.25$.

selects τ to be the time delay where the first local minimum in the autocorrelation function occurs.

Choosing N If N is not known, it can be found as follows. Let n_r be the dimension of the reconstruction space. Set $n_r = 1$ and calculate the dimension D_{cap}, say, of the reconstructed attractor. Repeat this calculation, incrementing n_r each time, until D_{cap} does not change any more. This final D_{cap} is the proper value of the correlation dimension, and the lowest n_r yielding this value is the minimum dimension of the reconstruction space. Grassberger and Procaccia [1983] have used this technique to reconstruct a finite-dimensional attractor of an infinite-order system.

The discrete-time case Reconstruction is the same in the discrete-time case except that the time delays in the definition of F are replaced with iteration delays. Let $P: \mathbb{R}^p \to \mathbb{R}^p$ be a diffeomorphism with an attractor A that lies on an N-dimensional manifold M. The reconstruction function $F: M \to \mathbb{R}^{2N+1}$ is defined by

$$F(x) := [\, x^{(j)} \quad P^m(x)^{(j)} \quad \cdots \quad P^{2Nm}(x)^{(j)} \,]^T \tag{7.53}$$

where m is any positive integer and the superscript (j) indicates the jth component of the state. If P is twice continuously differentiable, then, generically, F is an embedding and can be used to reconstruct the attractor A.

7.3 Summary

- *Dimension*: The dimension of an attractor is a lower bound on the number of state variables needed to describe the steady-state behavior. The dimension of a non-chaotic attractor is an integer, and the dimension of a chaotic attractor is almost always a non-integer.

- *Fractal*: A set with non-integer dimension.

- *Fractal dimension*: Generic term for a dimension that allows non-integer values.

- *Scenario for definitions*: Cover an attractor $A \subset \mathbb{R}^n$ with $N(\epsilon)$ n-dimensional volume elements (e.g., n-dimensional hyperspheres) of radius ϵ. Let P_i be the relative frequency of visitation of a typical trajectory to the ith volume element of this cover.

- *Capacity*:

$$D_{cap} := \lim_{\epsilon \to 0} \frac{\ln N(\epsilon)}{\ln(1/\epsilon)}.$$

  Alternatively, $N(\epsilon) \propto (1/\epsilon)^{D_{cap}}$ for some $k > 0$.

- *Information dimension*:

$$D_I := \lim_{\epsilon \to 0} \frac{H(\epsilon)}{\ln(1/\epsilon)}$$

  where $H(\epsilon) := -\sum_{i=1}^{N(\epsilon)} P_i \ln P_i$.

  Alternatively, $H(\epsilon) = k\epsilon^{-D_I}$ for some $k > 0$.

- *Correlation dimension:*

$$D_C := \lim_{\epsilon \to 0} \frac{\ln \sum_{i=1}^{N(\epsilon)} P_i^2}{\ln \epsilon}.$$

Alternatively,

$$D_C = \lim_{\epsilon \to 0} \frac{\ln C(\epsilon)}{\ln \epsilon}.$$

where

$$C(\epsilon) := \lim_{N \to \infty} \frac{1}{N^2} \{ \textit{the number of pairs of points } (x_i, x_j) \textit{ such that } \|x_i - x_j\| < \epsilon \}.$$

Alternatively, $C(\epsilon) = k\epsilon^{D_{cap}}$ for some $k > 0$.

- *kth nearest-neighbor dimension:*

$$D_{nn} := \frac{\ln k + c(x_1, \ldots, x_N)}{g(k, D_{nn}) + \ln \bar{r}(k)}$$

where $\bar{r}(k)$ is the mean over the sample points of the distance between a point and its kth nearest neighbor. Since g is small and c is constant given the set of sample points, this equation can be rewritten as

$$\begin{aligned} \bar{r}(k) &\approx e^{c(x_1, \ldots, x_N)} k^{1/D_{nn}} \\ &= c' k^{1/D_{nn}} \end{aligned}$$

where $c' \in \mathbb{R}$ is a constant.

- *Lyapunov dimension:* Let j be the largest integer such that $\lambda_1 + \cdots + \lambda_j \geq 0$. Then

$$D_L := j + \frac{\lambda_1 + \cdots + \lambda_j}{|\lambda_{j+1}|}.$$

If no such j exists, D_L is defined to be 0.

- *Relationships between dimensions:* In general, $D_C \leq D_I \leq D_{cap}$. For uniform probability densities, $D_C = D_I = D_{cap}$.

- *Reconstruction*: Let $A \subset \mathbb{R}^n$ be an attractor that lies in an N-dimensional manifold. Given a sequence $\{y_k\} := \{\phi_{k\tau}^{(j)}(x)\}_{k=1}^{K}$ that corresponds to a uniformly time-sampled component of a trajectory that lies on an attractor A, then, generically, the sequence of points

$$
\begin{gathered}
[\,y_0 \;\; y_1 \;\; \cdots \;\; y_{2N}\,]^T \\
[\,y_1 \;\; y_2 \;\; \cdots \;\; y_{2N+1}\,]^T \\
\vdots \\
[\,y_i \;\; y_{i+1} \;\; \cdots \;\; y_{i+2N}\,]^T \\
\vdots \\
[\,y_{K-2N} \;\; y_{K-2N+1} \;\; \cdots \;\; y_K\,]^T
\end{gathered}
$$

lies on a diffeomorphic copy of A and can be used to calculate the dimension of A.

Chapter 8

Bifurcation Diagrams

8.1 Definitions and Theory

Consider an nth-order continuous-time system

$$\dot{x} = f(x, \alpha) \tag{8.1}$$

with a parameter $\alpha \in \mathbb{R}$. As α changes, the limit sets of the system also change. Typically, a small change in α produces small quantitative changes in a limit set. For instance, perturbing α could change the position of a limit set slightly, and if the limit set is not an equilibrium point, its shape or size could also change. There is also the possibility that a small change in α can cause a limit set to undergo a qualitative change. Such a qualitative change is called a *bifurcation* and the value of α at which a bifurcation occurs is called a *bifurcation value.*

Remark: A qualitative change can occur only when the system is structurally unstable. Thus, the set of bifurcation values is the set of parameter values at which the system is structurally unstable.

Examples of bifurcations are the disappearance or creation of a limit set and a change in the stability type of a limit set (e.g., stable to non-stable). The definition of bifurcation also includes the case of a non-stable limit set that remains non-stable but undergoes a change in the number of positive Lyapunov exponents.

Example 8.1 Saddle-node bifurcation:
Consider the first-order system

$$\dot{x} = \alpha - x^2. \tag{8.2}$$

Figure 8.1: Phase portraits for a saddle-node bifurcation. (a) $\alpha < 0$; (b) $\alpha > 0$.

For $\alpha < 0$, (8.2) has no equilibrium points (Fig. 8.1(a)). At $\alpha = 0$, there is a single equilibrium point at the origin with eigenvalue 0. For $\alpha > 0$, there is a stable equilibrium point at $\sqrt{\alpha}$ with eigenvalue $-2\sqrt{\alpha}$, and an unstable equilibrium point at $-\sqrt{\alpha}$ with eigenvalue $2\sqrt{\alpha}$ (Fig. 8.1(b)). Since two equilibrium points are created as α passes through 0, $\alpha = 0$ is a bifurcation value for this system.

The bifurcation in Example 8.1 is depicted graphically in Fig. 8.2. This *bifurcation diagram* is a plot of the position of the equilibrium points versus the bifurcation parameter α. A solid line is often used to indicate a stable limit set and a dashed line, to indicate an unstable limit set.

Remarks:

1. The bifurcation diagram is composed of continuous curves. This is another way of saying that a small change in α results in a small change in the position of the limit set.
2. Since hyperbolic limit sets are structurally stable, a limit set must be non-hyperbolic at a bifurcation value. In Example 8.1, there is one equilibrium point for $\alpha = 0$. This equilibrium point, located at the origin, has eigenvalue 0 and is, therefore, non-hyperbolic.

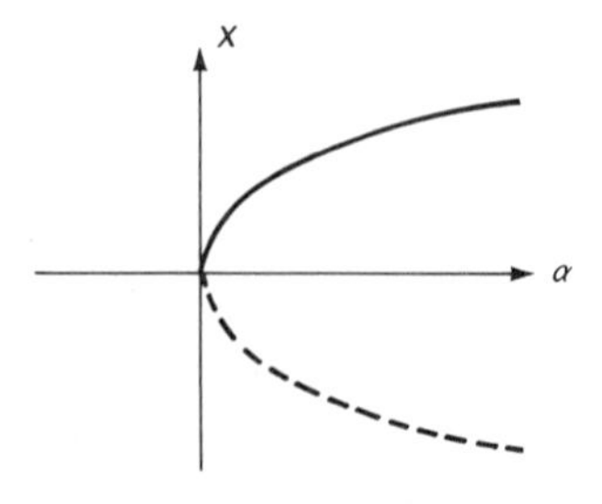

Figure 8.2: A bifurcation diagram for a saddle-node bifurcation. A solid curve indicates a stable equilibrium point; a dashed curve indicates an unstable equilibrium point.

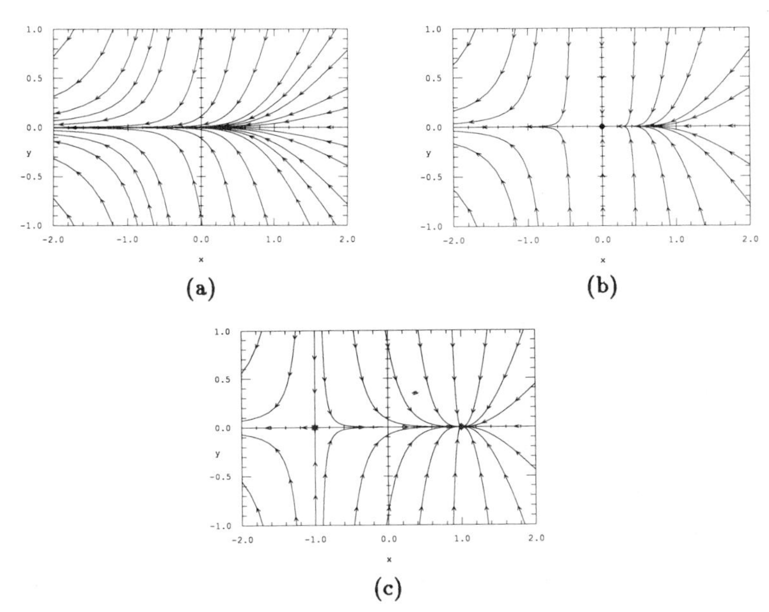

Figure 8.3: Saddle-node bifurcation in two dimensions. The system is $\dot{x} = \alpha - x^2$, $\dot{y} = -5y$. (a) $\alpha = -1$; (b) $\alpha = 0$; (c) $\alpha = 1$.

In Fig. 8.1(b), there is no saddle point, so why is this called a "saddle-node" bifurcation? The reason is clear when this bifurcation occurs in a higher-order system. Fig. 8.3 shows a saddle-node bifurcation in a two-dimensional system. For $\alpha > 0$, one equilibrium is a node and the other is a saddle.

Example 8.2 Pitchfork bifurcation:
Consider the first-order system

$$\dot{x} = \alpha x - x^3. \tag{8.3}$$

For any value of α, there exists an equilibrium point at the origin. Its eigenvalue is equal to α, so it is stable for $\alpha < 0$ and unstable for $\alpha > 0$. For $\alpha > 0$, there are two additional equilibrium points at $\pm\sqrt{\alpha}$. Both these equilibrium points have eigenvalue $-2/\alpha$, so the equilibrium points are stable. The position and stability of the equilibrium points are shown in Fig. 8.4. A bifurcation occurs at $\alpha = 0$ since at this bifurcation value, the equilibrium point at the origin changes stability type and two new equilibrium points are created. The bifurcation diagram for a pitchfork bifurcation is shown in Fig. 8.5.

Figure 8.4: Phase portraits for a pitchfork bifurcation. (a) $\alpha < 0$; (b) $\alpha > 0$.

In the previous two examples, a bifurcation created a pair of equilibrium points and, in addition, in Example 8.2, an equilibrium point changed stability type at the bifurcation. In the next example, a bifurcation creates a limit cycle.

Example 8.3 Hopf bifurcation in a continuous-time system: Consider the second-order system

$$\begin{aligned} \dot{x} &= y - x(x^2 + y^2 - \alpha) \\ \dot{y} &= -x - y(x^2 + y^2 - \alpha) \end{aligned} \tag{8.4}$$

This system has an equilibrium point at the origin with eigenvalues $\alpha \pm i$ where $i = \sqrt{-1}$. For $\alpha < 0$, the equilibrium point is stable. When α is increased to $\alpha = 0$, the equilibrium point becomes nonhyperbolic (it has pure imaginary eigenvalues), and for $\alpha > 0$, the equilibrium point is unstable. Furthermore, a stable limit cycle, given by the solution set of $x^2 + y^2 = \alpha$, exists for $\alpha > 0$. Since the equilibrium point changes stability type at $\alpha = 0$ and a new limit set is created, $\alpha = 0$ is a bifurcation value. This process of a pair of complex conjugate eigenvalues passing through the imaginary axis thereby spawning a limit cycle is called a *Hopf bifurcation.*

Phase portraits for the Hopf bifurcation are shown in Fig. 8.6 and a bifurcation diagram is shown in Fig. 8.7. This diagram is plotted in α-x-y space. For each value of $\alpha = \alpha_0$, the limit sets of the system

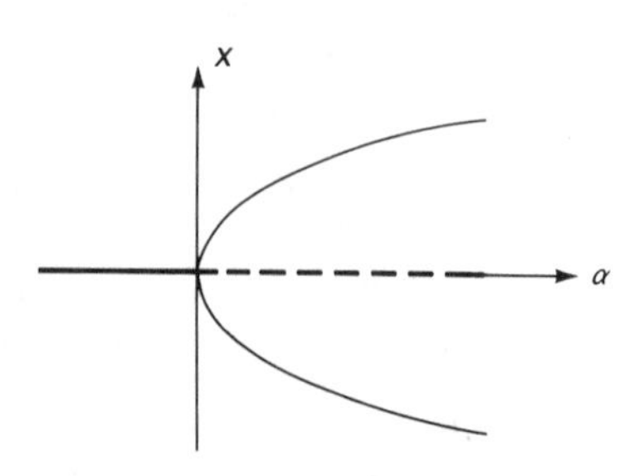

Figure 8.5: A bifurcation diagram for a pitchfork bifurcation.

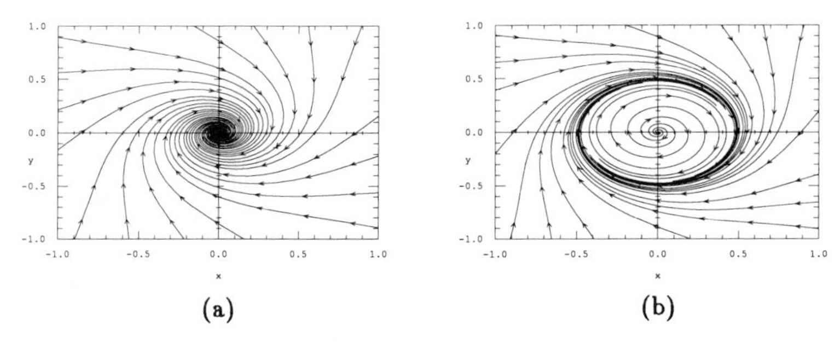

Figure 8.6: Phase portraits for a Hopf bifurcation in system (8.4). (a) $\alpha = -0.2$; (b) $\alpha = 0.2$.

are drawn in the plane defined by $\alpha = \alpha_0$. The parabolic surface indicates the limit cycle.

Example 8.4 Period doubling:
Consider the logistic equation, a one-dimensional discrete-time system defined by

$$x_{k+1} = f(x_k, \alpha) := 4\alpha x_k(1 - x_k). \tag{8.5}$$

For $0 \leq \alpha \leq 1$, f maps the unit interval to itself and this is the range of α that interests us.

The origin is a fixed point of the logistic equation with characteristic multiplier 4α. Thus, it is stable for $0 \leq \alpha < 0.25$. A second fixed point, $x^* = 1 - 1/4\alpha$, exists for $\alpha > 0.25$. The characteristic multiplier for x^* is $2 - 4\alpha$, so x^* is stable for $0.25 < \alpha < 0.75$. As α passes through 0.75, x^* becomes unstable and spawns a stable period-two closed orbit. Observe that the period-one fixed point still exists after the period-two orbit is created, though it has become unstable.

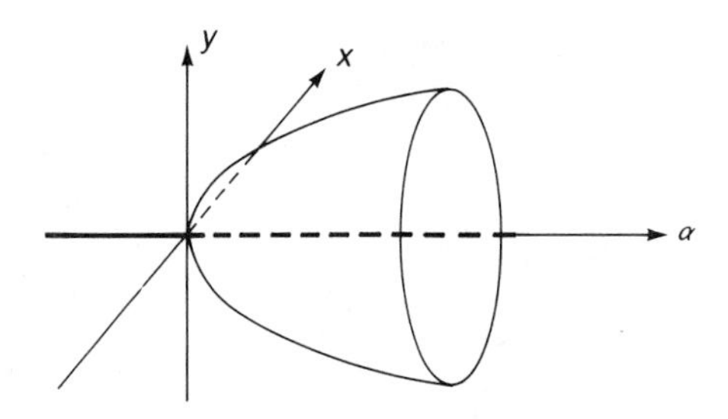

Figure 8.7: Bifurcation diagram for the Hopf bifurcation.

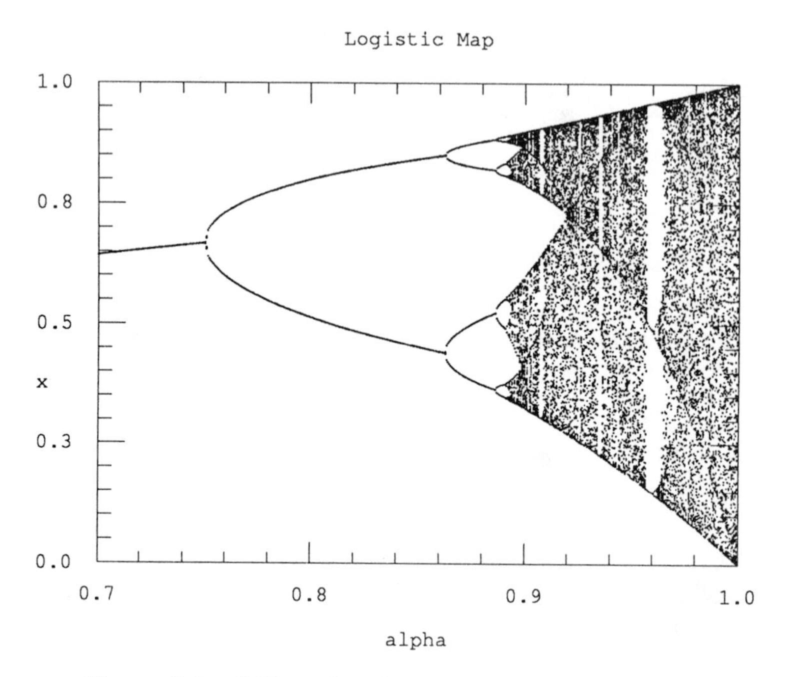

Figure 8.8: Bifurcation diagram for the logistic map.

As α is increased further, the period-two closed orbit becomes unstable and spawns a stable period-four closed orbit. This period-four orbit spawns a period-eight orbit and so on. These period-doubling bifurcations accumulate at a bifurcation value α^* at which the system becomes chaotic. This *period-doubling route to chaos* is shown in the bifurcation diagram of Fig. 8.8. Only the stable (i.e., observable) limit sets are plotted in Fig. 8.8. A period-K closed orbit is indicated by plotting K points for a particular value of α. A smear of points for a particular value of α indicates chaotic behavior.

As Fig. 8.8 shows, there are regions of periodic behavior between regions of chaotic behavior. Such *periodic windows* occur frequently in chaotic systems.

The family of maps on the interval, of which the logistic map is the prime example, has been studied extensively. These maps display quite a bit of surprising behavior. For example, let α_k be the bifurcation value for the bifurcation from a period-2^k to a period-2^{k+1} orbit. It can be shown (Feigenbaum [1978]) that for sufficiently

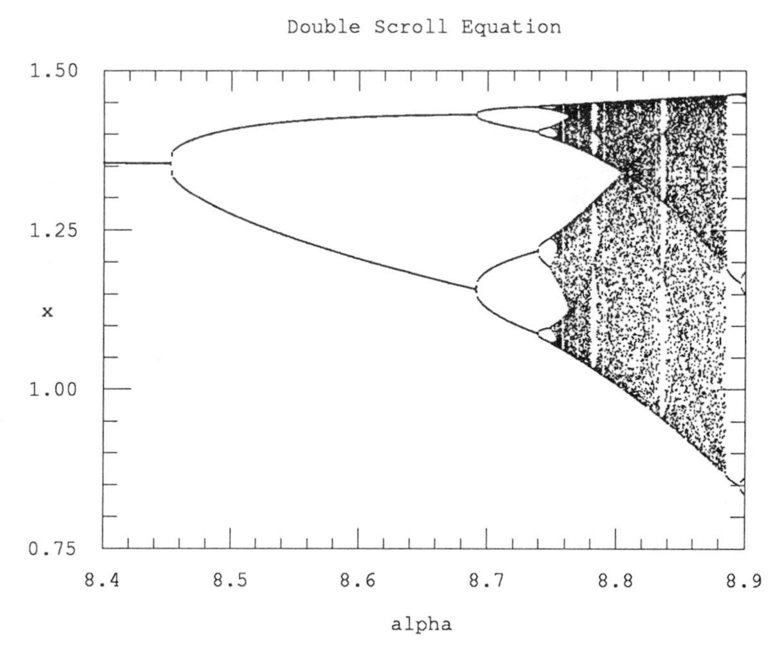

Figure 8.9: Bifurcation diagram for the double scroll equation with $\beta = 15$.

smooth maps with a quadratic maximum, the number

$$\delta := \lim_{k \to \infty} \frac{\alpha_k - \alpha_{k-1}}{\alpha_{k+1} - \alpha_k} \tag{8.6}$$

is a constant that is independent of the exact shape of the map. The constant δ is called *Feigenbaum's number* and is approximately 4.669. Readers who wish to learn more about maps of the interval are referred to Collet and Eckmann [1980].

Period-doubling also occurs in continuous-time systems. Fig. 8.9 shows a bifurcation diagram for the double scroll equation (1.24). Details of how Fig. 8.9 was calculated will be presented in Section 8.2. For now, think of it as the bifurcation diagram of the Poincaré map of the double scroll system. Thus, for a given parameter value, a single point indicates a period-one limit cycle, K points indicate a period-K limit cycle, and a smear of points indicates chaos.

The similarity between Figs. 8.8 and 8.9 is remarkable. The logistic map is not invertible and, therefore, cannot be embedded in the Poincaré map of a continuous-time system. It must be that a

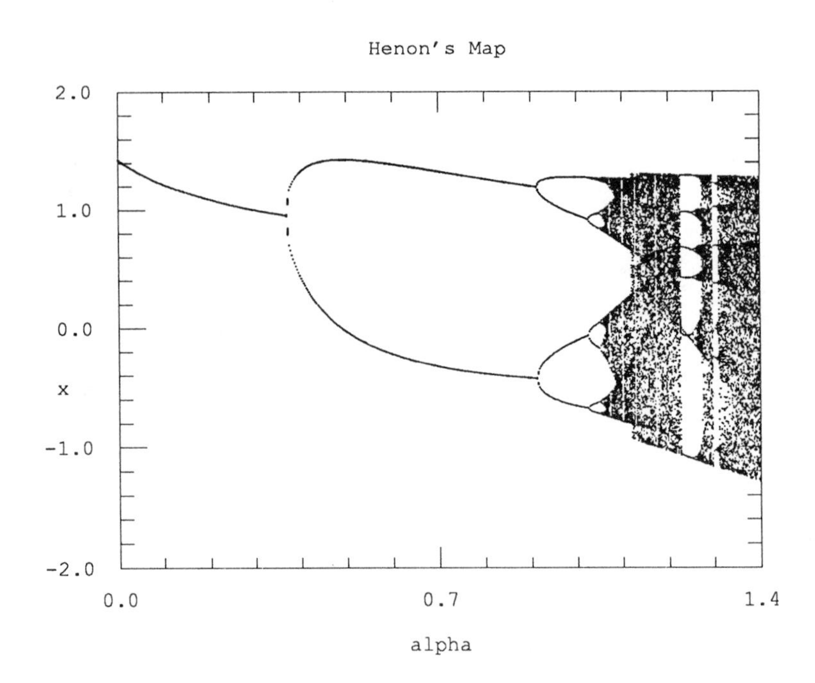

Figure 8.10: Bifurcation diagram for Henon's map for $\beta = 0.3$.

similar but invertible map is embedded in the Poincaré map of the double scroll. One candidate is Henon's map

$$\begin{aligned} x_{k+1} &= y_k + 1 - \alpha x_k^2 \\ y_{k+1} &= \beta x_k. \end{aligned} \tag{8.7}$$

A bifurcation diagram for Henon's map is shown in Fig. 8.10.

Example 8.5 Hopf bifurcation in a discrete-time system:
As a final example of the different types of bifurcation, we present the Hopf bifurcation of a fixed point of a map. This bifurcation is the discrete-time analogue of the Hopf bifurcation of an equilibrium point. In a discrete-time Hopf bifurcation, an invariant closed curve is created as a stable fixed point loses stability when the real parts of its (complex conjugate pair of) characteristic multipliers pass through the unit circle.

If the map is a Poincaré map, this Hopf bifurcation can be interpreted as a stable limit cycle which loses stability and spawns a stable two-torus in the process. Precisely this state of affairs occurs

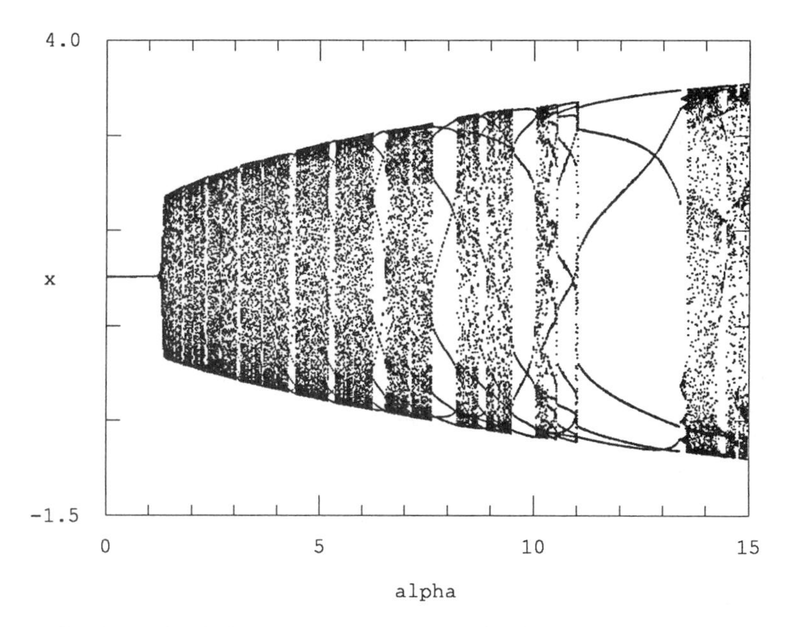

Figure 8.11: The bifurcation diagram for (8.8) with $a = 0.07$, $b = 0.1$. The bifurcation diagram was calculated using the Poincaré map defined by the plane $y = 0$.

in the third-order piecewise-linear system

$$\begin{aligned} \dot{x} &= -\alpha f(y-x) \\ \dot{y} &= -f(y-x) - z \\ \dot{z} &= y \end{aligned} \tag{8.8}$$

where the piecewise-linear function f is defined by

$$f(u) := -au + 0.5(a+b)(|u+1| - |u-1|). \tag{8.9}$$

A bifurcation diagram of this system is shown in Fig. 8.11. For the given values of a and b, the Hopf bifurcation occurs at $\alpha \approx 1.0$. The continuous-time limit sets, both before and after the bifurcation, are shown in Fig. 8.12.

It is apparent from Fig. 8.11 that (8.8) undergoes a variety of bifurcations. Within the region where the torus exists, periodic windows occur where the steady state phase-locks onto a subharmonic. It is hard to see in Fig. 8.11, but for several values of K there exists a period-$(2K+1)$ window between the period-K and period-$(K+1)$ windows. For instance, between the period-6 and period-7 windows,

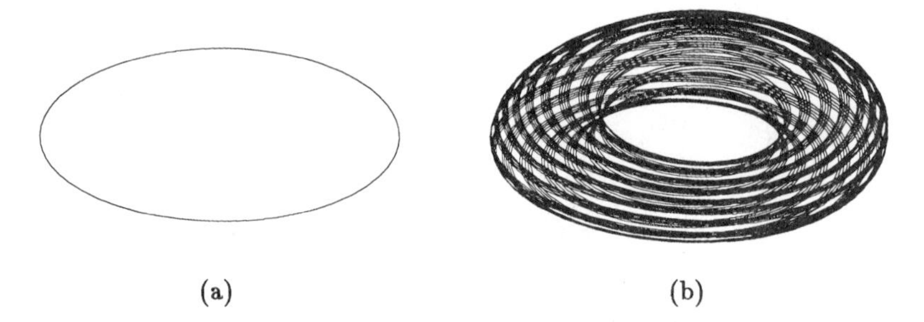

Figure 8.12: Limit sets of (8.8) with $a = 0.07$ and $b = 0.1$ (a) before the discrete-time Hopf bifurcation, at $\alpha = 0.5$; (b) after the Hopf bifurcation, at $\alpha = 2.0$.

there is a period-13 window. This type of bifurcation sequence is called *period-adding.*

A detailed study of the bifurcation behavior of this system is presented in Matsumoto *et al.* [1987].

8.2 Algorithms

The calculation of bifurcation diagrams comes down to one basic but difficult task: for a given parameter α, find all the steady-state solutions.

We present two algorithms. The brute-force algorithm (sometimes referred to as parameter stepping) tackles the problem in a straightforward but inefficient manner. The continuation method is more efficient, but cannot handle chaotic behavior. Besides its computational benefits, we will see that the theory behind the continuation method yields valuable insight into the structure and interpretation of bifurcation diagrams.

8.2.1 Brute force

The *brute-force method* is a general method for calculating bifurcation diagrams of stable steady-state solutions of arbitrary systems. Its advantages are generality and simplicity. Its drawbacks are long simulation times and the inability to locate unstable or non-stable steady-state behavior.[1]

[1]Brute force can find unstable limit sets of a system that is invertible by running the system in reverse time.

```
begin brute_bif
      choose x0[], α_min, α_max, K, i_d, N_t, N_ss
      set x[] = x0[]
      for k from 1 to K
          set α = α_min + (k − 1)(α_max − α_min)/(K − 1)
          for i from 1 to N_t do
              set x[] = P_b(x[], α)
          endfor
          for i from 1 to N_ss do
              set x[] = P_b(x[], α)
              plot α versus x[i_d]
          endfor
      endfor
end brute_bif
```

Figure 8.13: Pseudo-code for brute_bif.

Definition

Given a map, $P_b: \mathbb{R}^p \times \mathbb{R} \to \mathbb{R}^p$, that depends on a parameter $\alpha \in \mathbb{R}$, we would like to calculate the bifurcation diagram of P_b over the parameter range $\alpha_{min} \leq \alpha \leq \alpha_{max}$.

Pseudo-code for the brute-force technique is shown in Fig. 8.13. Choose an integer i_d such that $1 \leq i_d \leq p$. i_d is the index of the state component that will be displayed. Choose an integer $K > 0$. The system is simulated for K evenly spaced parameter values between α_{min} and α_{max},

$$\alpha_k := \alpha_{min} + (k-1)\frac{\alpha_{max} - \alpha_{min}}{K-1}, \qquad k = 1, \ldots, K. \tag{8.10}$$

For the kth simulation, the system is iterated for a fixed number of iterations. The first N_t iterations are assumed to be the transient and are discarded. After these N_t iterations, the system is assumed to be in the steady state and $x[i_d]$ of the next N_{ss} iterations is plotted on the bifurcation diagram.

Definition of P_b for continuous-time systems

Non-autonomous systems For a non-autonomous continuous-time system, P_b is chosen to be the Poincaré map.

Autonomous systems One way to define P_b is as the Poincaré map (Chapter 2). This approach has the drawback that one must

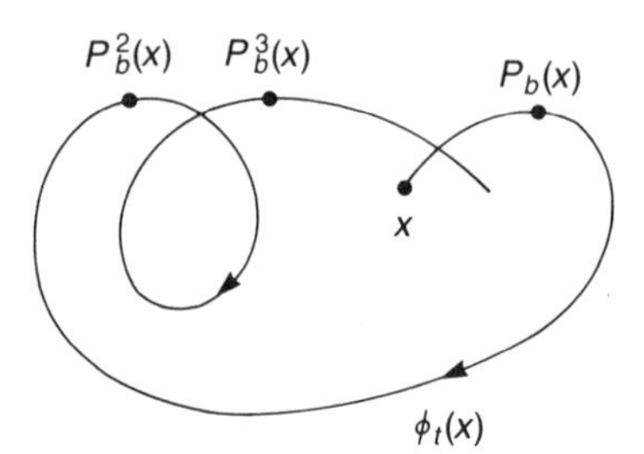

Figure 8.14: The map P_b for an arbitrary trajectory of an autonomous system.

know in advance a cross-section Σ that the steady state intersects for $\alpha_{min} \leq \alpha \leq \alpha_{max}$. Since the limit set might move as α is changed, it may be impossible to find a suitable Σ for some systems.

A more robust definition for P_b is as the next point on $\phi_t^\alpha(x)$ such that the i_dth component of the state is at a local maximum. Here ϕ_t^α denotes the flow for parameter value α. This idea is illustrated in Fig. 8.14.

Note that P_b is undefined when $\phi_t^\alpha(x)$ has no local maximum for $t > 0$. This can occur when the trajectory is unbounded or when x lies in the basin of attraction of an equilibrium point that has no complex eigenvalues. Equilibrium points with complex eigenvalues pose no difficulty because complex eigenvalues indicate rotational behavior and imply an infinity of local maxima.

Remark: If a local maximum for the i_dth component of the state occurs at x, then the i_dth component of the vector field at x is zero. Thus, using local maxima to define P_b is equivalent to using the Poincaré map with a (nonlinear) cross-section defined by the manifold $\{x : f_{i_d}(x) = 0\}$ where $f_{i_d}(x)$ is the i_dth component of $f(x)$.

Practical considerations

Calculating P_b in the autonomous case If P_b is defined as the Poincaré map, then the algorithm `find_crossing`, presented in Section 2.4.2, can be used to calculate it.

A similar algorithm can be used to calculate P_b when it is defined as the next local maximum of the i_dth component of $\phi_t^\alpha(x)$. The difference from the hyperplane crossing case is that instead of finding the time t where the function

$$H(\phi_t(x)) := \langle h, \phi_t(x) - x_\Sigma \rangle \tag{8.11}$$

equals 0, we need t such that the i_dth component of $f(\phi_t^\alpha(x), \alpha)$ is 0. Thus, `find_crossing` can be used to calculate $P_b(x, \alpha)$ as long as H is redefined as $H(x) := f_{i_d}(x, \alpha)$ where f_{i_d} is the the i_dth component of the vector field f.

Transient considerations The brute-force approach makes one major assumption: the steady state is reached during the first N_t iterations. To ensure that this assumption is satisfied, a large value of N_t can be used, but this results in long simulation times especially when simulating continuous-time systems.

To reduce the duration of the transients, `brute_bif` performs the simulations in order from α_1 to α_K and uses the final condition of the kth simulation as the initial condition for the $(k+1)$th simulation. The initial condition for the first simulation is provided by the user. If K is reasonably large (i.e., α_k is close to α_{k+1}) and if no bifurcations occur between α_k and α_{k+1}, then the steady-state solutions at α_k and α_{k+1} are approximately the same. Thus, if the kth simulation achieves steady state, then the initial condition for the $(k+1)$th simulation lies close to the steady-state solution and, therefore, the duration of the transient is reduced considerably. Even if the simulation at α_k has not achieved steady state—this is often the case for $k = 1$ or directly after a bifurcation occurs—α_k is still a good choice of initial condition because it causes the $(k+1)$th simulation to resume calculation of the transient where the kth simulation left off.

With this technique, the value of N_t can be made fairly small. A small N_t, however, leads to an interesting artifact that we call a *bifurcation transient.* Bifurcation transients appear on a brute-force bifurcation diagram as unusual behavior just after a bifurcation. They occur when the simulation at a particular parameter value does not achieve the steady state in N_t iterations.

An example of a bifurcation transient is shown in Fig. 8.15. This figure shows a period-doubling bifurcation calculated with two different values of N_t. Fig. 8.15(a) is calculated with a value of N_t large enough to ensure that the steady state is achieved at each value of α_k. The value of N_t used to calculate Fig. 8.15(b) is smaller and leads to a bifurcation transient.

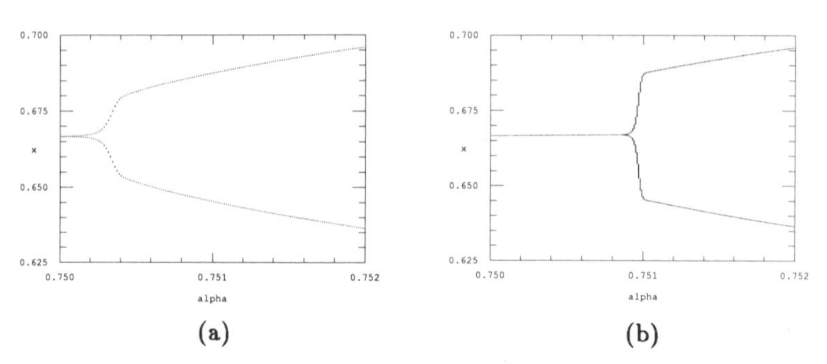

Figure 8.15: A period-doubling bifurcation calculated with (a) $N_t = 200$; (b) $N_t = 100$.

To explain the bifurcation transient in Fig. 8.15(b), assume that the period-doubling bifurcation occurs for a values of $\hat{\alpha}$ between $\alpha_{\hat{k}}$ and $\alpha_{\hat{k}+1}$. Figs. 8.15(a,b) agree for $\alpha_k \leq \alpha_{\hat{k}}$ so it is clear that in this parameter range, Fig. 8.15(b) achieves the steady state. When Fig. 8.15(a) period-doubles, however, Fig. 8.15(b) appears to still be following a period-one solution. Recall that after a period-doubling bifurcation, the period-one solution does not vanish, but does become unstable. It is this unstable period-one solution that is followed in Fig. 8.15(b).

It may seem odd that an unstable solution appears in the bifurcation diagram, but its appearance is explained by several factors. First, just after the bifurcation, the unstable period-one orbit is only weakly repelling. This is due to the fact that its one unstable characteristic multiplier has just passed through the unit circle and must have a magnitude only slightly greater than 1. Second, since the simulation for $k = \hat{k}$ achieved the steady state, the initial condition for the $(\hat{k}+1)$th simulation lies nearly on the period-one solution. These two facts plus the fact that N_t is small imply that for $\alpha_{\hat{k}+1}$, too few iterations are taken for the orbit to move significantly far from the weakly repelling period-one solution. Only after α increases well past the bifurcation value and the iterations of subsequent simulations accumulate does the period-two solution appear on the bifurcation diagram.

Hysteresis Fig. 8.16 shows an example of hysteresis in a brute-force bifurcation diagram. As α is increased from point A, the path from $AB'C'C''D$ is followed with a jump from C' to C''. When α is

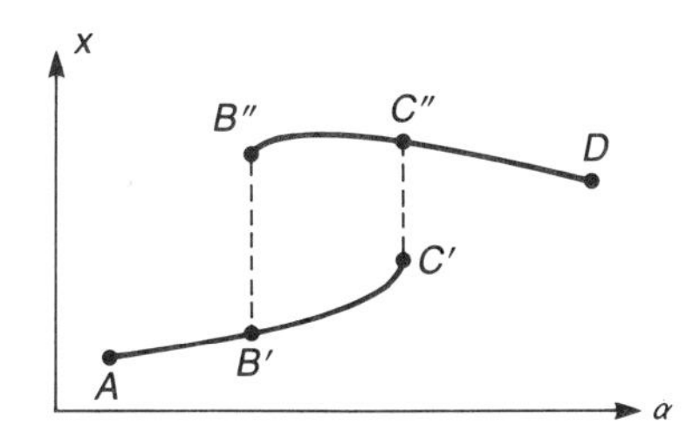

Figure 8.16: Hysteresis in a brute-force bifurcation diagram.

decreased from point D, the system follows the path $DC''B''B'A$ with a jump from B'' to B'.

Hysteresis is displayed by calculating the bifurcation diagram twice. The first time, α is varied from α_1 to α_K; the second time, α is varied backwards from α_K to α_1.

To explain the phenomenon, note that Fig. 8.16 was calculated using the brute-force technique and, therefore, only stable limit sets are displayed. In the section on the continuation method, we will see that points B'' and C' correspond to bifurcation points where the limit set loses stability and that there is a smooth curve connecting points B'' and C'. This curve corresponds to a limit set that is not stable and which is, therefore, not observable.

Newton-Raphson enhancements Brute force has no way of detecting whether the steady state has been achieved. It blindly assumes that the steady state is reached after N_t iterations. One of the advantages of the Newton-Raphson techniques for finding steady-state solutions (see Chapter 5) is that convergence of the Newton-Raphson iterations is a positive indication that the steady state has been found. Thus, if Newton-Raphson could be incorporated into the brute-force technique, then brute force would have a more reliable indicator that the steady state has been reached. A hybrid Newton-Raphson/brute-force algorithm should also be more efficient because the Newton-Raphson algorithm eliminates the calculation of the transient response.

The hybrid algorithm starts the calculation of the bifurcation diagram in the brute-force mode. If the output of the brute-force algorithm suggests that a periodic steady state has been reached, the hybrid algorithm switches to the Newton-Raphson mode for the next parameter value. The Newton-Raphson mode is used as long as

```
global variables:  period, x[][]

begin hybrid_bif
      choose x0[], α_min, α_max, K, i_d, N_t, N_ss
      set brute_mode = TRUE
      for k from 1 to K
          set α = α_min + (k - 1)(α_max - α_min)/(K - 1)
          if (brute_mode = TRUE) then
             call do_brute_force(α, N_T, N_ss, x0[])
          else
             call do_Newton_Raphson_mode(α, x0[])
             if (do_Newton_Raphson did not converge to
                    a stable limit set) then
                set brute_mode = TRUE
                call do_brute_force(α, N_T, N_ss, x0[])
             endif
          endif
          if (brute_mode = TRUE) then
             for i from 1 to N_ss do
                 output x[i][i_d]
             endfor
             set x0[] = x[N_ss][]
             call find_period
             if (period ≠ 0) then
                set brute_mode = FALSE
             endif
          else
             for i from 1 to period do
                 output x[i][i_d]
             endfor
             set x0[] = x[1][]
          endif
      endfor
end hybrid_bif
```

Figure 8.17: Pseudo-code for hybrid_bif.

it converges to a stable limit set. If it fails to converge or converges to a limit set that is not stable, the brute-force mode is resumed. Pseudo-code is presented in Figs. 8.17 and 8.18.

Remarks:

1. find_period checks for periodicity in $x[\,][\,]$. It is similar in nature to the minimum period finding routine presented in Sec-

```
begin do_brute_force(α, N_t, N_ss, x_0[])
      set x[1][] = x_0[]
      for i from 1 to (N_t + 1) do
          set x[1][] = P_b(x[1][], α)
      endfor
      for i from 2 to N_ss do
          set x[i][] = P_b(x[i - 1][], α)
      endfor
end do_brute_force

begin do_Newton_Raphson(α, x_0[])
      perform Newton-Raphson algorithm
      if (Newton-Raphson converged to
              a stable periodic orbit) then
         place the result in x[][]
         set period = minimum period of the orbit
      endif
end do_Newton_Raphson

begin find_period
      choose E_r and E_a
      for K from 1 to (N_ss/2 + 1) do
          if (||x[1][] - x[K + 1][]|| <
                  E_r ||x[1][]|| + E_a) then
             set period = K
             for j from 2 to K do
                 if (||x[j][] - x[j + K][]|| >
                         E_r ||x[j][]|| + E_a) then
                    set period = 0
                    exit the j for loop
                 endif
             endfor
             if (period = K) then
                return
             endif
          endif
      endfor
      set period = 0
end find_period
```

Figure 8.18: Pseudo-code for do_brute_force, do_Newton_Raphson, and find_period.

tion 5.4. It attempts to find a K such that

$$x[i][\,] \approx x[i+K][\,], \qquad i = 1, \ldots, K-1. \tag{8.12}$$

The test should be fairly lax—it is used to indicate that a periodic orbit may exist, not that one definitely exists—so the relative and absolute error tolerances should be chosen relatively large.

2. **do_Newton_Raphson** requires a stability test whenever the Newton-Raphson algorithm converges. See Section 5.4 for details.

Interpreting brute-force bifurcation diagrams

The brute-force approach introduces a few ambiguities into a bifurcation diagram so we now turn to a discussion of how to interpret a brute-force bifurcation diagram.

The following discussion is restricted to bifurcation diagrams of continuous-time systems. The interpretation for discrete-time systems is analogous.

Equilibrium points An equilibrium point appears as a single point above each value of α for which it is stable. Since the vector field is assumed to vary smoothly with respect to α, these points form a curve that extends over the interval of α for which the equilibrium point is stable.

Periodic solutions—non-autonomous case A period-one periodic solution of a non-autonomous system is represented by a single curve that spans the interval of α for which the periodic solution is stable.

Periodic solutions—autonomous case Regardless of whether a limit cycle is displayed using the Poincaré map or by plotting the local maxima of one of the states, a limit cycle is represented by one or more curves that span the interval of α for which the limit cycle is stable (see Fig. 8.19).

A limit cycle that is represented by a single curve cannot be distinguished from an equilibrium point by looking only at the bifurcation diagram. This is rarely a problem because the user typically

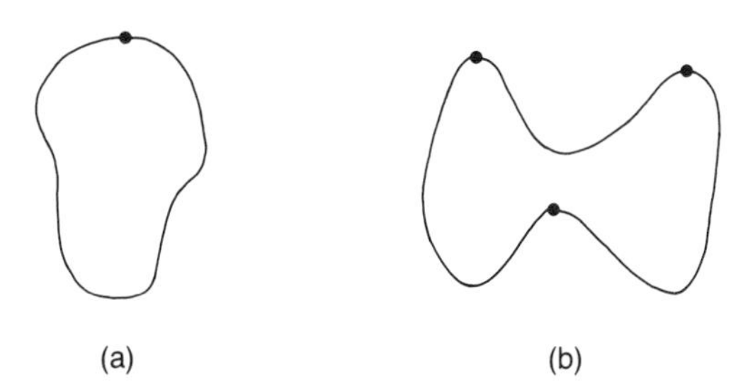

Figure 8.19: (a) This limit cycle is represented by one point on a bifurcation diagram; (b) this limit cycle, by three.

knows where the equilibrium points of the system are. If, in a particular application, it is necessary to distinguish equilibrium points from limit cycles, control logic can be incorporated into the bifurcation diagram program to check for equilibrium points and the program can draw equilibrium points in a different color.

Subharmonics—non-autonomous case A Kth-order subharmonic of a non-autonomous system is represented by K curves on the bifurcation diagram.

Subharmonics—autonomous case Subharmonics do not usually make sense in the autonomous case—there is no well-defined fundamental period—but as Fig. 8.9 demonstrates, when period-doubling occurs, the concept of autonomous subharmonics *is* meaningful. Suppose Γ is a limit cycle that for a particular value of α is represented by K points on the bifurcation diagram. Due to continuity with respect to the initial condition, when Γ period doubles, the new limit cycle is composed of two pieces, each of which is nearly identical to Γ (see Fig. 8.20). Hence, the newly created limit cycle is represented by a family of $2K$ curves that span the interval of α for which the period-$2K$ limit cycle is stable. Observe that after period-doubling, Γ still exists; it is, however, not stable and does not appear on the brute-force bifurcation diagram (provided there are no bifurcation transients).

Two-periodic solutions—non-autonomous case A two-periodic solution appears on a bifurcation diagram not as a curve but as a region that contains a smear of points.

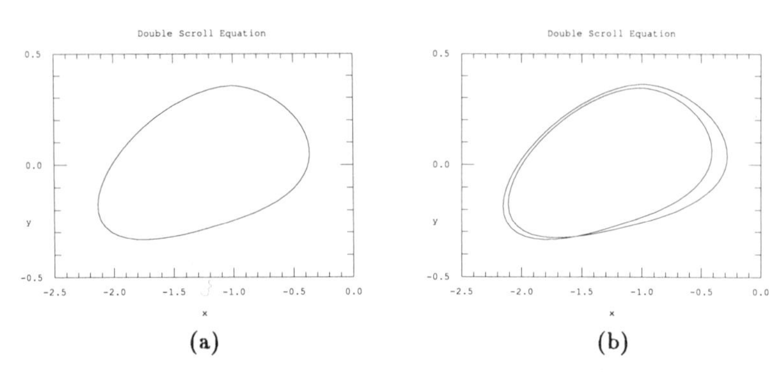

(a) (b)

Figure 8.20: Autonomous subharmonics for system (1.24) with $\beta = 15$. (a) The original limit cycle ($\alpha = 8.4$); (b) the newly created period-two limit cycle ($\alpha = 8.47$).

To understand why, recall that a Poincaré orbit corresponding to a two-periodic solution of the underlying flow lies on a diffeomorphic copy of the circle. Let x_i be the state that is plotted on the y-axis of the bifurcation diagram. What is displayed in the bifurcation diagram for each value of α is the projection of a steady-state orbit onto the x_i-axis. If the orbit is uniformly distributed on the circle, then the result of the projection is a smeared line segment that is darkened at the two ends.[2] If, on the other hand, the orbit favors certain regions of the circle, then the result of the projection is darker near those regions. If the two-periodic behavior is stable over a range of α and the steady-state orbits are not uniformly distributed over the circle, then dark horizontal bands appear in the bifurcation diagram (see Fig. 8.11).

A Kth-order two-periodic subharmonic is represented by a set of K smears.

Two-periodic solutions—autonomous case For autonomous systems, two-periodic solutions appear much the same as in the non-autonomous case. They are smears of points with darker regions indicating the portions of the limit set favored by a typical orbit. Observe that autonomous two-periodic behavior can appear as a set of K isolated smears. This analogous to the fact that a limit cycle can appear as a set of K curves.

[2]One may think that the projection of the limit set should be a uniformly dark line segment and not a smear of points. This is true if it is the limit set that is being projected; however, in this situation, the set that is being projected is a truncated orbit that lies on the limit set, not the limit set itself.

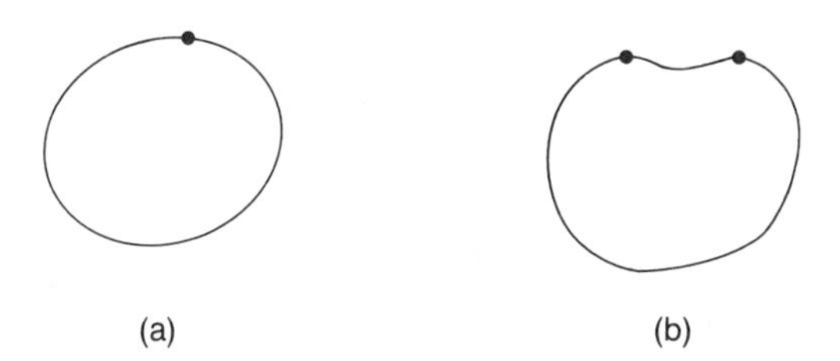

Figure 8.21: A false bifurcation using local maxima. (a) The original limit cycle; (b) after a slight perturbation, a second local maximum is created but no bifurcation has occurred.

Quasi-periodic behavior On a brute-force bifurcation diagram, higher-order quasi-periodic behavior looks identical to two-periodic behavior. It is impossible to distinguish the order of the quasi-periodic behavior from the bifurcation diagram alone.

Chaotic solutions Chaotic solutions appear much the same as quasi-periodic solutions. There is no robust way to distinguish chaotic from quasi-periodic solutions on the basis of the bifurcation diagram alone. The smear representing a chaotic solution does tend to be less uniformly distributed than the smear of a quasi-periodic solution, but this characterization is unreliable. Trajectories, spectra, Lyapunov exponents, and, whenever possible, theory should be used to distinguish chaotic from quasi-periodic behavior.

False bifurcations When the brute-force technique is applied to autonomous systems, it can generate false bifurcations. Fig. 8.21(a) shows a limit cycle with a single local maximum. This limit cycle appears on the bifurcation diagram as a single branch. If, at a particular value of α, a second local maximum appears on the limit cycle (Fig. 8.21(b)), then at that value of α, a second branch appears on the bifurcation diagram. The appearance of this second branch could be interpreted as a bifurcation but, in fact, no bifurcation has occurred because the limit cycle has undergone no real qualitative change. Fig. 8.22 is a bifurcation diagram of the same system as Fig. 8.11. Instead of using the Poincaré map, Fig. 8.22 was calculated using the local maxima of the trajectory. Overall, the bifurcation diagrams are remarkably similar. There are several differences, however. For instance, where Fig. 8.11 shows a period-6 orbit, Fig. 8.22 indicates a period-8 orbit. This is not contradictory; it means that when the limit cycle intersects the plane that defines

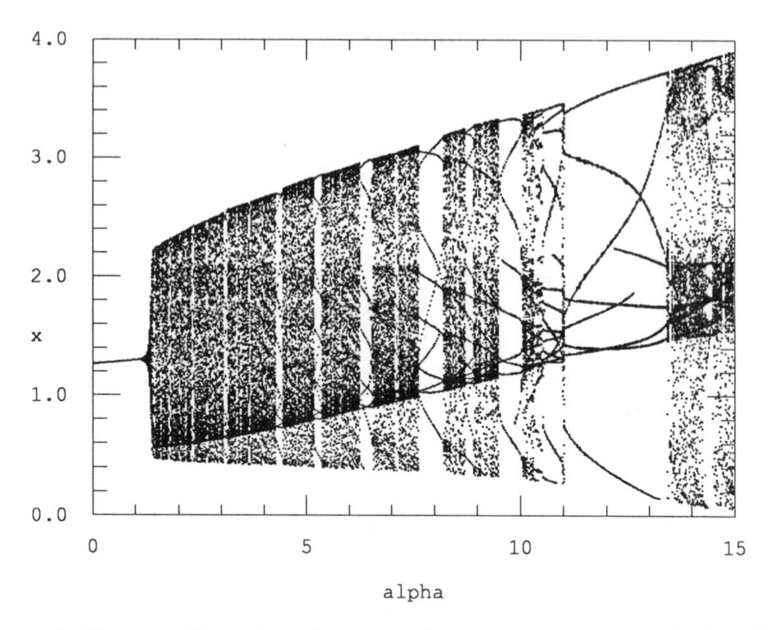

Figure 8.22: A bifurcation diagram of the same system as in Fig. 8.11. This diagram was calculated using the local maxima of x.

the Poincaré map six times, the limit cycle has eight local maxima. Also note that in the range of $12 < \alpha < 13$, there are two branches that end abruptly—one coming in from the left and one from the right. These branches are false bifurcations.

False bifurcations are not restricted to the case of using local maxima to calculate a bifurcation diagram; they can also occur when the Poincaré map is used. If, for example, the position of the limit cycle changes such that the number of intersections with the hyperplane increases by one (see Fig. 2.4), then a new branch will appear on the bifurcation diagram even though no bifurcation has occurred.

8.2.2 Continuation

Definitions

We (loosely) define a *branch* of a bifurcation diagram as a single curve of the bifurcation diagram that *i)* intersects no other curves and *ii)* cannot be extended in either direction without violating *i)*.

For example, Fig. 8.2 contains two branches and Fig. 8.5 contains four.

A more complicated example is a period-doubling bifurcation. It has one branch going in and three branches going out, two for the stable period-two solution and one for the unstable period-one solution. Branches are not well-defined when the steady state is chaotic because the bifurcation diagram of chaos is not made of curves.

Remark: The endpoints of a branch are always bifurcation points; if they were not, the branch could be extended. We shall see, however, that a branch can contain other bifurcation points as well.

The general bifurcation problem can be defined as follows. Given a map $F: \mathbb{R}^p \times \mathbb{R} \to \mathbb{R}^p$, find a branch $B \subset \mathbb{R}^p \times \mathbb{R}$ such that

$$F(x, \alpha) = 0 \tag{8.13}$$

is satisfied for every $(x, \alpha) \in B$.

The continuation function In our application, the map F is defined as follows. If we wish to find the equilibrium points of the parameterized system $\dot{x} = f(x, \alpha)$, then

$$F(x, \alpha) := f(x, \alpha). \tag{8.14}$$

To find the fixed points of a parameterized map $P(x, \alpha)$, define

$$F(x, \alpha) := P(x, \alpha) - x. \tag{8.15}$$

If P is a Poincaré map, (8.15) can be used to locate periodic solutions of the underlying continuous-time system. F defined by

$$F(x, \alpha) := P^K(x, \alpha) - x. \tag{8.16}$$

is used to locate period-K closed orbits of a map and, via the Poincaré map, can be used to find Kth-order subharmonics of a continuous-time system.

In theory, continuation can be applied to find branches corresponding to quasi-periodic solutions by finding a function F such that $F(x, \alpha) = 0$ when x lies on the quasi-periodic solution. We know of no work that has been done in this area.

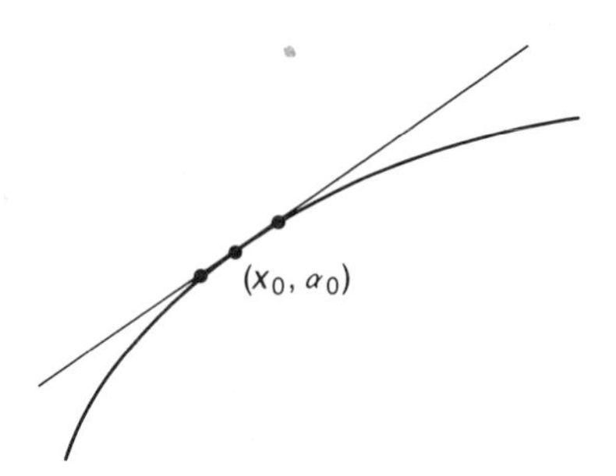

Figure 8.23: The slope at (x_0, α_0) can be used to estimate points on the branch near (x_0, α_0).

Continuation Let (x_0, α_0) be a point on a particular branch of a bifurcation diagram. If the slope is known at (x_0, α_0), it can be used to estimate two more points on the branch, one on either side of (x_0, α_0) (see Fig. 8.23). If the slopes at these points are known, they can be used to extend the branch further in each direction. The branch can be extended in this manner until a point is reached on the branch where the slope is not well defined. This is the basis of the continuation algorithm. Note the similarity to numerical integration algorithms.

To calculate the slope at a point on a branch, differentiate (8.13) to obtain

$$D_x F(x, \alpha)\, dx + D_\alpha F(x, \alpha)\, d\alpha = 0 \tag{8.17}$$

Note that $D_x F(x, \alpha)$ is a $p \times p$ matrix, $D_\alpha F(x, \alpha)$ and dx are p-vectors, and $d\alpha$ is a real number.

Equation (8.17) is a linear system of p equations in the $p+1$ unknowns, $[dx^T\ d\alpha]^T$. Define $DF(x, \alpha) \in \mathbb{R}^{p \times (p+1)}$ by

$$DF(x, \alpha) := [D_x F(x, \alpha) \quad D_\alpha F(x, \alpha)] \tag{8.18}$$

and rewrite (8.17) as

$$DF(x, \alpha) \begin{bmatrix} dx \\ d\alpha \end{bmatrix} = 0. \tag{8.19}$$

If DF has full rank, then (8.19) constrains $[dx^T\ d\alpha]^T$ to lie in a one dimensional sub-space of $\mathbb{R}^{p+1}$ that is tangent to the branch at (x, α) (see Fig. 8.24).

Append to (8.19) the normalization constraint

$$\|\, [dx^T\ d\alpha]^T \| = 1. \tag{8.20}$$

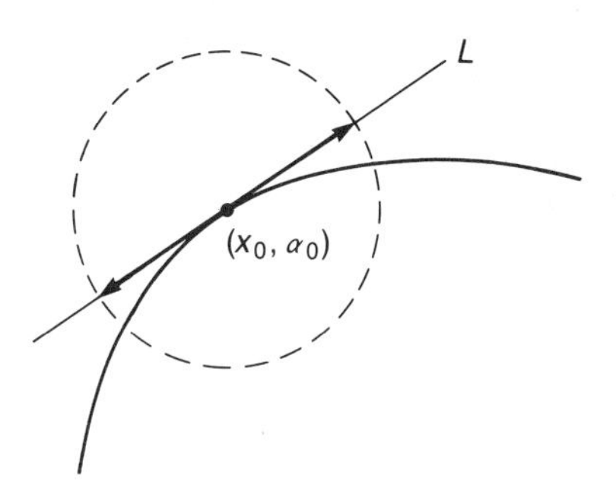

Figure 8.24: If DF has full rank, the solution space of (8.19) is a one-dimensional subspace L that is tangent to the branch at (x, α). The normalization constraint (8.20) further reduces the solution space to the two points on L that have magnitude 1.

Equation (8.20) constrains $[dx^T \ d\alpha]^T$ to be one of the two unit vectors in the one-dimensional sub-space defined by (8.19). By consistently choosing one of these two unit vectors to be $[dx^T \ d\alpha]^T$ (more on this later), equations (8.19) and (8.20) generate a vector field $\mathcal{F}: \mathbb{R}^{p+1} \to \mathbb{R}^{p+1}$,

$$\mathcal{F}(x, \alpha) := [dx^T \ d\alpha]^T \tag{8.21}$$

defined by

$$\begin{aligned} DF(x,\alpha)\begin{bmatrix} dx \\ d\alpha \end{bmatrix} &= 0 \\ \| [dx^T \ d\alpha]^T \| &= 1. \end{aligned} \tag{8.22}$$

Trajectories of $\mathcal{F}$ lie on level curves of $F(x, \alpha)$. Specifically, if the initial condition (x_0, α_0) satisfies $F(x_0, \alpha_0) = 0$, then the trajectory emanating from (x_0, α_0) lies on a branch of the bifurcation diagram. Thus, the branches of the bifurcation diagram can be calculated by integrating (8.21).

Remarks:

1. The system (8.22) is not in the standard form for differential equations because the independent variable (which we have always assumed to be time) has been suppressed. The normalization constraint (8.20) implies that the independent variable is arc-length, that is, one time unit corresponds to integrating over one unit of arc-length.

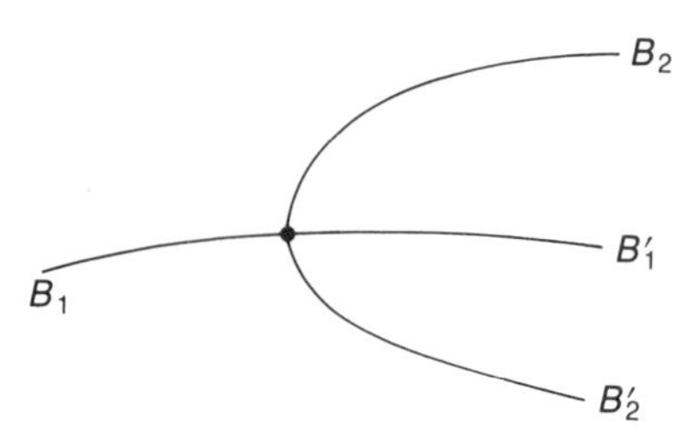

Figure 8.25: Four branches meet at a period-doubling bifurcation. B_1 and B_1' represent the period-one solution; B_2 and B_2' represent the period-two solution.

2. Assuming DF has full rank, (8.22) specifies the magnitude and direction of $[dx^T \; d\alpha]^T$, but not its sign, that is, if $[dx'^T \; d\alpha']^T$ satisfies (8.22), then so does $-[dx'^T \; d\alpha']^T$. For $\mathcal{F}$ to be well defined, the sign must be chosen such that $\mathcal{F}$ is continuous along a branch. It follows that the choice of sign of $[dx^T \; d\alpha]^T$ at one point of a branch determines the sign at all the other points of the branch. The initial choice of sign determines the direction of the flow, that is, whether integration in forward time proceeds to the left or to the right along the branch.

This sounds all well and good, and it does work, but there are several complications. The main difficulty is that $\mathcal{F}$ can be a poorly behaved vector field at bifurcation points. As an example, consider the period-doubling bifurcation shown in Fig. 8.25. The branch B_1 represents the period-one solution and is a trajectory of $\mathcal{F}_1$ defined for

$$F_1(x,\alpha) := P(x,\alpha) - x. \tag{8.23}$$

The other branch B_2 represents the period-two solution and is a trajectory of $\mathcal{F}_2$ defined for

$$F_2(x,\alpha) := P^2(x,\alpha) - x. \tag{8.24}$$

Since fixed points of P are also fixed points of P^2, branch B_1 is also a trajectory for $\mathcal{F}_2$ and, therefore, $\mathcal{F}_2$ possesses two transversally intersecting trajectories. The implication is that the vector field at the bifurcation point is not uniquely defined. Thus, it is worthwhile examining under what conditions (8.19) is well defined.

Equation (8.19) is an under-specified system and, therefore, the dimension of the space of solutions always has dimension greater than

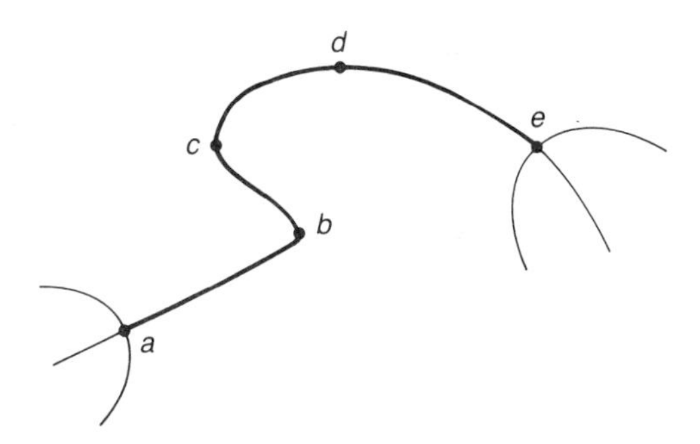

Figure 8.26: Four different types of points on a branch. Points a and e are the end-points of the branch; they are multi-branch bifurcation points (case 4). Points b and c are turning points (case 3). Point d is a local maximum (case 2). All other points on the branch correspond to case 1.

or equal to one. We consider $\mathcal{F}$ well defined if the dimension of the solution space of (8.19) is one. If so, the normalization constraint (8.20) reduces the solution set to two vectors of equal magnitude and opposite direction. As discussed earlier, the direction can then be chosen to create a consistent (i.e., continuous) vector field.

Note that the solution of (8.19) is the null space of the $p \times (p+1)$ matrix $DF(x, \alpha)$. Thus, the vector field is well-defined at $[x^T \ \alpha]^T$ if and only if the null space of $DF(x, \alpha)$ has dimension one, that is, if and only if $DF(x, \alpha)$ has full rank. There are four cases (see Fig. 8.26).

1. *D_xF and $D_\alpha F$ are of full rank:* Clearly, DF has full rank so the vector field is well-defined. Also note that $D_\alpha F$ is in the range of D_xF. It follows that (8.22) can be solved by setting $d\alpha = 1$, solving

$$D_xF(x, \alpha)\, dx' = -D_\alpha F(x, \alpha) \tag{8.25}$$

   for dx', and then normalizing $[dx'^T \ 1]^T$ to have magnitude 1.

2. *D_xF has full rank and $D_\alpha F$ does not:* Again, DF has full rank so the vector field is well-defined. $D_\alpha F$ is a vector so the fact that it does not have full rank means it is the zero vector. Thus, (8.19) becomes

$$D_xF(x, \alpha)\, dx = 0 \tag{8.26}$$

   with normalized solution $[dx^T \ d\alpha]^T = \pm [0 \cdots 0 \ 1]^T$. In this case, the vector field is parallel to the α-axis.

3. *D_xF has rank $p-1$, $D_\alpha F$ has full rank, and $D_\alpha F$ does not lie in the range of D_xF:* In this case, the matrix D_xF is singular. Since $D_\alpha F$ does not lie in the range of D_xF, DF possesses p linearly independent columns and is, therefore, of full rank. Hence, the vector field is well-defined.

   For further insight, observe that since $D_\alpha F$ is not in the range of D_xF, (8.25) admits no solution. This implies that any solution of (8.19) has $d\alpha = 0$. Thus, points of this type occur when a branch changes direction with respect to α. Such points are called *turning points*.

4(a). *D_xF has rank $p-1$, $D_\alpha F$ has full rank, and $D_\alpha F$ lies in the range of D_xF:* In this case, DF has rank $p-1$ and, therefore, the vector field is not well-defined. Since the vector field is not well-defined, trajectories (i.e., bifurcation branches) can intersect transversally at such a point. An example is the period-doubling bifurcation in Fig. 8.25.

4(b). *D_xF has rank less than $p-1$:* In this case the rank of DF is necessarily less than p and the vector field is not well-defined. As in case 4(a) above, bifurcation branches can intersect transversally at such a point.

Remarks:

1. For case 1, $dx \neq 0$ and $d\alpha \neq 0$.
2. For case 2, $dx = 0$ and $d\alpha \neq 0$.
3. For case 3, $dx \neq 0$ and $d\alpha = 0$.
4. The $(p+1) \times (p+1)$ matrix

$$M(x,\alpha) := \begin{bmatrix} DF(x,\alpha) \\ \\ [dx^T \;\; d\alpha] \end{bmatrix} \tag{8.27}$$

   is useful for distinguishing cases 4(a) and 4(b) from the other three. Since $[dx^T \; d\alpha]^T$ is in the null space of $DF(x,\alpha)$, $M(x,\alpha)$ is non-singular if and only if $DF(x,\alpha)$ has full rank.[3] It follows that for cases 1 through 3, the determinant of $M(x,\alpha)$ is non-zero and for cases 4(a) and 4(b), the determinant is zero.

[3]This follows from the fact that for any matrix A, the null space of A is orthogonal to the range of A^T.

Turning points and stability

A *turning point* $[\hat{x}^T \ \hat{\alpha}]^T$ of a branch is a point where the branch changes direction with respect to α (case 3(a) above). At a turning point, the matrix $D_xF(\hat{x},\hat{\alpha})$ is necessarily singular with rank $p-1$. The determinant of a singular matrix is 0 and since the determinant of a matrix is equal to the product of the eigenvalues of the matrix, it follows that at a turning point, at least one of the eigenvalues of $D_xF(\hat{x},\hat{\alpha})$ is 0.

Hence, in the generic case, a turning point indicates that an eigenvalue passes through the origin either from the left half-plane to the right half-plane or from the right half-plane to the left half-plane. We now show that for our application, this migration of an eigenvalue implies that turning points are always bifurcation points. We also discuss the effect of turning points on the stability of the associated limit set.

Equilibrium points: In the case of a branch corresponding to an equilibrium point of the system $\dot{x} = f(x,\alpha)$, the function F is identical to f and, therefore, the eigenvalues of the matrix $D_xF(x,\alpha)$ are the eigenvalues of the equilibrium point. Since the flow near an equilibrium point is structurally unstable when an eigenvalue of the equilibrium point lies on the imaginary axis, a turning point is necessarily a bifurcation point.

Fixed points: In the case of a branch corresponding to a fixed point of a map $P(x,\alpha)$, the function F is defined by

$$F(x,\alpha) := P(x,\alpha) - x. \tag{8.28}$$

Hence,

$$D_xF(x,\alpha) := D_xP(x,\alpha) - I. \tag{8.29}$$

Let the eigenvalues of $D_xF(x,\alpha)$ be $\lambda_i(x,\alpha)$ and the characteristic multipliers of $D_xP(x,\alpha)$ be $m_i(x,\alpha)$. From (8.29) it follows that $\lambda_i(x,\alpha) = m_i(x,\alpha) - 1$. In other words, a fixed point x^* of $P(x,\alpha)$ is stable if the eigenvalues of $D_xP(x^*,\alpha)$ lie in the unit circle centered at -1. Since a fixed point is structurally unstable when one of its characteristic multipliers lies on the unit circle, a turning point is necessarily a bifurcation point.

Periodic solutions: As usual, the previous discussion of fixed points applies also to periodic solutions when they are considered as fixed

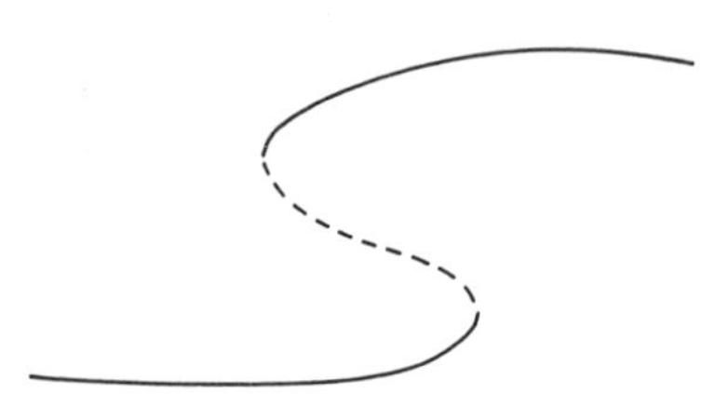

Figure 8.27: Hysteresis in a branch is associated with two turning points. The turning points are connected by a sub-branch corresponding to a limit set that is not stable.

points of the Poincaré map. Thus, a turning point is a bifurcation point for a periodic solution.

Remarks:

1. In the generic case, a turning point always indicates a change in the stability of a stable or unstable limit set. For example, at a turning point, a stable fixed point either becomes non-stable or unstable but cannot, in the generic case, remain stable.

2. A turning point does not necessarily imply a change in the stability type of a non-stable limit set. At a turning point, a non-stable fixed point can become either stable or unstable but can also remain non-stable but with a different number of characteristic multipliers inside the unit circle.

3. If a pair of complex conjugate eigenvalues passes through the imaginary axis (as in a Hopf bifurcation), then the stability type of a limit set may change but the matrix D_xF will not be singular. Thus, there is no geometrical feature (like a turning point) of the branch that indicates such a bifurcation. The eigenvalues of D_xF must be calculated explicitly, and checked to see whether the real part of any of the complex eigenvalues has changed sign.

Hysteresis

When hysteresis occurs in a branch, there are two associated turning points. The limit set loses stability at the first turning point and regains stability at the second (see Fig. 8.27). Thus the portion of the branch connecting the two turning points corresponds to a limit set that is not stable. This portion of the branch cannot be calculated by the brute-force algorithm, but is calculated easily using continuation.

Integration of the continuation equation

The system (8.22) can be integrated using any integration algorithm, but explicit algorithms are favored over implicit ones. There are two reasons.

First, implicit algorithms require that an implicit equation involving $\mathcal{F}$ be solved at every step. Since $\mathcal{F}$ is costly to calculate and its derivative is not readily available, implicit algorithms are inefficient for this application.

Second, it is known that the trajectory $\phi_t^\alpha(x)$ satisfies

$$F(\phi_t^\alpha(x), \alpha) = 0. \tag{8.30}$$

After every step of an integration algorithm, (8.30) can be used to correct any error introduced by the integration algorithm. Thus, step-sizes that are larger than those usually allowed by explicit algorithms can be taken.

In particular, continuation is one of the few situations where the forward Euler algorithm can be used. The problems of inaccuracy and numerical instability that usually plague the forward Euler algorithm are significantly reduced by using (8.30) as a corrector after each step.

The forward Euler algorithm is

$$x_{k+1} = x_k + h\mathcal{F}(x_k) \tag{8.31}$$

which, using the continuation notation, can be rewritten as

$$\begin{bmatrix} \bar{x}_{k+1} \\ \bar{\alpha}_{k+1} \end{bmatrix} = \begin{bmatrix} x_k \\ \alpha_k \end{bmatrix} + h_k \begin{bmatrix} dx_k \\ d\alpha_k \end{bmatrix}. \tag{8.32}$$

where h_k is the step-size at the kth step.

The point $(\bar{x}_{k+1}, \bar{\alpha}_{k+1})$ will not in general satisfy (8.30). To get the corrected value (x_{k+1}, α_{k+1}) use the Newton-Raphson algorithm on (8.30) with $(\bar{x}_{k+1}, \bar{\alpha}_{k+1})$ as the initial guess. The Newton-Raphson iteration is

$$\begin{bmatrix} x_{k+1}^{(i+1)} \\ \alpha_{k+1}^{(i+1)} \end{bmatrix} = \begin{bmatrix} x_{k+1}^{(i)} \\ \alpha_{k+1}^{(i)} \end{bmatrix} + \begin{bmatrix} \delta x_{k+1}^{(i)} \\ \delta \alpha_{k+1}^{(i)} \end{bmatrix}. \tag{8.33}$$

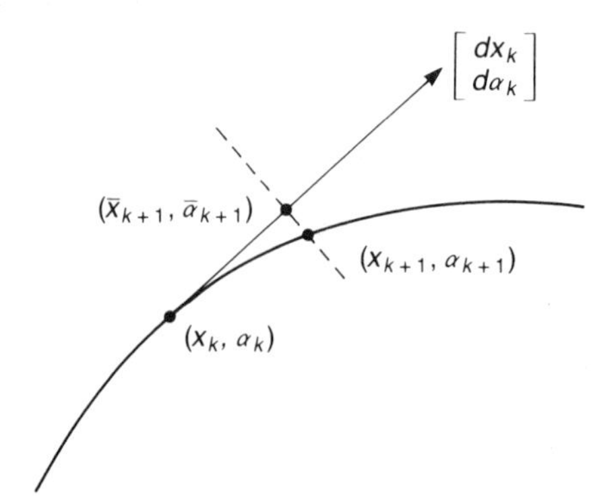

Figure 8.28: One step in the continuation method.

where

$$DF(x_{k+1}^{(i)}, \alpha_{k+1}^{(i)}) \begin{bmatrix} \delta x_{k+1}^{(i)} \\ \\ \delta \alpha_{k+1}^{(i)} \end{bmatrix} = -F(x_{k+1}^{(i)}, \alpha_{k+1}^{(i)}). \tag{8.34}$$

Equation (8.34) is an under-specified linear system of equations. To give it a unique solution, append the orthogonality constraint

$$\begin{bmatrix} dx_k^T \; d\alpha_k \end{bmatrix}^T \begin{bmatrix} \delta x_{k+1}^{(i)} \\ \\ \delta \alpha_{k+1}^{(i)} \end{bmatrix} = 0. \tag{8.35}$$

Equation (8.35) forces the Newton-Raphson correction vector to be orthogonal to the vector field at (x_k, α_k) (see Fig. 8.28).

The step-size h_k at the kth step is chosen using a simple step-size halving/doubling scheme. If the Newton-Raphson algorithm converges at the kth step, then set $h_{k+1} = h_k$; otherwise, halve h_k and repeat the kth step.

Automation

It is possible to automate the continuation method so that it calculates all the branches emanating from a bifurcation point. We briefly outline the procedure. Refer to Doedel [1986] and Kubíček and Marek [1983] for more details.

Define a *branch point* as a bifurcation point where two curves of a bifurcation diagram intersect transversally (see Fig. 8.25). Note that these two curves actually comprise four branches, B_1, B_1', B_2, and B_2', because by our definition, a branch ends where it intersects any other curve. We are faced with two tasks. First, we need to be able to locate a branch point. Second, given a branch B_1 that ends at a branch point, we need to calculate the initial direction of the remaining three branches so that we can use the continuation algorithm to calculate them. Since the directions of the primed branches, B_1' and B_2', are opposite those of the corresponding unprimed branches, B_1 and B_2, the second task reduces to finding the direction of B_2.

In theory, (8.22) becomes ill-conditioned at a branch point, so difficulty in solving (8.22) indicates that a branch point is close by. In practice, however, the step-size is usually such that the continuation algorithm steps over a branch point with no complaint and starts calculating branch B_1'. To take account of this possibility, use the matrix $M(x, \alpha)$ defined by (8.27). If (x, α) is a branch point, then the determinant of $M(x, \alpha)$ is zero. Thus, to locate a possible branch point, calculate the determinant of $M(x, \alpha)$ at each step in the continuation algorithm. When the determinant changes sign, use a time-step halving algorithm to locate the branch point more precisely.

Given a possible branch point, how do we calculate the direction of branch B_2? At a branch point $(\hat{x}, \hat{\alpha})$, the matrix $DF(\hat{x}, \hat{\alpha})$ has rank $p-1$. Thus, the null space of $DF(\hat{x}, \hat{\alpha})$ has dimension two. Let ξ_1 be the direction of B_1 and ξ_2 be the direction of B_2. Since B_1 and B_2 are assumed transversal, ξ_1 and ξ_2 must span the null space of $DF(\hat{x}, \hat{\alpha})$. There is a formula for calculating ξ_2 (see Kubíček and Marek [1983]), but it depends on second-order derivatives of F which are usually not directly available. A more practical approach is to calculate the null space of $DF(\hat{x})$ and then choose $\bar{\xi}_2$ as a vector that lies in the null space and that is orthogonal to ξ_1. This is only an approximation to ξ_2, but the Newton-Raphson step in the continuation algorithm will remove any error introduced by this approach. If the Newton-Raphson algorithm fails to converge using direction $\bar{\xi}_2$, then either $(\hat{x}, \hat{\alpha})$ is not a branch point, or ξ_1 is nearly tangent to ξ_2 indicating that $\bar{\xi}_2$ is a bad approximation to ξ_2.

8.3 Summary

- *Bifurcation*: A qualitative change in a limit set as a parameter is infinitesimally perturbed. Examples are the creation or disappearance of a limit set and the change in stability type of a limit set.

- *Bifurcation value*: A parameter value at which a bifurcation occurs. A system is structurally unstable at a bifurcation value.

- *Saddle-node bifurcation*: Two equilibrium points are created.

- *Pitchfork bifurcation*: An equilibrium point spawns two additional equilibrium points.

- *Hopf bifurcation—continuous-time case*: A complex-conjugate pair of eigenvalues of an equilibrium point pass through the imaginary axis, thereby creating a limit cycle.

- *Period doubling*: A stable period-K solution of a map becomes unstable and spawns a stable period-$2K$ solution. Period doubling also occurs in continuous-time systems.

- *Period-doubling route to chaos*: A sequence of period-doubling bifurcations (period-two to -four to -eight, etc.). The bifurcation values accumulate at a particular parameter value for which the system becomes chaotic. For maps of the interval that have a quadratic maximum, the rate of convergence of the bifurcation values is a constant independent of the exact shape of the map.

- *Hopf bifurcation—discrete-time case*: A complex-conjugate pair of characteristic multipliers of a fixed point pass through the unit circle thereby creating an invariant closed curve. If the map is a Poincaré map, this bifurcation corresponds to a limit cycle spawning a two-periodic steady state.

- *Bifurcation diagram*: A plot that indicates the steady-state behavior of a system over a range of parameter values.

- *Brute-force bifurcation diagram*: For each parameter value, the map is iterated for $N_t + N_{ss}$ iterations. The first N_t iterations are discarded and the last N_{ss} iterations are displayed on the bifurcation diagram.

- *Bifurcation transient*: An artifact that occurs in a bifurcation diagram when the brute-force algorithm does not reach the steady state for one or more parameter values.

- *Branch of a bifurcation diagram*: A single curve of a bifurcation diagram that *i)* intersects no other curves and *ii)* cannot be extended in either direction without violating *i)*.

- *Turning point*: A point on a bifurcation diagram where a branch changes direction with respect to α. At a turning point one of the Lyapunov exponents of the associated limit set passes through 0. In the generic case, a stable limit set loses stability at a turning point.

- *Continuation method*: Calculates a single branch on a bifurcation diagram by integrating an associated vector field. The continuation method cannot display chaotic behavior.

Chapter 9

Programming

A great deal of thought must go into the selection and development of the numerical algorithms used in a simulation program. It is also important to realize that an excellent numerical algorithm is useless if it is incorporated in a program that is awkward to use.

This chapter addresses the important issues of the user interface and how to write maintainable and portable code. These topics are approached from the viewpoint of writing simulation programs for dynamical systems. The material for this chapter is drawn from our experience writing INSITE—a collection of interactive, graphically-based programs for the simulation of nonlinear dynamics—and many of the examples are taken directly from the INSITE code.

9.1 The user interface

The user interface is the link between the user and the numerical algorithms. It determines how the user controls the various aspects of a simulation.

There is always a compromise between helpfulness and ease-of-use in the design of a user interface. An interface that is friendly enough for the inexperienced user is often a hindrance to someone familiar with the program and, conversely, an interface that lets an experienced user perform complex tasks quickly often leaves a novice feeling frustrated and defeated.

For the interactive simulation programs in INSITE, there are three components of the user interface.

1. *The dynamical system interface:* Governs how the user specifies the state equation of the system to be simulated.

2. *The program initialization interface:* Governs how the simulation options are set when a simulation program begins execution.

3. *The interactive interface:* Governs how the user controls the simulation while it is running.

9.1.1 The dynamical system interface

The dynamical system interface determines the way in which the user specifies to a simulation program the state equation of the system under study.

The routines

In INSITE, the dynamical system interface is implemented as a set of routines (see Fig. 9.1). There is a routine that sets parameter values, one that evaluates the state equation, one that evaluates the derivative of the state equation, and several others.

Remarks:

1. The routine `get_parameter_names` returns an array of strings. The ith entry in this array contains the name of the ith parameter. This information allows the user of a simulation program to refer to parameters by name. Also, since the size of the array is determined solely by the number of parameters, there is no limit to the number of parameters allowed.

2. There is a routine `get_system_name` that supplies the name of the system and a routine `get_equations` that supplies the state equations in string form. None of this information is required for a simulation. It is supplied so that a simulation program can document the simulation data and inform the user exactly what system is being simulated.

One of the major decisions in the design of INSITE was how the user would create the dynamical system interface for a particular dynamical system. From the point of view of the person writing the simulation package, the simplest method is to require the user to

get_dimension() Fetch the system dimension
Return value: the dimension of the dynamical system

is_discrete() Fetch the type of the system
Return value: TRUE for discrete-time systems

get_period(p) Fetch the system period
Upon return: p—if the system is time-periodic, the period of the system
Return value: TRUE for time-periodic systems

get_system_name() Fetch the system name
Return value: a string containing the name of the dynamical system

get_equations() Fetch the system equations
Return value: an array of strings. The ith string contains the ith entry in the right-hand side of the state equation.

get_parameter_names() Fetch the parameter names
Return value: an array of strings that contains the parameter names. The final entry in the array is an empty string.

set_parameter_value(n, x) Set a parameter value
Upon entry: n—the number of the parameter to be set. This number is the index into the array returned by **get_parameter_names**.
x—the value

get_parameter_value(n) Fetch a parameter value
Upon entry: n—the number of the parameter to be fetched
Return value: the value of the parameter

f(z[], x[], t) Evaluate the state equation
Upon entry: x[]—the state
t—the time
Upon return: z[]—the value of the state equation at x[] and t

df(z[][], x[], t) Evaluate the derivative of the state equation
Upon entry: x[]—the state
t—the time
Upon return: z[][]—the value of the derivative of the state equation at x[] and t

Figure 9.1: Routines of the dynamical system interface.

write source code to implement the routines of Fig. 9.1. This solution is obviously unsatisfactory as far as the user is concerned. Writing the routines is tedious, time consuming, and error prone.

An alternative method, ideal from the user's viewpoint, is to require the user to enter only the state equation describing the dynamical system and have the simulation program write the dynamical system interface. There are several ways to accomplish this.

1. Compile an interpreter into each simulation program.

2. Write a small compiler that reads the state equation and writes object code that implements the dynamical system interface.

```
continuous
2
Duffing's Equation
x[1]
gamma*cos(w*t) + x[1]*(1.0 - x[1]*x[1]) - delta*x[2]
```

Figure 9.2: An input file for `makef`.

3. Write a program that reads the state equation and writes source code that implements the dynamical system interface.

Method 1 requires no separate compilation or linking phase and has the desirable feature that dynamical systems can be changed quickly. The major drawback of this method is the computational speed of evaluating the state equation—interpreted code is often an order of magnitude slower than compiled code.

Since method 2 uses a compiler, it does not suffer from the slow computation speeds of method 1. In fact, since the compiler need not support all the features of a high-level language, it can generate efficient object code quite quickly. Method 2 does, however, suffer the serious disadvantage of poor portability. A new compiler must be written for each new type of host machine. This is not a simple task.

Method 3 is the method used by INSITE. Its drawback is that it requires three phases to change the dynamical system—a program must write the source code which is then compiled and finally linked into the simulation program. Its advantages, however, are enormous. In particular, there is no portability problem and compilation guarantees quick computation times. Furthermore, all the features of the high-level language are available to implement the routines of the dynamical system interface. These features include control structure (e.g., `if-else` blocks, `while` loops) and file input/output routines.

Simulation without programming—`makef`

The INSITE program that writes the source code to implement the dynamical system interface is called `makef`. An example of the input to `makef` is given in Fig. 9.2. This input file is for Duffing's equation (1.13). The first line indicates that the system is a continuous-time system. The second line is the dimension of the system, and the third line contains the system name. The remaining n lines are the right-hand side of the state equation where n is the dimension of the system. The state variable is an array `x[]` with index starting at 1 and continuing to n. In these equations, `t` has the special meaning of time and `w` always denotes the parameter ω which is radian frequency. An unrecognized variable is assumed to be a parameter. Thus, in the example given, there are three parameters: `w`, `delta`, and `gamma`. Since the state equation depends on `t` and `w`, `makef` assumes the system is periodic with period $2\pi/$`w`.

In most situations, `makef` frees the user from performing any programming. However, in situations where the code for the dynamical system is too complex for `makef` to write by itself, the user can easily edit the source code produced by `makef` to fit the problem at hand. Since a high-level language is used to implement the routines of the dynamical system interface, sophisticated simulations can be performed. For instance, consider a dynamical system whose state equation cannot be described analytically, but can be approximated using data gathered from an experiment. The source code produced by running `makef` on a dummy dynamical system with the same dimension and same parameter names as the system to be studied can be modified so that the experimental data is read from a file and then used to implement the desired state equation.

Symbolic differentiation

The input to `makef` is the state equation and its output contains not only the state equation but the derivative of the state equation with respect to the state (routine `df` of Fig. 9.1). The derivative is not approximated numerically; `makef` produces the exact analytical expression for each entry of the derivative. It finds these entries through symbolic differentiation.

To perform symbolic differentiation, a character string containing an algebraic expression is parsed and converted into binary tree form as shown in Fig. 9.3(a). Once in this form, the expression is easily differentiated (Fig. 9.3(b)) using a recursive differentiation routine. After differentiation, the binary tree is simplified to remove useless operations (multiply by one, add to zero, etc.) and to minimize computations (e.g., $x + x + x + x$ is converted to $4x$). Once simplified, the binary tree is converted back into a character string.

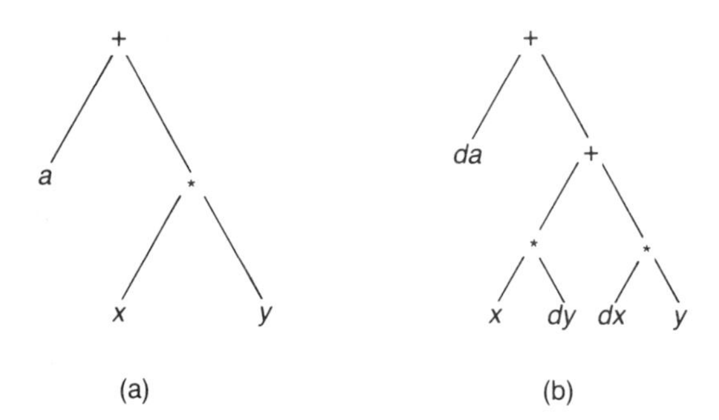

Figure 9.3: (a) The binary tree for $a + xy$; (b) the tree for its derivative.

9.1.2 The program initialization interface

Once the code implementing the dynamical system is compiled and linked into a simulation program, the simulation program is ready to run. Each program, however, requires a certain amount of information before a simulation can begin. These initialization options typically include parameter values, integration tolerances, and the initial condition. The program initialization interface of INSITE is specifically designed to make the entry of this data easy and self-documenting.

There are three methods to specify the initialization options: interrogative mode, an options file, and command line arguments.

Interrogative mode When the user enters the program name alone on the command line (i.e., with no command line arguments), interrogative mode is entered. The program explicitly asks the user to enter a value for each initialization option. To make life easier for the user, default values are offered whenever reasonable. When all the options have been entered, the user is asked whether the options should be written to a file. If the user consents, an options file is written to disk and the simulation proceeds.

Options file To initialize a simulation program using an options file, the command line argument `file` is used followed by the name of the options file. For example,

```
traj file traj.opt
```

runs `traj` using the options contained in the file `traj.opt`. Once the options are read from the file, the simulation begins.

The exact format of the options file varies from program to program, but there is always one option per line. Each option consists of a keyword followed by one or more arguments. For example,

```
graphics yes
```

specifies that graphical output is desired. The line

```
parameter alpha 0.6
```

sets the parameter named `alpha` to 0.6.

The options file is particularly useful since the user can easily alter any of the options using a text editor and then re-run the simulation program without the bother of reentering the unchanged options.

Option files can be nested by including the `file` option in an options file. Also the full option name need not be given; the programs need just enough of the name to identify it uniquely.

Command-line arguments Any option which can appear in an options file can also appear on the command line (and vice versa). For example, if a user usually runs the simulation program `traj` with graphics, but wants to run it once without graphics, she can type

```
traj file traj.opt graphics no
```

Arguments are read from left to right so the `graphics no` option overrides any `graphics yes` option in `traj.opt`.

9.1.3 The interactive interface

The interactive interface governs how a simulation program interacts with the user once the simulation is running.

As an example, consider a program `poin` that finds fixed points and invariant manifolds of a Poincaré map. Such a program typically has options that adjust the graphics parameters: erase the screen, change the graphics bounds, change which state is mapped to which axis. There are also options that control the Newton-Raphson algorithm: set the tolerances, set the initial condition, set the period, start the algorithm, abort the algorithm, restart the algorithm from a new initial condition. There are also program control options: find fixed points, find invariant manifolds, suspend the simulation, quit the simulation.

Menus

Menus are advantageous because if the option names that appear on the menu are chosen properly, the menu is self-documenting. Also, by using sub-menus, a tremendous number of options can be represented compactly in a logical, hierarchical manner.

There are two main approaches to menus: mouse-oriented and keyboard-oriented.

Mouse-oriented input is typically used with window-based graphics systems. One common approach is pop-up menus. When the user wants to select an option, she clicks the mouse and a menu appears on the display with a list of the available options. By pointing and clicking, the user chooses which options she wants to execute.

The advantages of the pop-up menu are that the user does not have to type any menu names and that the menu appears only when it is needed. The main drawback of the pop-up menu approach is that it does not allow the user to enter data. For example, if the user chooses the `new parameter value` option, she must enter the new value of the parameter. Hence, any user interface that uses pop-up menus must also use a keyboard-oriented interface.

INSITE uses a keyboard-oriented interface based on the pop-up menu idea. When the user hits a carriage return while a simulation is in progress, the simulation is suspended and the simulation program prompts

```
Enter option (? for help):
```

At this point an option is entered by specifying a keyword. Two keywords that are always available are `?` which lists all applicable keywords and `quit` which exits the program. The other options are program dependent. To simplify matters for the user, only enough of the keyword must be specified so as to be unambiguous.

For our applications, we have found the keyword approach easier to use than a Mouse-driven menu system for several reasons.

1. Keywords allow an experienced user to execute options more quickly and without taking his hands from the keyboard. For example, in the trajectory plotting program, there is an `erase` option—useful for erasing transients—which clears the graphics screen and resumes the integration from the last point. With a pop-up menu interface, the user must search the menu for

the **erase** entry and then position the cursor over the entry. This is fine for a novice user who is unsure what options are available, but is tedious for an experienced user. With the keyword approach, an experienced user just has to type an `e<CR>` followed by a carriage return.

2. Since many of the options in a simulation program require text input, the keyword approach frees the user from continuously switching from the mouse to the keyboard.

3. The keyword approach is more portable—it does not require the simulation program to run in a windowing environment.

Program feedback

It is important that an interactive program provide feedback to the user about the progress of the simulation. For example, in a program that locates periodic trajectories using the shooting method, the trajectory that is calculated during each iteration can be displayed on the graphics screen. This feedback allows the user to see immediately whether the algorithm is converging and if it is not, provides insight as to a better initial guess.

A program should also provide immediate feedback to the user whenever the user enters invalid data. To this end, the program can read each line of input as a character string. The character string is then checked to see whether the input is valid. For instance, if the program is expecting a floating-point number, it can check that the characters in the string form a valid floating-point number. If they do, a further check can be performed to see whether the value lies within the limits of the floating-point representation of the computer. If so, the string is converted to a floating-point variable. If not, the program prompts the user to enter a new value. The formatted input routines supplied with languages like C and FORTRAN do not perform reliable type checking and should not be used. Definitions for several useful parsing routines are presented in Section 9.3.2.

History files

In INSITE, each simulation program stores all the data it calculates in a data file called a *history file.* History files have many uses.

1. The graphics display can be redrawn without recalculating any data. For instance, a useful command for graphically oriented

programs is `center` which automatically alters the graphics bounds such that the data is centered in the graphics box. Without the history file, such an operation is prohibitively slow.

2. The history file allows stand-alone programs to post-process the data calculated by a simulation program. For example, a program that performs the fast Fourier transform can be used to calculate the spectrum using data from a simulation program that calculates trajectories.

3. The user may exit any of the simulation programs at any point in the simulation with no loss of data. When the user exits a program, the program writes its state to the history file. The program's state is defined as whatever information is needed so that the program can resume the simulation at a later time. For example, the program state of a trajectory plotting program might include the current parameter values, the current graphics bounds, the current integration time, the current (dynamical system) state, and the integration tolerances.

   In most applications, it is unrealistic (and unnecessary) for the state to contain enough information for the simulation program to resume calculations *exactly* where they were suspended. Instead, the state contains just enough information to resume the simulation with only a small number of duplicate calculations. For instance, when a trajectory plotting program is suspended in the middle of an integration step, the integration should be resumed at the beginning of the step, not in the middle. Thus, the entire state of the integration routine does not need to be stored in the history file.

4. The history mechanism protects against loss of data caused by operating system crashes. If simulation data is written to the history file as it is calculated, and if the program periodically writes its state to the history file as well, then the simulation can be resumed after an operating system crash.

In INSITE, data is stored in a history file in binary form to reduce the size of the file and to eliminate the costly conversion to and from the ASCII format. A simple program can be written to translate the data in a history file to ASCII format so that the data can be viewed by the user or sent to programs that require ASCII input.

9.2 Languages

One of the most important decisions in writing simulation software is which programming language to use. There are a dizzying number of programming languages to choose from, but the two most widely used languages for numerical simulations are FORTRAN and C. Each language has advantages and disadvantages, and their relative merits are the topic of this section.

Since the two main operating systems that are used for interactive numerical programs are UNIX and PC-DOS, the discussion that follows is restricted to these two operating systems.

Advantages of C

String manipulation Information is entered by the user as characters, so a language for interactive programs must have flexible string manipulation routines. C was explicitly designed to write interactive programs and has flexible and powerful string handling capabilities.

Interaction with the operating system There are a few occasions when a program needs access to operating system routines. One example is an interactive program that relies on system interrupts to indicate when the user has entered text from the keyboard. UNIX is written in C so on UNIX machines, it is easier to interact with the operating system when programming in C than when programming in FORTRAN. On non-UNIX machines, the number of operating system functions that is available to the programmer depends on the implementation of the compiler.

Parsing assistance There are two standard UNIX programs—`lex`, a lexical analyzer, and `yacc`, a parser compiler—that considerably reduce the amount of effort required to write an input parser. Though, these programs must be run on UNIX, their output is C source code that can be compiled by any C compiler under any operating system.

Dynamic memory allocation Dynamic memory allocation gives a program the ability to allocate memory at run time. It is used to create open-ended data structures such as linked lists, stacks, and binary trees. Dynamic memory allocation is also used to minimize the memory used by arrays and matrices. The standard C library supports dynamic memory allocation; FORTRAN does not.

Structured language By nature, C is a more structured language than FORTRAN. The C control statements (`if-else`, `do`, `while`, `for`, `switch`) are more flexible, more powerful, and cleaner than those of FORTRAN. This limitation of FORTRAN has been somewhat alleviated by preprocessors like RATFOR.

Data structures With one exception we will discuss shortly, C and FORTRAN have similar simple data types such as character and integer. Unlike FORTRAN, however, C allows the programmer to build compound data types called *structures* out of these simpler data elements. Structures allow the programmer to group into a single entity variables that are logically related but that may be of different types. This capability promotes modularity and allows the programmer to create flexible data structures such as linked lists and binary trees.

Advantages of FORTRAN

Computation speed FORTRAN was originally written for numerical algorithms—FORTRAN comes from *for*mula *tran*slation—and is typically faster than C in performing floating-point operations. The main reason is that all floating-point operations in C are performed in double precision, even for single-precision operands.[1] FORTRAN, on the other hand, performs single-precision operations on single-precision operands.

Another reason is that FORTRAN is a simpler language than C and, therefore, a compiler that produces efficient object code is easier to write. Related to this is the fact that since FORTRAN was designed for numerical applications, a great deal of effort has gone into optimizing the floating-point performance of FORTRAN compilers. Many FORTRAN installations support array processors for even faster processing.

Complex data type FORTRAN has a simple data type for complex numbers. C does not. A complex data type can be implemented

[1] Extensions proposed in the new ANSI C standard allow single-precision operations on single-precision operands.

in C using structures, and complex arithmetic operations can be implemented using macros or function calls, but this approach is computationally slower than the direct implementation available in FORTRAN.

Ease of learning FORTRAN has fewer operators and control structures than C and is easier to learn and to debug.

Numerical libraries Most of the standard numerical libraries like LINPACK and EISPACK are written in FORTRAN.

Discussion

The comparison of C and FORTRAN can be summed up by saying that C is better at control and data manipulation, and FORTRAN is better at numerically intensive tasks. Since a simulation program requires all three types of processing, the question, "Which language is better?" still remains unanswered.

The ideal solution would be to write the numerically intensive parts of the program in FORTRAN and the rest in C and somehow combine the two parts into a single executable program. The UNIX operating system has this capability. In UNIX, FORTRAN subroutines can be linked into C programs and vice-versa.

If other operating systems are used, then mixing C and FORTRAN may or may not be possible. For example, there are compiler companies that sell both a FORTRAN and a C compiler that run under PC-DOS and whose output can be linked into a single executable program, but it may not be advisable to lock a project into using one company's product.

If the capability of mixing C and FORTRAN code is not available, then, by necessity, C is the language of choice. FORTRAN simply does not have the features to support a user-friendly, interactive program.

9.2.1 Modular design

Interactive simulation programs are often quite large. Due to their size, they are difficult to write and once written, are difficult to debug, to modify, and to document. There needs to be some way of breaking the program into smaller, more easily understood parts. Ideally, each part is written, debugged, and documented separately, and then all the parts are integrated to form a single program. This approach is called *modular design*, and though the ideal is rarely reached in practice—the interconnectivity of programs usually makes it impossible to break a program into completely independent parts—modular design is still a powerful technique for organizing large programs.

There is no formula for breaking a program into modules—experience is the best teacher—but there are two simple approaches that are quite useful: top-down design and libraries. Top-down design organizes a single program in a simple hierarchical manner. A library is a collection of routines that are used by several different programs.

Top-down design

Top-down design breaks a single program into a handful of tasks. These tasks, in turn, are broken into a few sub-tasks which are broken into smaller tasks, and so on. This breaking-down process is continued until the lowest-level tasks can be implemented in ten or twenty lines of code.

For example, consider a program `traj` that integrates and plots trajectories. The main routine might be divided into four tasks:

```
call initialize
call process_initialization_options
call find_trajectory
call uninitialize.
```

`find_trajectory` might be

```
call initialize_trajectory
repeat
       call get_next_point
       call plot_point
until the final time is reached
call uninitialize_trajectory.
```

The routine `get_next_point` might be

```
set t to the current time plus the time-step
if (t is greater than the final time) then
   set t to the final time
endif
integrate until t.
```

Observe that each routine performs a well-defined task and contains only a few lines of code. The simplicity of each routine makes it easy to understand and to debug.

Another advantage of the top-down approach is that it forces the programmer to think through the overall control structure before any low-level code is written. The process of breaking the problem into smaller tasks often suggests efficient and modular ways of organizing the program structure.

Also, many programs have similar high-level structures. For instance, both a trajectory plotting program and a bifurcation diagram plotting program are initialized, read program initialization options, perform a simulation, etc. With top-down design, a fair amount of programming time can be saved when writing a new program by using an existing program with a similar structure as a skeleton.

Libraries

A library is simply a collection of related routines. Libraries are used to share code among several programs. For instance, many simulation programs use numerical integration. It would be silly to write a numerical integration routine into each program. Instead, a numerical integration library is written and this library is used by programs that require numerical integration.

A library always contains one or more routines that can be called from outside the library. These routines are called *entry points*. Additionally, most libraries contain routines that are used by the library internally and that are not available to users.

The functionality of a library is completely defined by the *library interface*, that is, by the library's entry points and by the task that each entry point performs. This is an important point for modular design. The library interface distinguishes *what* a library does from *how* the library does it. For example, it is useful to have several implementations of a numerical integration library. If each implementation uses the same library interface but a different integration algorithm, then which integration algorithm a program uses can be changed by simply linking into the program the library that implements the desired algorithm. The code for the program never has to be changed.

A library interface should be as clean as is practical. The number of entry points to a library should be small. The number of parameters passed to each entry point should also be small. The names of the entry points should be descriptive.

Entry point names often have a prefix unique to the library. For instance, the entry points for a symbolic algebra library might be `sym_initialize`, `sym_differentiate`, etc. The prefix serves two purposes. It is self-documenting in that it readily identifies the origin of a library routine. It also considerably reduces the chance of a name conflict with a routine external to the library.

9.3 Library definitions

There is no recipe for defining a library interface in a logical, modular fashion. The best ways to learn are by seeing how others do it and by doing it yourself. For INSITE, we have defined library interfaces for a variety of libraries that are useful for interactive simulation programs. By no means do we claim that these are the best or most complete definitions. We present them simply to let the reader see how we did it. We hope that the knowledge gained from looking at these definitions will protect a novice programmer from some of the pitfalls we encountered and give the experienced programmer some fresh ideas.

There are two basic types of libraries. The first type of library is a collection of tightly coupled routines that perform a single complex task. An example is RKF, a library that performs numerical integration using the Runge-Kutta-Fehlberg algorithm. The second type of library is a collection of independent routines that perform a variety of independent but related tasks. An example of this type of library is PARSE, a library of routines for parsing input.

When a library of the first type is used, there are typically three stages: initialize, perform one or more tasks, uninitialize. A common example is the standard C routines for reading and writing files. Initialization is performed by calling `fopen` which returns a pointer to an open file. An open file can be read from or written to by passing its pointer to the routines `fscanf` and `fprintf`. The third stage, uninitialization, is performed by passing the file pointer to `fclose` which closes the file associated with the pointer.

We have found this open-close approach to be a useful model for defining interfaces for libraries of the first type. Thus most of our libraries have `open` and `close` routines.

We now present the interface definitions for several libraries.

```
rkf (f, t, x[], tf, rel_tol, abs_tol,
          n, iflag, ws[], iws[])
```
Perform RKF integration

Upon entry:

- `f`—the function to be integrated
- `t`—the initial time
- `x[]`—the initial condition
- `tf`—the desired output time
- `rel_tol`—relative error tolerance
- `abs_tol`—absolute error tolerance
- `iflag`—integration mode flag
  - If integration is being initialized:
    - −1 for ONE-STEP mode: return after every step but don't go past `tf`.
    - 1 for END-POINT mode: return only upon error or when `tf` is reached.
  - If integration is continuing:
    - Use the value of `iflag` returned by the previous call to *rkf* unless 6 or 7 is returned in which case set iflag to
      - −2 to continue in ONE-STEP mode
      - 2 to continue in END-POINT mode
- `n`—the system dimension
- `ws[]`—an array of reals for working storage. Should have dimension 8n.
- `iws[]`—an array of integers for working storage. Should have dimension `n`.

Upon return:

- `t`—the output time; not `tf` in ONE-STEP mode or if an error occurs
- `x[]`—the output point
- `iflag`—integration status flag. Takes on values −2, 2, 3, ..., 8.

Figure 9.4: The library interface for RKF—FORTRAN version.

9.3.1 RKF—Runge-Kutta-Fehlberg integration

RKF is an integration library that performs numerical integration using the Runge-Kutta-Fehlberg algorithm. We initially received the library written in FORTRAN. It had the single entry point shown in Fig. 9.4. This single entry point is inadequate for several reasons.

1. Information that is used only for initialization (`f`, `rel_tol`, `abs_tol`, and `n`) is passed to `rkf` even when the integration is continuing and needs no initialization.

2. The programmer is burdened with the responsibility of allocating enough memory for the working arrays, `ws` and `iws`. If insufficient memory is allocated, `rkf` malfunctions.

3. There are two working arrays instead of one. The reason is that in FORTRAN, the real and integer arrays must be stored separately.

4. t and x[] play a dual role. Upon entry, they are initial conditions; upon return, they are final conditions. This change of meaning can be confusing. Also, if the user wants to save the initial values, he is forced to copy them to a safe place.

5. The iflag control scheme may make sense to the computer, but it is extremely cumbersome. First, iflag carries two separate pieces of information: the mode (ONE-STEP or END-POINT) and whether the integration is being initialized or is continuing. Second, the program is forced to carry iflag around and pass it to rkf the next time rkf is called.

6. The meaning of certain combinations of arguments is unclear. For example, what will rkf do when confronted with a completely new value of x[] but with a value of iflag that indicates that the integration is continuing? Will it ignore the new value of x[]? Will it restart the integration using x[] as the initial condition?

Most of these complaints arise from the fact that everything is performed with a single function call. The rest are due to the library being written in FORTRAN.

A much cleaner interface, shown in Fig. 9.5, is obtained using the open-close model and the features available in C. Every integration is associated with a pointer obtained by calling rkf_open. The pointer points to a compound data structure which contains working storage that is allocated by rkf_open using dynamic memory allocation. Thus the programmer is freed from the burden of allocating the working storage.

Note that just as more than one file may be open at any time, more than one integration may be open at a given time. This feature might be useful, for instance, in a phase portrait program where it is desired to display the evolution of several trajectories simultaneously.

The initial condition for the integration is set by calling rkf_initial_condition. The routine rkf_tolerances sets the error tolerances and rkf_mode sets the integration mode. The routine that actually performs the integration is rkf. It takes a single step when in ONE-STEP mode and integrates all the way to time tf when in

rkf_open(n, f) Allocate an integration structure

Upon entry: n—the dimension of the system
f—the function to be integrated

Return value: an integration pointer; NIL if an error occurs

rkf_tolerances(rel_tol, abs_tol, ip) Set error tolerances

Upon entry: rel_tol—the relative error tolerance
abs_tol—the absolute error tolerance
ip—the integration pointer

Return value: status code

rkf_initial_condition(t0, x0[], ip) Set the initial conditions

Upon entry: t0—the initial time
x0[]—the initial state
ip—the integration pointer

Return value: status code

rkf_mode(mode, ip) Set the integration mode

Upon entry: mode—The desired mode of operation, ONE-STEP or END-POINT
ip—the integration pointer

Return value: status code

rkf(tf, t, x[], ip) Perform the integration

Upon entry: tf—the final time
ip—the integration pointer

Upon return: t—the output time. Not equal to tf if an error occurs or if in ONE-STEP mode
x[]—the output point

Return value: status code

rkf_close(ip) Dispose of an integration structure

Upon entry: ip—an integration pointer

Figure 9.5: The library interface for RKF—C version.

END-POINT mode. The routine rkf_close is called to dispose of the integration structure once the integration is complete.

Each of these routines performs a single, well-defined task and each routine requires only the information that is necessary to perform its task. In particular, the routine that performs the integration requires just four arguments. No argument is used for both input and output information. Each routine returns a status code that indicates whether an error occurred, except rkf_open which returns a NIL pointer to indicate an error condition.

One important consequence of the way the RKF interface is defined is that the programmer is freed from the burden of keeping track of the state of the integration. For example, when rkf_initial_condition is called, it sets a flag in the working storage that indicates that the initial condition has been changed. When rkf is next called, it checks this flag to see whether the integration should

rkf_interrupt(mode, f_poll, ip) Set interrupt mode

Upon entry: **mode**—the interrupt mode: TRUE to enable interrupt mode; FALSE to disable interrupt mode

f_poll—the polling routine. At certain points in an integration step, **rkf** calls this routine to poll the interrupt status. **f_poll** returns TRUE to indicate that an interrupt occurred; FALSE otherwise.

ip—an integration pointer

Figure 9.6: Interrupt mode entry point for RKF.

be continued or restarted. Thus, the question posed earlier of what happens if the programmer continues an integration from a new initial condition is rendered nonsensical simply by the way the library interface is defined.

Remark: Under an operating system like UNIX that allows the mixing of FORTRAN and C code in a single program, there is no need to rewrite the FORTRAN `rkf` routine in C. The C entry points simply serve as an interface between the main program and the FORTRAN routine.

Adding features Another advantage of this interface definition is that new features can be added to the library without modifying currently existing programs.

For example, suppose it is desired to add a feature to RKF that allows a program to force `rkf` to exit prematurely, even in the middle of an integration step. This feature would be used by an interactive program that needs to suspend an integration upon receipt of a keyboard interrupt.

This feature can be added to RKF by adding the entry point shown in Fig. 9.6. To use this interrupt facility, the program's interrupt handler sets a flag in program memory whenever a keyboard interrupt occurs. Then, when `rkf` calls `f_poll`, `f_poll` checks this flag and returns TRUE if it is set or FALSE if it is not. With this approach, `rkf` controls the points at which it can be interrupted, making it easier to continue the integration after an interrupt.

The interrupt mode defaults to no interrupts. Thus, any program that uses RKF and that was written prior to the addition of `rkf_interrupt` runs properly without any modifications. Compare this with the original FORTRAN routine where incorporating the interrupt capability requires the addition of a new argument to `rkf` which, in turn, requires modification of every program that calls `rkf`.

rkf_copy(ip) Copy an integration structure

Upon entry: ip—an integration pointer

Return value: a pointer to a new integration structure that is a copy of ip. NIL is returned if an error occurs.

rkf_read(code, result, ip) Fetch RKF data

Upon entry: code—a code representing what internal variable is to be read from ip

ip—an integration pointer

Upon return: result—the value of the internal variable

rkf_message(status) Fetch a text version of a status code

Upon entry: status—an RKF status code

Return value: a string that holds a text string briefly explaining the meaning of status

Figure 9.7: Additional entry points for RKF—C version.

Additional entry points Several additional entry points for RKF are given in Fig. 9.7.

rkf_copy returns a copy of an integration structure and is useful for restarting integrations from a previous point. It is used by the hyperplane locating routine time_half as discussed on page 51.

rkf_read fetches data that is internal to an integration structure (e.g., the current step-size, the current error). rkf_message returns a string containing a brief description of the status code. These two routines are used mostly for debugging purposes.

9.3.2 PARSE—input parsing routines

PARSE is a library of input parsing routines. One reason for writing specialized input routines is to perform error checking. If the program requests an integer and the user enters a real, the user should be politely reminded to enter an integer value. The standard C formatted input routine fscanf does not have this capability. The second reason for these routines is to standardize the user interface. When several related programs are distributed in a single simulation package, they should each interact with the user in a similar fashion. Writing a single input parsing library that is used by all the programs in the package is a major step toward this goal.

The entry points for PARSE are listed in Figs. 9.8–9.10. The routines, is_integer and is_real, test whether a character string contains an integer or a real number.

The get_* and the read_* routines are used by interactive programs for reading input entered from the keyboard. get_character reads characters from the terminal until the user enters one that is in

is_integer(s) Checks for an integer

Upon entry: s—a character string

Return value: TRUE if s contains an integer; FALSE otherwise

is_real(s) Checks for a real

Upon entry: s—a character string

Return value: TRUE if s contains a real number; FALSE otherwise

get_character(list) Read a character

Upon entry: list—a character string of valid characters

Return value: the first valid character entered by the user. Will keep prompting for input until a valid character is entered.

get_integer(min, max) Read an integer

Upon entry: min—the minimum value allowed

max—the maximum value allowed

Return value: the first integer the user enters that lies between min and max. Will keep prompting for input until a valid integer in the proper range is entered.

get_real(min, max) Read a real

Upon entry: min—the minimum value allowed

max—the maximum value allowed

Return value: the first real the user enters that lies between min and max. Will keep prompting for input until a valid real in the proper range is entered.

Figure 9.8: The entry points for PARSE.

list. The routines get_integer and get_real read from the terminal until the user enters an integer or a real number that lies within the specified bounds.

The read_* routines are similar to the get_* routines except that they print a prompt and provide for a default value. For example

```
read_character("Plot trajectories? [%c]: ",
                                    "yYnN", 'y');
```

prints to the terminal

```
Plot trajectories? [y]:
```

If a carriage return is hit, read_character returns the character y. If the user enters an upper- or lower-case y or n, that character is returned. If any other character is entered, the prompt is redisplayed and read_character waits for more input.

The scan_* routines are useful for parsing input read from an options file. Typically, an entire line is read from the file into a buffer. The scan_* routines are then used to process the buffer word by word.

The keyword matching routines implement the interactive interface. The valid[] array is used to mask out certain keywords that

read_character(prompt, list, default)
Read a character with prompt and default

Upon entry: **prompt**—a character string to be printed to the terminal. May contain a single formatting command (%c in C) for printing a character. **default** will be inserted at that point in the prompt.

list—a character string of valid characters

default—the default character

Return value: the first valid character entered by the user or, if the user enters just a carriage return, **default** is returned. Will keep prompting for input until a valid character or a line containing just a carriage return is entered.

read_integer(prompt, min, max, default)
Read an integer with prompt and default

Upon entry: **prompt**—a character string to be printed to the terminal. May contain a single formatting command (%d in C) for printing an integer. **default** will be inserted at that point in the prompt.

min—the minimum value allowed

max—the maximum value allowed

default—the default value

Return value: the first integer the user enters that lies between **min** and **max**, or, if the user enters just a carriage return, **default** is returned. Will keep prompting for input until a valid integer in the proper range or a line containing just a carriage return is entered.

read_real(prompt, min, max, default)
Read a real with prompt and default

same as **read_integer** but for reals. **prompt** may contain a single formatting command (%g in C) for printing a real.

scan_integer(s, n) Scan a string for an integer

Upon entry: **s**—the character string to be scanned

Upon return: **s**—unchanged unless the first white-space delimited word in **s** is an integer in which case that word is deleted from **s**

n—if the first white-space delimited word in **s** is an integer, the value of that integer

Return value: TRUE if an integer was found and decoded; FALSE otherwise. If FALSE, neither **s** nor **n** is modified.

scan_real(s, x) Scan a string for a real

same as **scan_integer** but for reals

scan_string(s, word)
Scan a string for a white-space delimited word.

same as **scan_integer** but returns the first white-space delimited word in **s**.

Figure 9.9: The entry points for PARSE (cont.).

match_keywords(s, key[]) Match keywords

Upon entry: **s**—character string containing the word to be matched

key[]—a NIL-terminated array of character strings; each entry is a legal keyword

Return value: the index (with respect to **key[]**) of the keyword that is matched. If no keyword is matched, −1 is returned. If more than one keyword is matched, −2 is returned. **s** matches **key[i]** if the strings are identical or if **s** is a prefix of **key[i]**.

print_keywords(fp, format, key[]) Print the entries in **key[]**

Upon entry: **fp**—a file pointer indicating where the output should be written; usually is the terminal

format—a character string indicating the format of the output

key[]—a NIL-terminated array of character strings; each entry is a legal keyword

match_valid_keywords(s, key[], valid[])
Match valid keywords

Upon entry: **s**—character string containing the word to be matched

key[]—a NIL-terminated array of character strings; each entry is a keyword

valid[]—a boolean array. If **valid[i]** is TRUE, then **key[i]** is a valid keyword.

Return value: same as **match_keywords** except only valid keywords are allowed to be matched

print_valid_keywords(fp, format, key[], valid[])
Print the valid entries in **key[]**

Is to **print_keywords** as **match_valid_keywords** is to **match_keywords**

Figure 9.10: The entry points for PARSE (cont.).

are not currently valid. For instance, in a Poincaré map program, the option to plot invariant manifolds could be masked out until a fixed point has been located.

9.3.3 BINFILE—binary data files

INSITE uses a library to create and maintain the binary history files. Each binary history file is composed of three parts (listed in the order in which they occur in the file):

> *The BINFILE header:* Contains information that is used internally by the BINFILE routines (e.g., the number of data vectors, the size of a data vector). This header has a fixed size which is the same in all BINFILE files.
>
> *The user-supplied header:* Contains information specific to the program that is using BINFILE. In INSITE, each program uses the user-supplied header to store enough information to resume

the simulation where it left off. The size of this header is determined by the program using BINFILE and varies from file to file.

The data vectors: A list of data vectors. In a given file, all the vectors must contain the same number of bytes, but the number of vectors is arbitrary. The size of this part of the binary file varies with the number of vectors that are stored.

Remark: As far as BINFILE is concerned the data stored in the user-supplied header and in the data vectors is arbitrary binary data.

The entry points for BINFILE are shown in Fig. 9.11. They are fairly self-explanatory. Data vectors and the user-supplied header may be read and written in any order. The `data_type` information is stored in the BINFILE header and can be used by a plotting program, for instance, so that it knows how to interpret the binary data. The BIN_STRUCT data type can be used to write structures or other compound data types into a binary file. Note that data types cannot be mixed in a single BINFILE file.

9.3.4 GRAF—graphics

The issue of graphics is mainly an issue of portability. If you know that a program will be used only on a particular model of graphics hardware, then it is reasonable to gear the capabilities of the program toward that hardware.

It is more common, however, for a program to be designed to run on a variety of different makes and models of graphics hardware. In this situation, a device-independent graphics library should be used. Device-independent graphics libraries rarely exploit the full capability of the graphics hardware on which they run—of necessity, they are geared toward the lowest common denominator—but they do provide a high degree of portability. For the UNIX version of INSITE, we chose the X windowing system. For the PC-DOS version, we used the Meta Windows graphics package.

Since INSITE uses two different graphics packages and the graphics packages have different characteristics and different interfaces, we wrote an intermediate graphics library called GRAF. GRAF has the capability of drawing a graphics box on the screen, labeling it, and displaying data in it. All the graphics calls in INSITE use the GRAF

bin_create(name, dim, data_type, data_size, hdr_size)
Create a file for writing

Upon entry: **name**—the name of the file
dim—the dimension of a data vector
data_type—The type of the data:
BIN_SIGNED—signed integer data
binunsigned—unsigned integer data
binreal—floating-point data
binstruct—none of the above; can be used with arbitrary data types
data_size—the number of bytes in one component of a data vector
hdr_size—the size of the user-supplied header in bytes

Return value: a pointer to the newly created binary file or NIL if an error occurs

bin_append(name)
Opens an existing file for writing at the end of the file

Upon entry: **name**—the name of the file
Return value: a pointer to the binary file or NIL if an error occurs

bin_write(vec, n, bfp) Write vectors to a binary file

Upon entry: **vec**—a pointer to the vectors
n—the number of vectors to write
bfp—a pointer to the file to write to
Return value: TRUE if all goes well; FALSE otherwise

bin_header_write(hdr, bfp) Write a user-supplied header

Upon entry: **hdr**—a pointer to the buffer containing the header
bfp—a pointer to the file to write to
Return value: TRUE if all goes well; FALSE otherwise

bin_open(name) Open an existing binary file for reading

Upon entry: **name**—the name of the file
Return value: a pointer to the binary file or NIL if an error occurs

bin_read(vec, n, bfp) Read vectors from a binary file

Upon entry: **vec**—a pointer to the buffer where data will be written
n—the number of vectors to read
bfp—a pointer to the file to read from
Return value: TRUE if all goes well; FALSE otherwise

bin_header_read(hdr, bfp) Read a user-supplied header

Upon entry: **hdr**—a pointer to the buffer where the header will be written
bfp—a pointer to the file to read from
Return value: TRUE if all goes well; FALSE otherwise

bin_close(bfp) Close a binary file

Upon entry: **bfp**—a pointer to the file to close
Return value: TRUE if all goes well; FALSE otherwise

Figure 9.11: The entry points for BINFILE.

graf_initialize() Initialize graphics
Called once, before any other GRAF commands.

graf_open() Allocate a GRAF structure
Return value: a pointer to a GRAF structure; NIL if an error occurs

graf_position(xmin, ymin, xmax, ymax, gp)
Specify the position of the graphics box on the graphics display
Upon entry: (xmin, ymin)—The position (in percent) of the lower left-hand corner of the graphics box
(xmax, ymax)—The position (in percent) of the upper right-hand corner of the graphics box
(xmin, ymin, xmax, ymax) = (0.0, 0.0, 100.0, 100.0) makes the graphics box the same size as the graphics area. This is not recommended because there will be no room for the axis labels which are normally positioned outside the graphics box.
(xmin, ymin, xmax, ymax) = (25.0, 25.0, 75.0, 75.0) places the graphics box in the middle of the graphics area and makes it one-fourth the size of the graphics area.
gp—a pointer to a GRAF structure

graf_real(xmin, ymin, xmax, ymax, gp)
Set the real coordinates of the graphics box
Upon entry: (xmin, ymin)—the real coordinates of the lower left-hand coordinate of the graphics box
(xmax, ymax)—the real coordinates of the upper right-hand coordinate of the graphics box
gp—a pointer to a GRAF structure

graf_close(gp) Dispose of a GRAF structure
Upon entry: gp—a pointer to a GRAF structure

Figure 9.12: The initialization entry points for GRAF.

library, and GRAF translates its graphics commands into X or Meta Windows calls. This extra level of insulation from the graphics libraries allowed INSITE to be written as if it were going to be run on only one graphics system.

GRAF—the graphics library

As far as graphics libraries go, GRAF is fairly primitive. The entry points are shown in Figs. 9.12–9.15.

The GRAF interface uses the open-close model, which has the desirable consequence that more than one graphics box can be open at a given time. This capability is useful, for instance, to display simultaneously, in separate graphics boxes, different projections of high-dimensional data. It also provides a simple technique for giving a single graphics box different scalings, which is useful when different sets of data are displayed in the same graphics box.

graf_move(x, y, gp) Move graphics cursor
Upon entry: (x, y)—real coordinates of the point to move to
gp—a pointer to a GRAF structure

graf_draw(x, y, gp)
Draw a line from the current graphics cursor position
Upon entry: (x, y)—real coordinates of the point to draw to.
After the draw, the graphics cursor is moved to (x, y).
gp—a pointer to a GRAF structure

graf_point(x, y, gp) Draw a point
Upon entry: (x, y)—real coordinates of where to draw the point
The graphics cursor is moved to (x, y).
gp—a pointer to a GRAF structure

graf_move_norm(x_norm, y_norm, gp)
Move graphics cursor—normalized
Upon entry: (x_norm, y_norm)—normalized coordinates of the point to move to
gp—a pointer to a GRAF structure

graf_draw_norm(x_norm, y_norm, gp)
Draw a line from the current graphics cursor position—normalized
Upon entry: (x_norm, y_norm)—normalized coordinates of the point to draw to.
After the draw, the graphics cursor is moved to (x_norm, y_norm).
gp—a pointer to a GRAF structure

graf_point_norm(x_norm, y_norm, gp)
Draw a point—normalized
Upon entry: (x_norm, y_norm)—normalized coordinates of where to draw the point
The graphics cursor is moved to (x_norm, y_norm).
gp—a pointer to a GRAF structure

graf_color(color) Change the current color
Upon entry: color—the color to change to
gp—a pointer to a GRAF structure

graf_erase() Erase the graphics screen

Figure 9.13: Drawing entry points for GRAF.

There are two coordinate systems used by GRAF. The more familar of the two are *real coordinates*, also called world coordinates, which are real numbers that correspond to the values of the data that are being displayed. Less familar are *normalized coordinates.* They, too, are real numbers and are used to draw graphical objects in a graphics box.

Normalized coordinates Many applications require that graphical objects like arrows or squares be drawn in the same graphics box as real data like trajectories or orbits. Since the size of a graphi-

graf_x_axis(n_labels, n_tics, tic_length, n_sig_figs, lbl_side, lbl_shift, axis_name, name_shift, gp)

Set formatting parameters for the x-axis

Upon entry:
- **n_labels**—the number of labels
- **n_tics**—the number of tic marks between labels
- **tic_length**—the length of a tic mark as a percentage of the height of the graphics box
- **n_sig_figs**—the number of significant figures for each label
- **lbl_side**—TRUE to position the labels at the bottom of the box; FALSE to position the labels at the top
- **lbl_shift**—a real number indicating a shift in the y direction for the numerical labels. The units are character height. A positive value shifts the labels away from the box; a negative value towards the box. A value of 0.0 places the labels just below the box (just above the box if **lbl_side** is FALSE).
- **axis_name**—a character string containing the name of the x-axis
- **name_shift**—similar to **lbl_shift** but alters the vertical positioning of **axis_name**. A value of 0.0 places the name just below the numerical labels (just above the labels if **lbl_side** is FALSE).
- **gp**—a pointer to a GRAF structure

graf_y_axis(n_labels, n_tics, tic_length, n_sig_figs, lbl_side, lbl_shift, axis_name, name_shift, gp)

Set formatting parameters for the y-axis. Similar to **graf_x_axis** except

Upon entry:
- **lbl_side**—TRUE to position the labels to the left of the box; FALSE to position the labels to the right

graf_title(title, size, offset, gp) Set the title parameters

Upon entry:
- **title**—a character string containing the title
- **size**—TRUE for emphasized text; FALSE for standard text. Emphasized text is either bold or a larger size.
- **offset**—a real number indicating the vertical shift (in units of character height) for the title. A positive value moves the title away from the graphics box; a negative value, towards the graphics box. A value of 0.0 positions the title just outside but touching the graphics box. The title is placed on the side of the box opposite from the x-axis labels.
- **gp**—a pointer to a GRAF structure

graf_bounds(box, labels, tics, axes, title, gp)

Draws the graphics box

Upon entry:
- **box**—TRUE to draw the outline of the box
- **labels**—TRUE to draw labels
- **tics**—TRUE to draw tic marks
- **axes**—TRUE to draw axes along $x = 0$ and $y = 0$
- **title**—TRUE to draw the title
- **gp**—a pointer to a GRAF structure

Figure 9.14: Formatting entry points for GRAF.

graf_screen_to_real(ix, iy, x, y, gp)
Convert screen coordinates to real coordinates
Upon entry: (ix, iy)—the screen coordinates
gp—a pointer to a GRAF structure
Upon return: (x, y)—the corresponding real coordinates

graf_real_to_screen(x, y, ix, iy, gp)
Convert real coordinates to screen coordinates
Upon entry: (x, y)—the real coordinates
gp—a pointer to a GRAF structure
Upon return: (ix, iy)—the corresponding screen coordinates

graf_screen_to_norm(ix, iy, x_norm, y_norm, gp)
Convert screen coordinates to normalized coordinates
Upon entry: (ix, iy)—the screen coordinates
gp—a pointer to a GRAF structure
Upon return: (x_norm, y_norm)—the corresponding normalized coordinates

graf_norm_to_screen(x_norm, y_norm, ix, iy, gp)
Convert normalized coordinates to screen coordinates
Upon entry: (x_norm, y_norm)—the normalized coordinates
gp—a pointer to a GRAF structure
Upon return: (ix, iy)—the corresponding screen coordinates

graf_real_to_norm(x, y, x_norm, y_norm, gp)
Convert real coordinates to normalized coordinates
Upon entry: (x, y)—the real coordinates
gp—a pointer to a GRAF structure
Upon return: (x_norm, y_norm)—the corresponding normalized coordinates

graf_norm_to_real(x_norm, y_norm, x, y, gp)
Convert normalized coordinates to real coordinates
Upon entry: (x_norm, y_norm)—the normalized coordinates
gp—a pointer to a GRAF structure
Upon return: (x, y)—the corresponding real coordinates

Figure 9.15: Coordinate conversion entry points for GRAF.

cal object should be independent of the scaling of the data, that is, the square or arrow should be the same size whether the x-axis is picoseconds or kilometers, a separate coordinate system is used for drawing graphical objects.

A natural candidate for this coordinate system is the screen coordinate system. Unfortunately, pixels are device dependent. A ten pixel by ten pixel square on a monitor with 1024 by 1024 resolution has one fourth the area of a ten by ten square on a monitor with 512 by 512 resolution. There is also the problem of aspect ratios: when a graphics device has a different number of pixels per inch in the x and y directions, a ten by ten square appears as a rectangle.

These problems can be solved by using *normalized coordinates.* Normalized coordinates are defined such that a line in the x direction that has unit length in normalized coordinates and a line in the y direction that also has unit length in normalized coordinates appear, on the graphics device, to have the same actual length. In addition, normalized coordinates are scaled such that the lower left-hand corner of the graphics box has normalized coordinates (0.0, 0.0), and such that at least one of the sides of the graphics box has length 1.0 (in normalized coordinates) and neither side has length greater than 1.0.

For example, if a graphics box appears on the screen to be five inches wide and three inches tall, then its normalized coordinates are $(x_{min}, y_{min}) = (0.0, 0.0)$ and $(x_{max}, y_{max}) = (1.0, 0.6)$. For a graphics box that is four inches wide and five inches tall, the normalized coordinates are $(x_{min}, y_{min}) = (0.0, 0.0)$ and $(x_{max}, y_{max}) = (0.8, 1.0)$.

Chapter 10

Phase Portraits

In this, the last chapter, we examine some of the difficulties encountered in combining several numerical algorithms into a usable, intelligent, simulation program. The example used throughout this chapter is the INSITE program `phase` which draws phase portraits of two-dimensional autonomous continuous-time systems. An added benefit of this example is that we get to discuss in detail some properties of basins of attraction and their boundaries.

A *phase portrait* is a graphical representation of the behavior of a continuous-time dynamical system. It consists of a collection of typical trajectories together with all the limit sets and basins of attraction. Though phase portraits may be drawn for any autonomous system, we restrict our attention to second-order systems. The main reason is that a second-order autonomous system is the most complicated system for which all possible types of behavior are known. Furthermore, all of its behavior can be displayed unambiguously in the plane. Phase portraits of two second-order systems are shown in Fig. 1.1.

Since structurally stable systems are the only kind that can be simulated accurately, they are all that concern us here. In particular, many of the algorithms presented in this chapter are unreliable for lossless systems.

A typical phase portrait program has three stages: calculate a family of trajectories, find the limit sets, and find the basins of attraction.

We discuss these three steps one at a time.

10.1 Trajectories

Drawing trajectories is broken into three tasks: choosing the initial condition, calculating the trajectory, and placing arrows on the trajectory to indicate the direction of the flow.

10.1.1 Selection of initial conditions

Initial conditions are selected by the user or generated automatically by the program.

Manual selection

The user specifies initial conditions by using a mouse to point to the desired initial condition on the screen or by entering the coordinates directly from the keyboard.

The initial condition can also be selected non-interactively using an options file. This approach is useful for prepared demonstrations.

Automatic selection

Typically the user wants the program to select the initial conditions automatically.

The naïve approach covers the display area with an imaginary rectangular grid. The center point of each rectangle in the grid is taken as an initial condition for a trajectory. The problem with this approach is that, as Fig. 10.1 shows, an evenly spaced set of initial conditions rarely yields an evenly spaced set of trajectories.

To obtain an evenly spaced family of trajectories, a subset of the grid points is used. Order the rectangles in the grid and associate a flag with each rectangle. Clear all the flags and repeat the following steps until all the flags are set. Choose the initial condition x_0 as the center point of the first grid rectangle whose flag is not set. Integrate the trajectory from x_0 in both forward and reverse time. During the integration of $\phi_t(x_0)$, set the flag of every grid rectangle that the trajectory enters.

This *grid algorithm* is simple and yields a more pleasing phase portrait than the naïve approach. Unfortunately, the aesthetic quality of the phase portrait depends on the ordering of the rectangles in the grid list. There is no best ordering—different orderings work better for different dynamical systems. We have found a few different orderings to be useful.

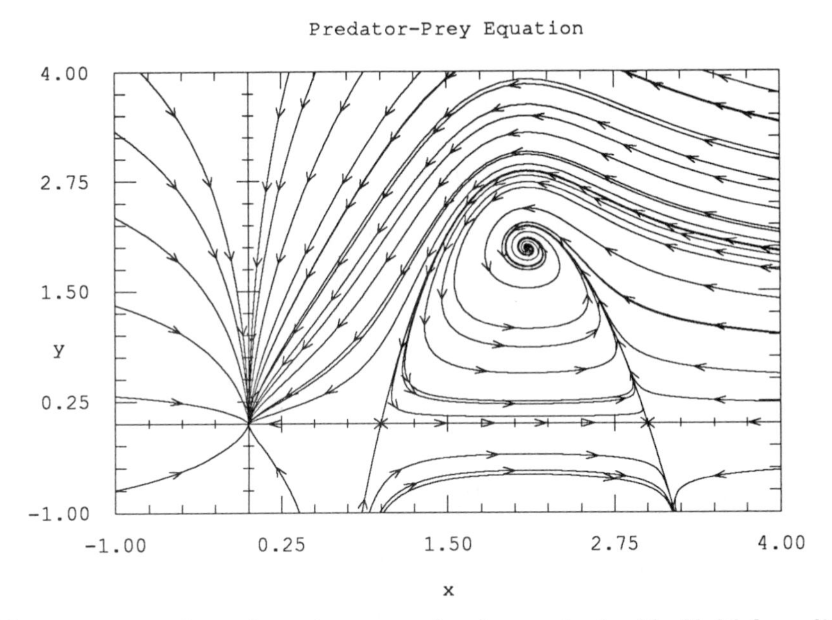

Figure 10.1: The trajectories emanating from a 6×6 grid of initial conditions for system (1.8). Compare this phase portrait with the one in Fig. 1.1, also calculated using a 6×6 grid, which uses the grid algorithm to obtain an evenly spaced family of trajectories.

One approach is to choose the order randomly using a pseudo-random number generator. Another ordering is the clockwise spiral shown in Fig. 10.2. This ordering gives priority to the outer rectangles where it is more important to have the trajectories evenly spaced.

Finally, we note that for an odd-symmetric system—a system is *odd-symmetric* if $f(-x) = -f(x)$ for all x—it is advantageous to choose the initial conditions in an odd-symmetric fashion, that is, if x_1 is the first initial condition, then $x_2 = -x_1$ is the second.

1	2	3	4
12	13	14	5
11	16	15	6
10	9	8	7

Figure 10.2: Clockwise ordering of the grid rectangles.

phase offers both random and clockwise ordering, and has an odd-symmetric option. Thus there are four different modes for choosing the initial conditions automatically.

10.1.2 Calculating the trajectory

Calculating the trajectory passing through a given point x_0 requires two integrations: one from x_0 in forward time, the other from x_0 in reverse time.

Each of these integrations proceeds until one of three conditions is satisfied:

1. The integration points converge.
2. The integration points travel out of bounds.
3. A time-out threshold is exceeded.

Convergence Convergence indicates that the trajectory is approaching a stable equilibrium point (or an unstable equilibrium point if the integration is in reverse time). To test for convergence, use a relative/absolute test (A.5). When convergence is detected, the program's equilibrium point list is checked to see whether the trajectory is near a known equilibrium point. If so, and if the equilibrium point is of the proper stability type—stable for forward-time integration, unstable for reverse time—then the integration is stopped.

If the trajectory is converging but is not near a known equilibrium point of the proper type, then the equilibrium point finding routine is called using the most recent integration point as the initial guess. If the routine converges to a new equilibrium point of the proper type, and if that equilibrium point is near the last integration point, then the integration is stopped. Otherwise, the integration continues and to avoid searching repeatedly for equilibrium points, the convergence testing is suspended temporarily. After a given number of integration steps, twenty, say, the convergence testing is resumed.

Out of bounds When the trajectory travels out of bounds, it is assumed that the trajectory is either unbounded or is attracted to a stable limit set that partially or totally lies out of bounds.

A better looking and more meaningful phase portrait is produced if, after the trajectory has passed out of the graphics area, the integration continues for a short time in the hope that the trajectory will

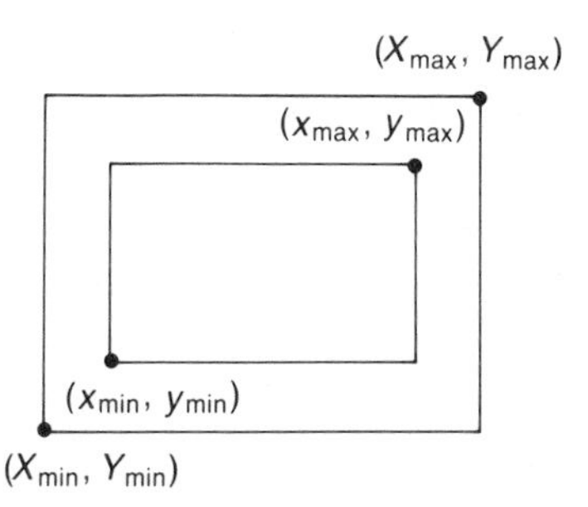

Figure 10.3: The graphics bounds, x_{min}, x_{max}, y_{min}, and y_{max}, determine what is displayed on the graphics screen. The calculation bounds, X_{min}, X_{max}, Y_{min}, and Y_{max}, are the values used by the program to judge whether the integration of a trajectory should be terminated.

reenter the graphics area. This is accomplished by using two different sets of bounds (see Fig. 10.3). The *graphics bounds* are the minimum and maximum x and y values for the graphics display. The *calculation bounds* are the minimum and maximum x and y values used in the out-of-bounds test. To achieve the desired effect of continuing the integration after the trajectory has passed out of the graphics region, the calculation bounds are chosen to frame the graphics region as in Fig. 10.3.

It is convenient to define the calculation bounds in terms of the graphics bounds by using a magnification factor $m > 1$. A value of $m = 1.5$ is typical. For example, let x_{min} and x_{max} be the x-axis graphics bounds, and define

$$\bar{x} := (x_{min} + x_{max})/2 \tag{10.1}$$

and

$$\Delta x := (x_{max} - x_{min})/2. \tag{10.2}$$

The x-axis calculation bounds are given by

$$X_{min} = \bar{x} - m\,\Delta x \tag{10.3}$$

and

$$X_{max} = \bar{x} + m\,\Delta x. \tag{10.4}$$

Similar formulas hold for the y-axis bounds.

Time-out A time-out test is needed to catch those cases that fall through the first two tests. Theoretically, the only situation that can slip past the first two tests is a trajectory that is attracted to a limit cycle that lies fully within the calculation bounds. It is difficult to devise a simple, efficient test that checks whether a trajectory is asymptotically approaching a limit cycle, but even if such a test were available, the time-out test would still be needed for practical reasons.

The major concern is that a large number of integration steps could be required before the integration points are close enough to satisfy the convergence test. This situation is likely to happen when the convergence tolerances are small and the attracting equilibrium point has complex eigenvalues.

There are several ways to implement the time-out. A maximum integration time t_{max} may be specified. If the integration proceeds past $t = t_{max}$, then it is stopped. An alternative, independent of time scale, is to limit the maximum number of integration steps. A third approach limits the total arc-length of each trajectory. This approach is quite useful because it is independent of the integration tolerances as well as the time scale and, therefore, the program can select the arc-length limit automatically—a value of

$$2(x_{max} - x_{min} + y_{max} - y_{min}) \qquad (10.5)$$

is reasonable. To avoid scaling dependencies, all length calculations are performed in normalized coordinates.

phase uses both arc-length and integration step limits. The integration step limit is included in the event a stiff system is being integrated by an integration algorithm that is not stiffly stable. In this case, without a limit on the number of integration steps, thousands of steps may be taken with little actual movement along the trajectory.

10.1.3 Arrowheads

The goal is to draw evenly spaced, properly oriented arrowheads along each trajectory. The arrows should be independent of the scaling of the x and y state variables, so normalized coordinates are used throughout this section.

The basic arrow template is shown in Fig. 10.4. The tip of the arrow is at the origin. The arrow's length is l and its base width is $2b$. Thus the arrow is defined by the three points: $(0,0)$, $(-l,b)$, and $(-l,-b)$.

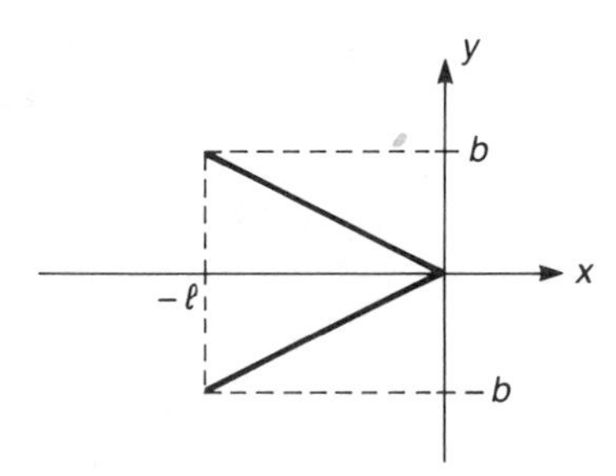

Figure 10.4: The basic arrow template has length l, base width $2b$, and is positioned with its tip at the origin.

Positioning an arrowhead requires two pieces of information: the position (x_t, y_t) of the tip of the arrow and the angle of rotation $\hat{\theta}$ from the basic position.[1] Given these values, the transformation

$$\begin{bmatrix} x' \\ y' \end{bmatrix} = \begin{bmatrix} \cos\hat{\theta} & -\sin\hat{\theta} \\ \sin\hat{\theta} & \cos\hat{\theta} \end{bmatrix} \begin{bmatrix} x \\ y \end{bmatrix} + \begin{bmatrix} x_t \\ y_t \end{bmatrix} \tag{10.6}$$

maps the arrow template to the desired position and orientation.

All that remains is specification of (x_t, y_t) and $\hat{\theta}$. One approach places an arrow every d units along the trajectory, where d is measured in terms of arc-length. The implementation is easy. Create a variable `length` which holds the arc-length since the last arrowhead. Whenever an integration is started or an arrow is drawn, set `length` to 0. Every time a point on the trajectory is found, calculate the distance from the last point, and add this distance to `length`. When `length` exceeds d, use linear interpolation to calculate the point (x_b, y_b) that is exactly d units of arc-length along the trajectory. This point is the mid-point of the base of the arrow. Similarly, calculate the point (x_t, y_t) that is $d + l$ units along the trajectory. This point is the tip of the arrow.

Simple geometry shows that $\cos(\theta) = (x_t - x_b)/l$ and $\sin(\theta) = (y_t - y_b)/l$. Using these values in (10.6), the two end-points of the base of the arrow are seen to map to $(x_b - \Delta y, y_b + \Delta x)$ and $(x_b + \Delta y, y_b - \Delta x)$ where $\Delta x := b(x_t - x_b)/l$ and $\Delta y := b(y_t - y_b)/l$. The tip of the arrow maps, of course, to (x_t, y_t).

There should always be an arrow at the initial condition of each trajectory. Since the trajectory is being integrated twice from each

[1] θ is measured in the usual manner—counter-clockwise with $\theta = 0$ indicating the positive x axis.

initial condition—once in forward time, once in reverse time—an arrow is placed at the beginning of every forward-time integration.

Trajectories accumulate on stable limit sets, and so do arrows. To alleviate this problem, use a grid algorithm similar to the one used to select initial conditions. The graphics region is divided into a rectangular grid. Whenever an arrow is drawn, the rectangle in which it lies is flagged and no other arrow is permitted in that rectangle. The one exception to this rule is that an arrow is always drawn at the initial condition of a forward-time integration. Thus each trajectory always has at least one arrow on it.

10.1.4 Drawing the vector field

The calculation of trajectories requires integration and, depending on the integration tolerances and the grid size, can be somewhat time consuming. A quicker alternative—useful for obtaining a rough idea of the system dynamics—is to display the vector field on the screen. Though this is not strictly within the purview of a phase portrait program, it is a simple task and can be added easily.

The vector field is drawn as follows. The graphics area is covered with an imaginary rectangular grid. An arrow is drawn based at the center point x_{ij} of each rectangle in the grid. The direction of the arrow is the direction of the vector field at x_{ij}.

`phase` has three options for the magnitude of the vector positioned at x_{ij} (see Fig. 10.5).

1. The magnitude is proportional to $\|f(x_{ij})\|$. The magnitudes are scaled such that the longest arrow is on the order of the distance between grid points. The maximum magnitude is found by calculating the value of the vector field at all of the center points before any arrows are drawn. To avoid recalculation, these values should be kept in an array. Once the scaling factor is known, values from the array are used to draw the arrows on the display.

2. The magnitude is proportional to $\log(\|f(x_{ij})\|)$. This option is useful when the vector field assumes a wide range of magnitudes. As in the first option, the length of each arrow is scaled such that the magnitude of the longest arrow is on the order of the spacing between grid points.

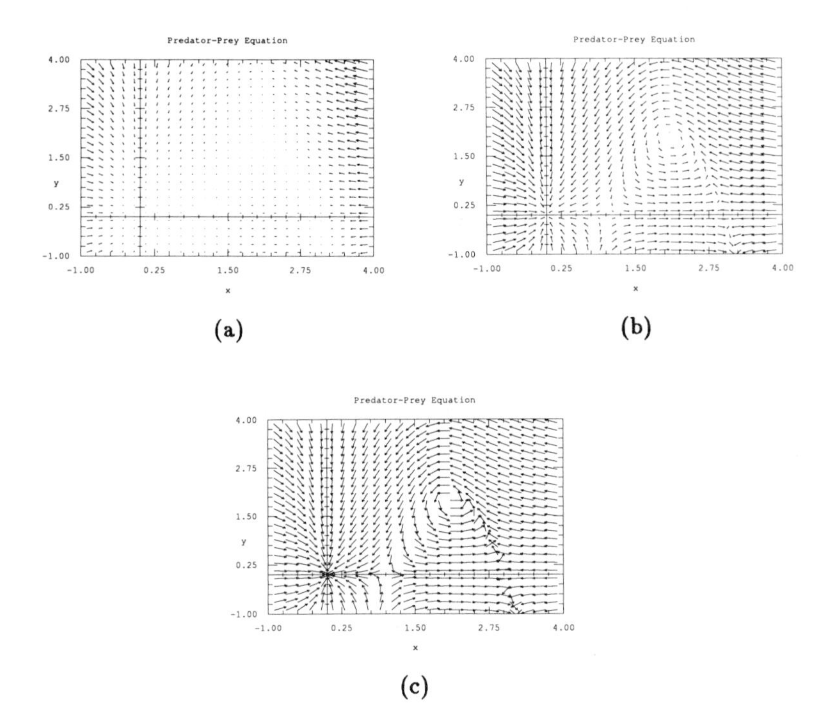

Figure 10.5: The vector field of the predator-prey system (1.8). (a) linear magnitudes; (b) log magnitudes; (c) constant magnitudes.

3. The magnitude of each arrow is fixed. This option provides the user with a good idea of the direction of the flow. To provide a pleasing display, the magnitude is chosen on the order of the distance between grid points.

10.2 Limit sets

Since we are interested only in structurally stable systems, all limit sets in this section are assumed hyperbolic.

The simplest possible limit set of a second-order autonomous system is an equilibrium point. The equilibrium point may be stable, unstable, or non-stable, as determined by the eigenvalues of $Df(x_{eq})$.

The next most complicated limit set that can occur in a second-order autonomous system is a limit cycle. A hyperbolic limit cycle is either stable or unstable depending on its single characteristic multiplier. A non-stable hyperbolic limit cycle in $\mathbb{R}^2$ is impossible.

Two- and higher-periodic behavior is also impossible in a second-order autonomous system because a K-periodic solution possesses at least $K+1$ Lyapunov exponents and, therefore, requires a state space of dimension at least $K+1$.

Chaos, too, is outlawed in second-order autonomous systems. For chaos to occur, one Lyapunov exponent must be positive. Since one Lyapunov exponent is always 0, and the sum of the exponents must be negative, chaos requires a third- or higher-order autonomous system.

To summarize, the only hyperbolic limit sets that occur in second-order autonomous systems are equilibrium points and limit cycles.

10.2.1 Equilibrium points

Equilibrium points are found using the Newton-Raphson algorithm presented in Section 5.2.

Since the Newton-Raphson algorithm requires an initial guess, the routine that finds equilibrium points cannot be fully automated. The one case where the program can provide a good initial guess is when the integration points of a trajectory converge. Theoretically, the only object to which the points can converge is an equilibrium point;[2] hence, the most recent integration point is an excellent initial guess for the Newton-Raphson algorithm. Under normal circumstances, the grid algorithm coupled with the Newton-Raphson algorithm can automatically locate all the stable and unstable equilibrium points that lie within the calculation bounds. Initial conditions for non-stable equilibrium points must be provided by the user.

phase maintains an equilibrium point list in program memory. Each entry of the list contains the coordinates, eigenvalues, and eigenvectors of the equilibrium point, and a code indicating its stability type. Whenever an equilibrium point is found, this list is checked to see whether the equilibrium point is a duplicate.[3] If not, the eigenvalues and eigenvectors of the equilibrium point are calculated and this information is stored in a new entry of the equilibrium point list.

When an equilibrium point is found, a symbol indicating its position and stability type is drawn on the graphics screen.

[2] In practice, the points can appear to converge for several other reasons. See Section 10.4.1.

[3] See Section 5.2 for details on the duplicate equilibrium point test.

10.2.2 Limit cycles

Limit cycles are located using the Poincaré map technique described in Section 5.5.2. This method is preferred over the autonomous shooting method because it frees the user from the difficult task of supplying an initial guess for the period. This feature is important because, from looking at the display created by the grid algorithm, it is easy to judge the position of a limit cycle but impossible to estimate its period.

The initial condition $x^{(0)}$ is specified interactively using a mouse or similar graphics pointing device, or by entering the coordinates from the keyboard. The cross-section defining the Poincaré map defaults to the line through $x^{(0)}$ that is orthogonal to $f(x^{(0)})$. The user may override the default by specifying a different value for the normal vector.

Recall that a limit cycle can correspond to a period-K closed orbit of P_+. The iteration order of the fixed point—the order is K if a fixed point is sought for P_+^K—can be specified by the user. It defaults to one.

A limit cycle in the plane has only one characteristic multiplier. It follows that a hyperbolic limit cycle must be stable or unstable; non-stable hyperbolic limit cycles are impossible in the plane. Since the convergence properties of the Poincaré map method are improved by integrating in reverse time for unstable limit cycles, the user may choose the integration direction. It defaults to forward time.

Note that it is simple for the user to determine K and the stability of the limit cycle, as well as $x^{(0)}$, by looking at the trajectories drawn by the grid algorithm.

It helps to display on the graphics screen the cross-section and, at each iteration, the trajectory calculated by the Poincaré map algorithm. This visual feedback allows the user to see at a glance whether the algorithm is converging and whether K should be changed.

phase maintains a limit cycle list in program memory. Each entry of the list contains the coordinates of a point x^* on the limit cycle, the minimum period T of the limit cycle, the characteristic multiplier, the normal vector of the cross-section that defines the Poincaré map, and the iteration order K of the Poincaré map. Whenever a limit cycle is found, the list is checked to see whether the limit cycle is a duplicate. If the limit cycle is not a duplicate, its minimum period is determined,[4] its characteristic multiplier is calculated, and a new entry is added to the limit cycle list.

When a limit cycle is located, it is drawn on the graphics display by integrating from x^* for T seconds.

10.2.3 Index

There are topological constraints on limit sets in the plane. For example, it is easy to draw a vector field with a single equilibrium point surrounded by a limit cycle. It is impossible to draw a smooth vector field for a system that has exactly two hyperbolic equilibrium points inside a limit cycle.

These constraints are quantified using the concept of index. In state space, choose a simple closed curve, called a *contour*, that intersects no equilibrium points. Traverse this contour in a counter-clockwise direction, noting the direction of the vector field at each point. The *index* is defined as the number of complete rotations the vector field makes in one complete journey around the contour. Counter-clockwise rotations add one to the count; clockwise rotations subtract one.

It can be shown that the index depends only on the equilibrium points interior to the contour. Thus, it makes sense to speak of *the* index of an equilibrium point. As Fig. 10.6 shows, the index of a stable hyperbolic equilibrium point is 1. The index of a hyperbolic saddle point[5] is -1. With a few quick sketches, the reader can show that the index of an unstable hyperbolic equilibrium point is one, and that the index using any limit cycle as a contour is also one. It can be shown that, if the contour surrounds more than one equilibrium point, its index is the sum of the indices of the equilibrium points. It follows that the sum of the indices of the equilibrium points inside a limit cycle is 1.

By calculating the index using the perimeter of the graphics box as a contour and comparing this value to the sum of the indices of all known equilibrium points, the phase portrait program may be able to let the user know if there are equilibrium points that are yet to be found. It is important to realize exactly what information this test provides. If the index of the graphics box does not equal the sum of the indices of the equilibrium points in the graphics box, then

[4] See Section 5.5.2 for details on the duplication and minimum period tests.

[5] Recall that a saddle point is a non-stable equilibrium point.

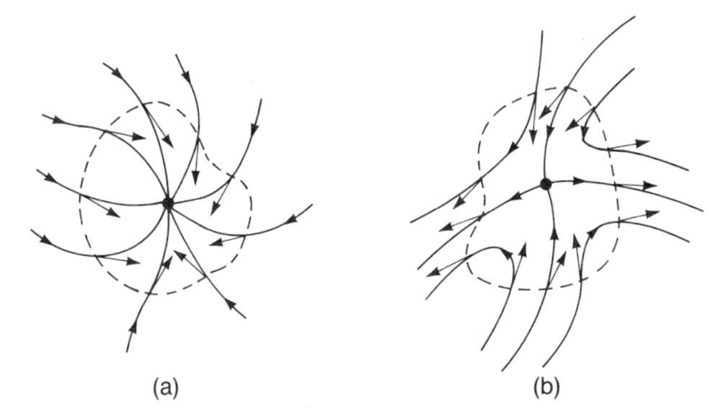

Figure 10.6: The index of (a) a stable equilibrium point; (b) a saddle point.

there are necessarily some undetected equilibrium points inside the graphics box. If, however, the two values agree, it does not follow that all the equilibrium points inside the graphics box have been detected—there may be two or more undetected equilibrium points whose indices sum to zero.

phase also checks that the indices of the equilibrium points interior to a limit cycle sum to one. To determine whether an equilibrium point is interior to a limit cycle, pick any ray emanating from the equilibrium point. If the ray crosses the limit cycle an even number of times, then the equilibrium point is exterior to the limit cycle; if there are an odd number of crossings, it is interior to the limit cycle.

For a more thorough treatment of indices and their properties, see Guillemin and Pollack [1974] and Krasnosel'skiy [1966].

10.3 Basins of attraction

10.3.1 Definitions

Given an attracting limit set A of a flow ϕ_t, its *basin of attraction* B_A is the set of all points x whose ω-limit set is A. It follows that the basin of attraction of an attracting set is open and invariant under the flow.

We wish to draw basins of attraction and are, therefore, interested in their boundaries. The *boundary* of a set B, denoted by $\text{bd}(B)$, is defined as the intersection of the closure[6] of B and the closure of the

[6]The *closure* of a set B is defined as the set of all points x such that every

complement of B, that is,

$$\mathrm{bd}(B) := \mathrm{cl}(B) \bigcap \mathrm{cl}(\overline{B}). \tag{10.7}$$

In the following facts, which we state without proof, the set B is a basin of attraction.

1. A point x lies on the boundary of B if and only if every open neighborhood of x contains at least one point in B and at least one point not in B.

2. No point is common to B and $\mathrm{bd}(B)$.

3. The boundary of B is invariant with respect to ϕ_t.

It follows from the last fact that the boundary of a basin of attraction is the union of one or more trajectories. These trajectories are called *boundary trajectories.* In the plane, the boundary of a basin of attraction is a curve, and can be visualized as one or more boundary trajectories linked end-to-end. In higher dimensions, a boundary is a hyper-surface and is the union of an uncountable set of boundary trajectories.

A boundary trajectory of a basin of attraction B is called *degenerate* if there exists an open neighborhood U of some point on the trajectory such that U is contained in $\mathrm{cl}(B)$. From invariance, it follows that every point on a degenerate boundary trajectory has such a neighborhood and, therefore, there exists an open neighborhood of the boundary trajectory that lies in $\mathrm{cl}(B)$. Thus, a degenerate boundary does not border on different basins of attraction and does not separate two different types of behavior.

A boundary trajectory that does separate one type of behavior from another is of special interest. A non-degenerate, non-constant boundary trajectory of a basin of attraction B is called a *separatrix for B.* Given any open neighborhood U of any point x on a separatrix, U contains at least one point not in $\mathrm{cl}(B)$.

The *boundary between two basins of attraction*, B_1 and B_2, is the intersection of $\mathrm{bd}(B_1)$ and $\mathrm{bd}(B_2)$. It is composed of the separatrices common to B_1 and B_2 together with any equilibrium points these separatrices approach.

---

open neighborhood of x contains at least one point of B. It follows that B is contained in $\mathrm{cl}(B)$, and that the closure of any set is closed.

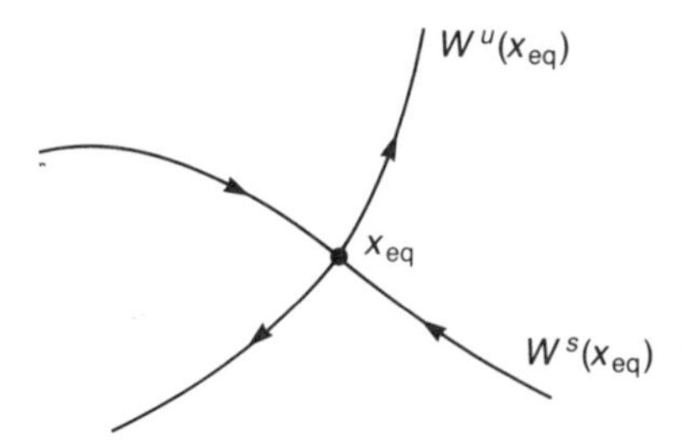

Figure 10.7: The four special trajectories associated with a saddle point.

Define B_∞ as the set of all points not in a basin of attraction or in the boundary of a basin of attraction. More precisely, if the family of attracting limit sets is $\{A_i\}$, then

$$B_\infty := \bigcap_i \overline{\mathrm{cl}(B_{A_i})}. \tag{10.8}$$

B_∞ is invariant under the flow, and every trajectory in B_∞ is unbounded, so, with care, one can think of B_∞ as the basin of attraction of an attracting limit set at infinity. There are two cautionary points, however. First, as we will see shortly, there can be unbounded trajectories not contained in B_∞. Second, the intersection of an infinite family of open sets is not necessarily open; therefore, if there are an infinity of attracting limit sets, there is no guarantee that B_∞ is an open set.

The following definitions will be helpful in the subsequent discussion of boundaries and separatrices.

In the plane, there are four special trajectories associated with a saddle point (Fig. 10.7). The two *stable saddle trajectories* approach the saddle point as $t \to \infty$; the two *unstable saddle trajectories* approach the saddle point as $t \to -\infty$.

The set of points that approach a limit set L as $t \to \infty$ is called the *stable manifold* of the limit set and is denoted by $W^s(L)$. The set of points that approach L in reverse time is called the *unstable manifold* and is denoted by $W^u(L)$. It follows immediately that the stable and unstable manifolds are invariant under the flow ϕ_t.

The stable manifold of a stable limit set is the basin of attraction and the unstable manifold is the limit set itself. For an unstable limit set, the stable manifold is the limit set itself; the unstable manifold is the basin of attraction of the limit set in reverse time. The stable manifold of a saddle point in the plane is the union of the saddle point and its two stable trajectories; the unstable manifold is the union of the saddle point and its two unstable trajectories.

10.3.2 Examples

We now present several examples of basins of attraction, boundary trajectories, and separatrices. The examples are divided into three groups.

The first set of examples is given in Fig. 10.8. These systems share the feature that all the boundaries of their basins of attraction are composed of stable manifolds. Furthermore, all the systems are structurally stable.

Fig. 10.8(a) shows the most common type of separatrix, a stable saddle trajectory. Both stable saddle trajectories are unbounded and, therefore, the stable manifold of the saddle point divides the plane into two regions, each of which is a basin of attraction.

The system of Fig. 10.8(b) possesses a bounded basin of attraction whose boundary is the unstable limit cycle.

The basin of attraction in Fig. 10.8(c) has an unusual boundary. It is clear that the stable manifold of the saddle point separates the two basins of attraction and should be on the boundary of each basin. What may be surprising is that the unstable limit cycle also lies on the boundary of each basin. Since every open neighborhood of the limit cycle contains points in $B_{x_1^s}$ and points in $B_{x_2^s}$, the limit cycle is a separatrix for both basins and, therefore, lies on the boundary "between" $B_{x_1^s}$ and $B_{x_2^s}$.

When just the basins of attraction are drawn, the system in Fig. 10.8(d) appears identical to the system in Fig. 10.8(a)—each system has two unbounded basins of attraction which share a common boundary. The systems are quite different though. The boundary in Fig. 10.8(d) is more complicated; it contains two stable manifolds and an unstable equilibrium point. To discover more differences between the two systems, the reader is invited to find the basins of attraction of each system in reverse time.

Figs. 10.8(b,e) also have identical basins of attraction. However, unlike Fig. 10.8(b), the boundary in Fig. 10.8(e) is not a limit cycle; it is the union of the unstable equilibrium point and the stable manifold of the saddle point.

Fig. 10.8(f) is an example of a boundary composed only of degenerate boundary trajectories, that is, there are no separatrices.

(a) stable limit sets: x_1^s, x_2^s;
unstable limit sets: none;
non-stable limit sets: $\hat{x}$;
separatrices: stable trajectories of $\hat{x}$;
degenerate boundary trajectories: none;
boundary of $B_{x_1^s}$: $W^s(\hat{x})$;
boundary of $B_{x_2^s}$: $W^s(\hat{x})$;

(b) stable limit sets: x^s
unstable limit sets: Γ^u;
non-stable limit sets: none;
separatrices: Γ^u;
degenerate boundary trajectories: none;
boundary of B_{x^s}: Γ^u;
boundary of B_∞: Γ^u;

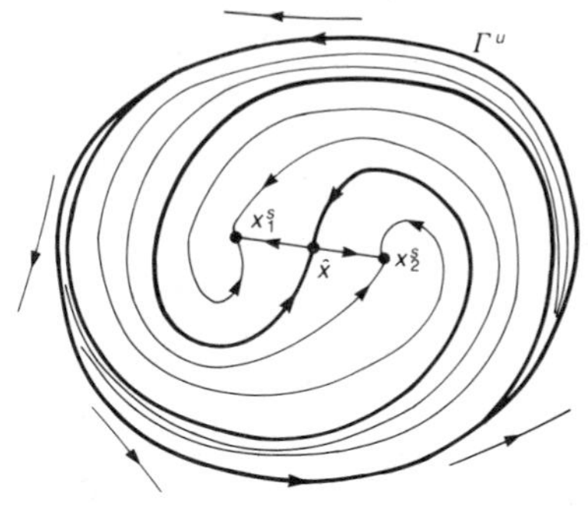

(c) stable limit sets: x_1^s, x_2^s;
unstable limit sets: Γ^u;
non-stable limit sets: $\hat{x}$;
separatrices: Γ^u, stable trajectories of $\hat{x}$;
degenerate boundary trajectories: none;
boundary of $B_{x_1^s}$: $\Gamma^u \bigcup W^s(\hat{x})$;
boundary of $B_{x_2^s}$: $\Gamma^u \bigcup W^s(\hat{x})$;
boundary of B_∞: Γ^u;

Figure 10.8: The boundaries of these basins of attraction are composed of stable manifolds.

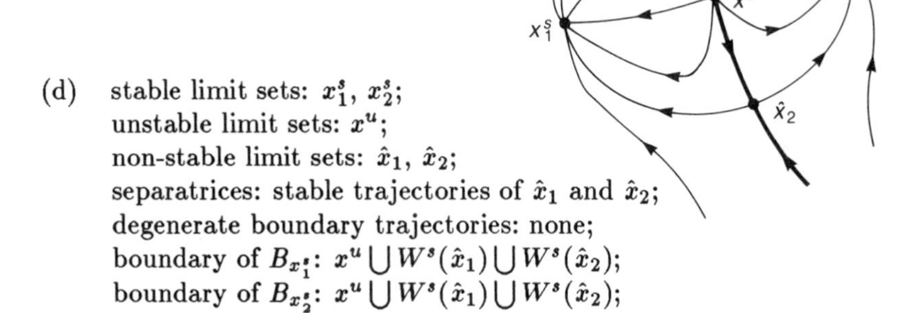

(d) stable limit sets: x_1^s, x_2^s;
unstable limit sets: x^u;
non-stable limit sets: $\hat{x}_1$, $\hat{x}_2$;
separatrices: stable trajectories of $\hat{x}_1$ and $\hat{x}_2$;
degenerate boundary trajectories: none;
boundary of $B_{x_1^s}$: $x^u \bigcup W^s(\hat{x}_1) \bigcup W^s(\hat{x}_2)$;
boundary of $B_{x_2^s}$: $x^u \bigcup W^s(\hat{x}_1) \bigcup W^s(\hat{x}_2)$;

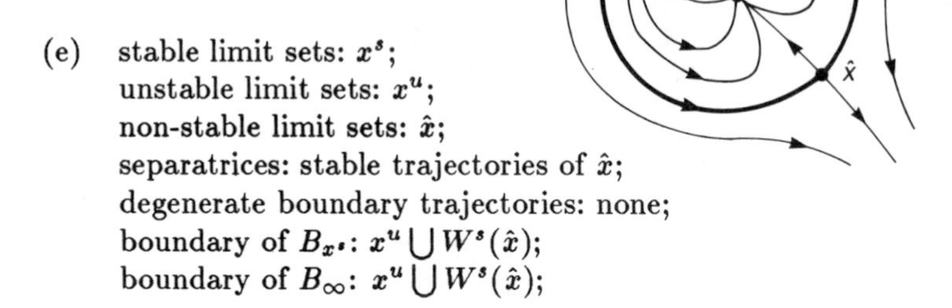

(e) stable limit sets: x^s;
unstable limit sets: x^u;
non-stable limit sets: $\hat{x}$;
separatrices: stable trajectories of $\hat{x}$;
degenerate boundary trajectories: none;
boundary of B_{x^s}: $x^u \bigcup W^s(\hat{x})$;
boundary of B_∞: $x^u \bigcup W^s(\hat{x})$;

(f) stable limit sets: x^s;
unstable limit sets: x^u;
non-stable limit sets: $\hat{x}$;
separatrices: none;
degenerate boundary trajectories: x^u, $\hat{x}$,
stable trajectories of $\hat{x}$;
boundary of B_{x^s}: $x^u \bigcup W^s(\hat{x})$.

Figure 10.8 (cont.)

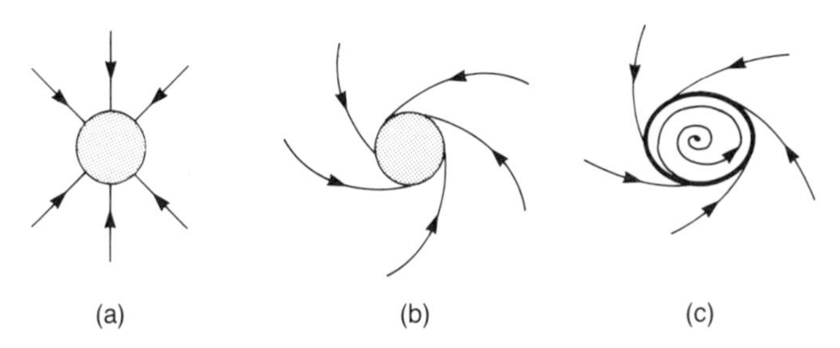

Figure 10.9: A stable equilibrium point can be replaced by a stable limit cycle in three steps: (a) stretch; (b) twist; (c) replace.

None of these examples has an attracting limit cycle. This is not an oversight. In the plane, any stable equilibrium point may be replaced by a stable limit cycle without affecting the global nature of the flow. Thus, all the systems given can be transformed easily into systems with stable limit cycles. The transformation, displayed in Fig. 10.9, may be visualized as follows. Take a stable equilibrium point and stretch it slightly in all directions so that it becomes a small disk. If the equilibrium point has real eigenvalues, twist the disk slightly so that the vector field spirals in around it. Replace the circumference of the disk with a stable limit cycle. Interior to the limit cycle, there can be any legal vector field.

Similarly, any unstable equilibrium point in the plane can be replaced by an unstable limit cycle. This substitution leads to some interesting boundaries. Try it on Figs. 10.8(d,e,f).

The second set of examples is shown in Fig. 10.10. Like Fig. 10.8, these systems are structurally stable. Unlike Fig. 10.8, however, the boundaries of the basins of attraction are not composed solely of stable manifolds of limit sets.

The boundary in Fig. 10.10(a) does not even lie on a manifold associated with any of the limit sets. It is a single trajectory that is unbounded in both forward and reverse time.

In Fig. 10.10(b), one of the *unstable* saddle trajectories is a separatrix. This system is an example where B_∞ is not the set of all initial conditions whose trajectory is unbounded as $t \to \infty$.

The basins of attraction of Fig. 10.10(c) are similar to those shown in Figs. 10.8(a,d). In this system, however, there exists a separatrix $\mathcal{S}$ that does not lie in the stable manifold of any limit set.

(a) stable limit sets: x_1^s, x_2^s;
unstable limit sets: none;
non-stable limit sets: none;
separatrices: the trajectory $\mathcal{S}$ that is unbounded as $t \to \infty$ and as $t \to -\infty$;
degenerate boundary trajectories: none;
boundary of $B_{x_1^s}$: $\mathcal{S}$;
boundary of $B_{x_2^s}$: $\mathcal{S}$;

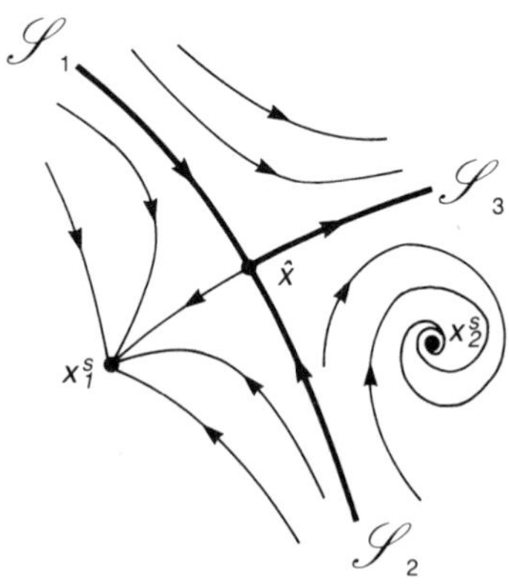

(b) stable limit sets: x_1^s, x_2^s;
unstable limit sets: none;
non-stable limit sets: $\hat{x}$;
separatrices: stable trajectories of $\hat{x}$, one unstable trajectory of $\hat{x}$;
degenerate boundary trajectories: none;
boundary of $B_{x_1^s}$: $\hat{x} \bigcup \mathcal{S}_1 \bigcup \mathcal{S}_2$;
boundary of $B_{x_2^s}$: $\hat{x} \bigcup \mathcal{S}_2 \bigcup \mathcal{S}_3$;
boundary of B_∞: $\hat{x} \bigcup \mathcal{S}_1 \bigcup \mathcal{S}_3$;

Figure 10.10: The boundaries of these basins of attraction are not composed solely of stable manifolds.

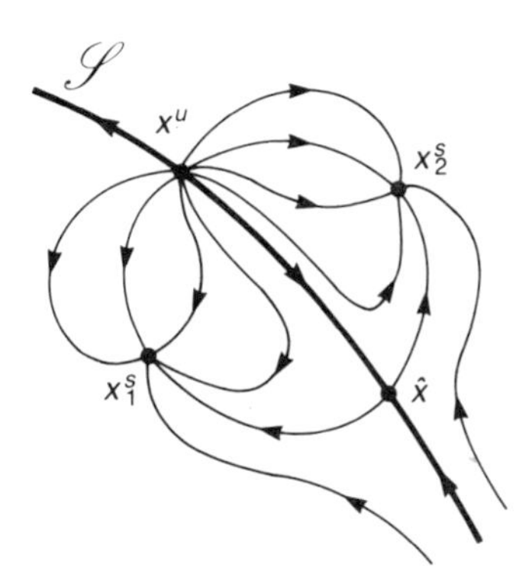

(c) stable limit sets: x_1^s, x_2^s;
unstable limit sets: x^u;
non-stable limit sets: $\hat{x}$;
separatrices: stable trajectories of $\hat{x}$,
the trajectory $\mathcal{S}$ that is unbounded as $t \to \infty$
and approaches x^u as $t \to -\infty$;
degenerate boundary trajectories: none;
boundary of $B_{x_1^s}$: $\mathcal{S} \bigcup W^s(\hat{x})$;
boundary of $B_{x_2^s}$: $\mathcal{S} \bigcup W^s(\hat{x})$;

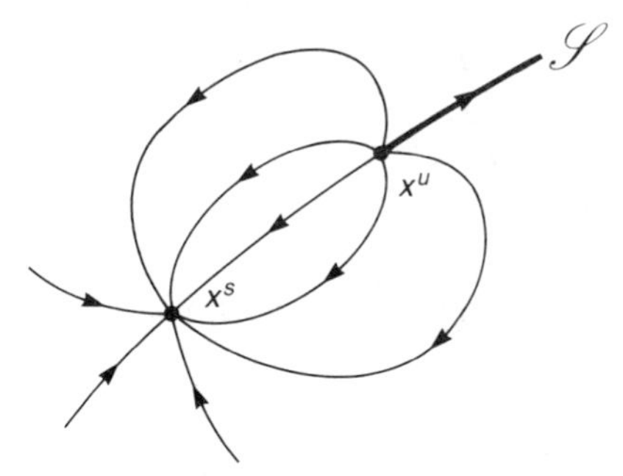

(d) stable limit sets: x^s;
unstable limit sets: x^u;
non-stable limit sets: none;
separatrices: none;
degenerate boundary trajectories: x^u,
the trajectory $\mathcal{S}$ that is unbounded as $t \to \infty$
and that approaches x^u as $t \to -\infty$;
boundary of B_{x^s}: $x^u \bigcup \mathcal{S}$.

Figure 10.10 (cont.)

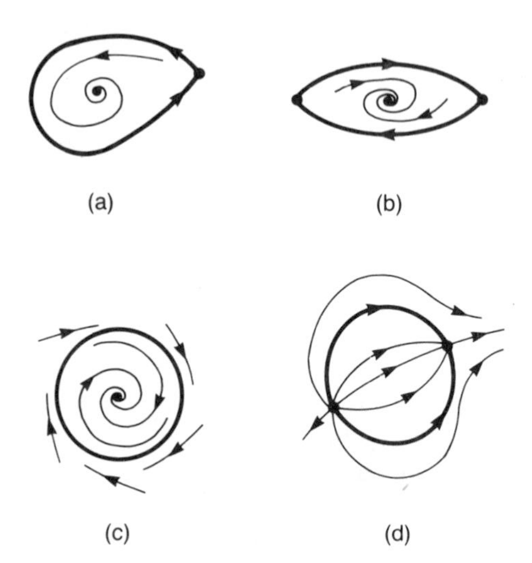

Figure 10.11: Separatrices in structurally unstable systems. (a) a homoclinic trajectory; (b) a heteroclinic trajectory; (c) a non-hyperbolic limit cycle; (d) a non-hyperbolic equilibrium point.

Fig. 10.10(d) has no separatrices, but does have a degenerate boundary trajectory $\mathcal{S}$. $\mathcal{S}$ does not lie in the stable manifold of any limit set.

The third and final set of examples is shown in Fig. 10.11. All the systems in this figure are structurally unstable. We reiterate that just because these systems are structurally unstable, it does not follow that they are of no interest—structurally unstable systems are of the utmost importance in the study of bifurcation phenomena. It does follow, however, that they cannot be simulated reliably using integration routines on a digital computer. The error introduced by the integration almost always ruins the delicate structure of these systems.

The separatrix shown in Fig. 10.11(a) is a *homoclinic trajectory*, that is, a non-constant trajectory that approaches the same equilibrium point in both forward and reverse time. A homoclinic trajectory lies in both the stable and unstable manifolds of some non-stable equilibrium point.

Each separatrix in Fig. 10.11(b) is a *heteroclinic trajectory*, that is, a trajectory that approaches one non-stable equilibrium point in forward time and another in reverse time. A homoclinic trajectory lies in the stable manifold of one non-stable equilibrium point and the unstable manifold of another.

The limit cycle in Fig. 10.11(c) is not hyperbolic. It attracts nearby trajectories that lie outside it, and repels nearby trajectories that lie inside it.

In the system of Fig. 10.11(d), there is a non-hyperbolic equilibrium point. It can be viewed as a combination of a saddle point and a stable node,[7] and is often called a *saddle-node*.

10.3.3 Calculating boundaries of basins of attraction

Theorems

Now that we know what boundaries and separatrices are, how can they be calculated? This is a difficult problem, and, currently, only a partial answer can be given. We rely on two theorems, both due to Chiang, *et al.* [1988]. The theorems apply to autonomous continuous-time dynamical systems of any order. Let A be an attracting set that satisfies the following three assumptions:

A.1 All the equilibrium points and limit cycles on $\mathrm{bd}(B_A)$ are hyperbolic.

A.2 The stable and unstable manifolds of the equilibrium points and limit cycles on $\mathrm{bd}(B_A)$ are transversal.

A.3 Every trajectory on $\mathrm{bd}(B_A)$ approaches one of the equilibrium points or limit cycles as $t \to \infty$.

The systems in Fig. 10.8 satisfy all three assumptions. Assumptions A.1 and A.2 rule out the structurally unstable systems shown in Fig. 10.11. Fortunately, since we are interested only in structurally stable systems, this is not a serious limitation. The third assumption, however, does rule out some structurally stable systems; in fact, all of the systems in Fig. 10.10 violate assumption A.3.

We now state the two theorems. Theorem 10.1 gives conditions for when an equilibrium point or limit cycle lies on the boundary of A. Given the limit sets on the boundary, Theorem 10.2 identifies the boundary of A.

[7] Recall that a *node* is a stable or unstable equilibrium point that has no complex eigenvalues.

Theorem 10.1 *Let Λ be either an equilibrium point or limit cycle. If Λ is distinct from the attracting set A, and if assumptions A.1–A.3 are satisfied, then*

$$\Lambda \text{ lies on } \mathrm{bd}(B_A) \text{ if and only if } W^u(\Lambda) \bigcap B_A \text{ is non-empty.}$$

Theorem 10.2 *Let $\{x_i\}$ and $\{\Gamma_j\}$ be the equilibrium points and limit cycles on the boundary of A. If assumptions A.1–A.3 are satisfied, then*

$$\mathrm{bd}(A) = \bigcup_i W^s(x_i) \bigcup_j W^s(\Gamma_j) \tag{10.9}$$

Remarks:

1. A stable equilibrium point or stable limit cycle cannot lie on the boundary of A.
2. Consider a two-dimensional autonomous system. Since an unstable limit cycle repels all nearby trajectories, any trajectory interior to the limit cycle must approach some attracting limit set in the interior of the limit cycle. It follows from Theorem 10.1 that, in the plane, an unstable limit cycle is always a boundary trajectory.
3. Furthermore, since trajectories exterior to an unstable limit cycle cannot be attracted to a limit set that is interior to the limit cycle, an unstable limit cycle in the plane is always a separatrix.

Algorithms

With the help of these two theorems, we now state an algorithm for locating the boundaries of the basins of attraction for a second-order system that satisfies assumptions A.1–A.3. For simplicity, we assume that there are no unstable limit cycles. The algorithm will be modified to incorporate unstable limit cycles shortly.

1. Locate all the equilibrium points and limit cycles of the system.
2. Using the technique presented shortly, calculate the stable and unstable trajectories of each saddle point. Record which unsta-

ble limit set each stable trajectory approaches in reverse time and each unstable trajectory approaches in forward time. If the trajectory is unbounded, record that fact instead.

3. For each stable limit set A_i:
   (a) Identify those saddle points that have only one unstable trajectory that approaches A_i. The stable trajectories of these saddle points are separatrices for B_{A_i}.
   (b) Identify those saddle points that have at least one unstable trajectory that approaches A_i. $\mathrm{bd}(B_{A_i})$ is the union of these saddle points, their stable trajectories, and any limit sets these stable trajectories approach.

Remarks:

1. Step 3(a) ignores any saddle point whose unstable trajectories both approach the same limit set. Thus, no degenerate boundary trajectories are found in this step.
2. If one of the unstable saddle trajectories of a particular saddle point is unbounded, the stable manifold of the saddle point lies on the boundary of B_∞. This case is handled nicely by the algorithm if ∞ is included as an attracting limit set.

In step 2, the limit set that a trajectory approaches must be identified. There are two types of tests. One tests for convergence to an equilibrium point; the other tests for convergence to a limit cycle.

The test for convergence to an equilibrium point is nearly the same as the trajectory termination algorithm used by the grid algorithm (Section 10.1.2). The difference is that the time-out limits are set extremely high. The reason is that the algorithm definitely needs to know which limit set the trajectory is approaching.

The test for convergence to a limit cycle is as follows. Every limit cycle Γ_i is associated with a cross-section Σ_i and a point on that cross-section x_i^*. Using the algorithm of Section 2.4.2, calculate, while the trajectory is being integrated, the intersection of the trajectory with each Σ_i. If the point of intersection agrees with x_i^*, as judged by a relative/absolute test, then the trajectory is approaching Γ_i. Only intersections with cross-sections of stable limit cycles need to be found.

Consistency check Since we assume that the system under study is structurally stable, assumptions A.1 and A.2 are always satisfied. It would be nice if the program could check whether the third assumption, A.3, holds.

There are two simple tests that can be incorporated into the algorithm.

Every stable limit set has a basin of attraction, and, therefore, unless the system possesses only one limit set, every stable limit set needs a boundary. If there is more than one stable limit set, the program checks that each stable limit set has a boundary; if not, assumption A.3 is not satisfied (see Figs. 10.10(a,b,d)).

The second test checks whether the non-degenerate boundary trajectories form meaningful boundaries. Separatrices should form closed curves as in Figs. 10.8(e,f) or curves unbounded at both ends as in Figs. 10.8(a,d). If they do not, assumption A.3 is not satisfied (see Fig. 10.10(c)).

In practice, failure of either test can also indicate that there are limit sets yet to be found.

Remark: If infinity is thought of as a point, then a curve unbounded at both ends is actually a closed curve. With this interpretation, there is really only one rule: the curves formed by the non-degenerate boundary trajectories should be closed.

Unstable limit cycles In the plane, an unstable limit cycle is always a separatrix, but the program must decide of which basins of attraction.

First, consider the dynamics of the system outside the limit cycle. To the rest of the flow, an unstable limit cycle behaves like a large unstable equilibrium point (this is the procedure of Fig. 10.9 in reverse). If A.1–A.3 are satisfied, it follows that the limit cycle is on $\text{bd}(B_A)$ if and only if there is a stable saddle trajectory that approaches it in reverse time. Furthermore, the limit cycle lies on the same boundaries as the stable saddle trajectory.

Now, consider the dynamics inside the limit cycle. Since a second limit cycle interior to the first may be considered an equilibrium point, we restrict our attention to the case of only equilibrium points inside the limit cycle. If there is a stable saddle trajectory that winds onto the limit cycle in reverse time, then the limit cycle is on the same boundaries as the stable trajectory (Fig. 10.8(c)). If there is no such trajectory, then there must be a single stable equilibrium point inside the limit cycle (e.g., Fig. 10.8(b)). The limit cycle is a separatrix for this limit set.

In the consistency check on the separatrices, an unstable limit cycle is treated as if it were an equilibrium point. Thus, its only role is to link two stable saddle trajectories. With the provision that infinity is treated as a single point, the rule for non-degenerate trajectories is as before: their union should form one or more closed curves.

Calculation of saddle trajectories Let $\hat{x}$ be a saddle point with eigenvalues, $\lambda_s < 0$ and $\lambda_u > 0$, and corresponding eigenvectors, η_s and η_u. At the saddle point, the stable manifold is tangent to the stable eigenvector η_s and the unstable manifold is tangent to the unstable eigenvector η_u. This fact leads immediately to an algorithm for finding the saddle trajectories.

The first unstable trajectory is calculated by integrating from an initial point lying on the unstable eigenvector,

$$\bar{x} := \hat{x} + \alpha\eta_u. \tag{10.10}$$

Here $\alpha > 0$ is some small number. The second unstable trajectory is found by integrating from $\bar{x} := \hat{x} - \alpha\eta_u$.

A suitable value for α may be determined by the method used in the invariant manifold algorithm for Poincaré maps (see Section 6.2). To use that algorithm in the current setting, define $P(x) := \phi_T(x)$ where $T := k/\lambda_u$ for some small integer k.

The stable trajectories are found by integrating in reverse time from initial conditions $\bar{x} := \hat{x} \pm \alpha\eta_s$. To find α, define $P(x) := \phi_T(x)$ where $T := -k/\lambda_s$ for some small integer k.

Drawing basins of attraction Once the separatrices and the limit sets whose basins of attraction they border are known, the basins of attraction can be displayed on the screen. Most graphics packages have a *flood* or *fill* command. These commands highlight a region on the graphics screen by flooding it with a particular color or by filling it with a stipple pattern. Depending on the graphics library, the boundary of the region is specified by either drawing it directly on the screen or by passing to the flood/fill routine an array containing points defining the boundary. The boundary must form a closed curve or the flood/fill routine will not work properly. If the basin of attraction extends outside the graphics region, it should be clipped, that is, the portion outside the graphics box should not be drawn. If the graphics library does not perform this clipping automatically, the program must include the relevant segments of the graphics box as part of the boundary.

Some flood/fill routines also require a *seed point* that lies within the region to be highlighted. Any point on a stable limit set can serve as the seed for the corresponding basin of attraction.

10.4 Programming tips

10.4.1 Consistency checking

When writing a simulation program, it is important to take into account practical as well as theoretical considerations. Computer programs are notorious for producing unforeseen results. Events that theory promises will never occur are often commonplace in a program.

For example, in theory, a trajectory converges to a point if and only if that point is a stable equilibrium point or a saddle point. Furthermore, due to integration error, theory also predicts that a simulated trajectory will almost never converge to a saddle point,[8] though it can spend a finite amount of time near the saddle point.

In practice, the integration points of a trajectory converge (according to the convergence test) if

1. the tolerances used in the convergence test are too large.

2. a stiff system is integrated by an integration algorithm that is not stiffly stable. This situation results in extremely small integration steps and virtually no progress along the trajectory.

3. the vector field changes rapidly, thereby forcing the integration routine to decrease the step-size.

4. the trajectory stays in a neighborhood of a saddle point long enough to appear to have converged.

5. the trajectory is approaching a stable equilibrium point.

[8] An example where an integrated trajectory will converge to a saddle point is the second-order system, $\dot{x} = -x$, $\dot{y} = (x+1)y$, using any initial condition with $y_0 = 0$.

As a check on the performance of the program and on the logic of the programmer, it is important for the program to verify the consistency of its conclusions. For instance, if the program thinks a trajectory is converging, it should check that there is a nearby stable equilibrium point. This consistency checking produces a more reliable program, and when inconsistencies arise, provides the user with an indication that tolerances need to be changed or that the program is being applied to a problem that it cannot simulate reliably.

10.4.2 History files

Format

The history file for the phase portrait program contains a widely varied assortment of data. The location and stability types of equilibrium points and limit cycles are stored in the history file as well as an unspecified number of trajectories each containing an unspecified number of points. It is not obvious how to store this data efficiently in a single file.

Recall that there are two user-defined segments of a history file: a header and a data segment. In `phase`, the data segment is used to record trajectory points only. The trajectories are stored consecutively, point by point, in the order that they are calculated. All other information is stored in the header. Specifically, the header contains the equilibrium point list, the limit cycle list, the trajectory list, the saddle trajectory list, and, of course, all the parameter and tolerance information used by the program. The header also contains the current number of entries in each of the lists.

Since the header must have a fixed format, a maximum number of entries for each of the lists is chosen and enough space is reserved in the header to accommodate that many items. The size (in bytes) of each list entry is small, so the maximum values can be chosen fairly large (e.g., 32 equilibrium points). If a list needs to be larger, the program can be recompiled using a different list size.

The contents of the equilibrium point and limit cycle lists are described in Section 10.2.

Each entry of the trajectory list corresponds to a single trajectory and contains three items: a code indicating whether the trajectory was integrated in forward or reverse time, an offset indicating the position in the history file of the initial condition of the trajectory, and the number of points stored in the history file for this particular trajectory.

Each entry of the saddle trajectory list describes one saddle trajectory. An entry contains the saddle point (specified by an index to the equilibrium point list), a code indicating whether the trajectory is a stable or unstable trajectory, the limit set that the saddle trajectory approaches (specified by a code indicating whether an equilibrium point or a limit cycle is reached together with an index to the equilibrium point or limit cycle list), an offset indicating the position in the history file of the initial condition of the saddle trajectory, and the number of points stored in the history file for this particular trajectory.

The use of an offset to identify the beginning of a trajectory in the data segment allows trajectory points to be retrieved quickly using random file access functions (e.g., `fseek` in C).

No explicit data on basins of attraction is stored in the history file because the basins of attraction can be reconstructed from the information in the separatrix list.

Erasure

Often the user wants to erase a phase portrait and start again, perhaps with different parameters. Or, sometimes, the user would like to erase only part of the phase portrait. For example, the user may want to erase the output of the grid algorithm, but retain the limit sets and separatrices.[9] When a total or partial erasure occurs, the history file must be updated accordingly.

As far as the history file is concerned, erasing a particular type of item corresponds to setting the number of entries in the corresponding list to zero. If, however, trajectories or separatrices are being erased, the data portion of the history file must be modified. This is accomplished using a temporary history file as follows. The history file is copied to the temporary file point by point, except that the trajectories being erased are not copied to the temporary file. When a trajectory is copied, the file offset that is stored in the corresponding

[9]Due to dependencies in the data, only certain partial erasures are allowed. For example, the limit sets cannot be erased without erasing the separatrices and basins of attraction at the same time.

list is updated to reflect the trajectory's position in the temporary history file. Finally, the first history file is deleted, and the name of the temporary history file is changed to the name of the first history file.

10.5 Summary

- *Phase portrait*: A phase portrait is a graphical display of a family of typical trajectories, the limit sets, and the basins of attraction, of an autonomous continuous-time system.

- *Index*: For a second-order autonomous continuous-time system, the index around a closed contour is the number of times the vector field makes a 360° rotation in one counter-clockwise traverse of the contour. A counter-clockwise rotation of the vector field adds 1 to the degree; a clockwise rotation adds -1. The contour is not allowed to pass through an equilibrium point.

- *Index of an equilibrium point*: The index of a stable or unstable hyperbolic equilibrium point is 1. The index of a hyperbolic saddle point is -1. The index of any contour is the sum of the indices of the equilibrium points inside the contour.

- *Index around a limit cycle*: The index around any limit cycle is 1 and, therefore, the sum of the indices of the equilibrium points inside the limit cycle must sum to 1.

- *Basin of attraction*: The basin of attraction of an attracting limit set A is the set of initial conditions whose trajectory approaches A as $t \to \infty$. The basin of attraction, denoted by B_A, is an open set and is invariant under the flow ϕ_t.

- *Boundary of a basin of attraction*: The boundary of B is defined as $\text{bd}(B) := \text{cl}(B) \bigcap \text{cl}(\overline{B})$. The boundary of a basin of attraction is closed, invariant under the flow, and does not intersect the basin.

- *Boundary trajectory*: A trajectory that lies on $\text{bd}(B)$ is called a boundary trajectory. If there exists a neighborhood of a boundary trajectory that lies in $\text{cl}(B)$, then the boundary trajectory is called degenerate.

- *Separatrix*: A non-constant, non-degenerate boundary trajectory of a basin B is called a separatrix for B.

- *Boundary between basins of attraction*: The boundary between two basins of attraction, B_1 and B_2, is $\text{bd}(B_1) \bigcap \text{bd}(B_2)$. It consists of the separatrices and equilibrium points common to the boundaries of B_1 and B_2.

- *Saddle trajectory*: Given a saddle point in the plane, the two trajectories that approach the saddle point as $t \to \infty$ are the stable saddle trajectories of the saddle point. The two trajectories that approach the saddle point as $t \to -\infty$ are called unstable saddle trajectories. At the saddle point, the stable saddle trajectories are tangent to the stable eigenvector (i.e., the eigenvector corresponding to the negative eigenvalue), and the unstable saddle trajectories are tangent to the unstable eigenvectors.

- *Invariant manifold*: The stable invariant manifold of a limit set L is the set of all points whose trajectory approaches L as $t \to \infty$. The unstable manifold is the set of all points that approach L as $t \to -\infty$. The stable invariant manifold of L is denoted by $W^s(L)$, and the unstable manifold by $W^u(L)$.

- *Limit sets on* $\text{bd}(B_A)$: Under assumptions A.1–A.3, an equilibrium point or limit cycle lies on $\text{bd}(B_A)$ if and only if its unstable manifold intersects B_A.

- *Structure of* $\text{bd}(B_A)$: If assumptions A.1–A.3 are satisfied, $\text{bd}(B_A)$ is the union of the stable manifolds of the equilibrium points and limit cycles that lie on $\text{bd}(B_A)$.

Appendix A

The Newton-Raphson Algorithm

The *Newton-Raphson algorithm* is a commonly used technique for locating zeros of a function.

Let $H: \mathbb{R}^n \to \mathbb{R}^n$ have a zero at x^*, that is, $H(x^*) = 0$. The Newton-Raphson algorithm calculates x^* by iterating from an initial guess $x^{(0)}$ using the relation

$$x^{(i+1)} = x^{(i)} - DH(x^{(i)})^{-1} H(x^{(i)}) \tag{A.1}$$

where the superscripts in parentheses indicate the iteration count.

Equation (A.1) is obtained by approximating the equation $y = H(x)$ by its linearization

$$\Delta y = DH(x)\, \Delta x. \tag{A.2}$$

Equation (A.2) predicts the change Δy in the value of $H(x)$ for a given change Δx of x. We are seeking a zero of H, so set $\Delta y = -H(x)$ and solve (A.2) for Δx to obtain

$$\Delta x = -DH(x)^{-1} H(x). \tag{A.3}$$

The iteration relation (A.1) is obtained from (A.3) by setting $x = x^{(i)}$ and $\Delta x = x^{(i+1)} - x^{(i)}$. Of course, since (A.2) is only an approximation, it is not expected that $H(x^{(i+1)}) = 0$, but it is hoped that successive iterations of (A.1) yield a better and better approximation to x^*.

Fig. A.1 shows a geometrical interpretation of the Newton-Raphson algorithm in the case of a real-valued function $H: \mathbb{R} \to \mathbb{R}$.

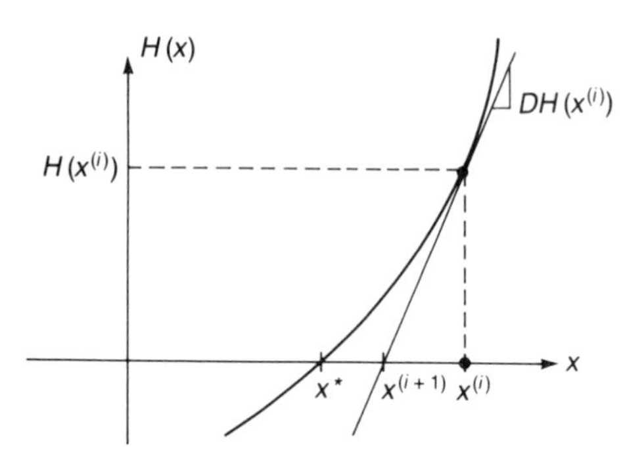

Figure A.1: Geometrical interpretation of the Newton-Raphson algorithm for a real-valued function H.

Equation (A.2) is a poor approximation when Δx or Δy is large. Thus, convergence is guaranteed only when the initial guess $x^{(0)}$ is sufficiently close to the actual zero x^*. If $x^{(0)}$ is not sufficiently close to x^*, then the Newton-Raphson algorithm may or may not converge. If it does converge, however, the convergence is quite rapid once $x^{(i)}$ is close to x^*. For more information on convergence properties, consult Chua and Lin [1975].

Pseudo-code for a practical implementation of the Newton-Raphson algorithm is presented in Fig. A.2. The routine exits if the algorithm converges, if an iteration limit is reached, or if $DH(x)$ is ill-conditioned.

Note that in `newton_raphson`, $DH(x)$ is never inverted. Instead, the system $DH(x)\Delta x = y$ is solved. The system can be solved using any reliable linear algebra library such as LINPACK (see Dongarra *et al.* [1979]).

Convergence test The convergence test that appears in the `until` statement of Fig. A.2 has two components. One checks whether $x^{(i)}$ is converging, and the other tests whether $H(x^{(i)})$ is near 0.

The obvious convergence test for $x^{(i)}$ is

$$\|x^{(i+1)} - x^{(i)}\| < \epsilon \tag{A.4}$$

for some $\epsilon > 0$. This test is sensitive to changes in the scaling of $x^{(i)}$ and should be replaced by a test that uses relative and absolute tolerances, E_r and E_a,

$$\|x^{(i+1)} - x^{(i)}\| < E_r \|x^{(i)}\| + E_a. \tag{A.5}$$

```
begin newton_raphson
      choose x0[], Er, Ea, ε, and kmax
      set k = 0
      set x[] = x0[]
      repeat
            set k = k + 1
            if (k = kmax) then
               exit--no convergence
            endif
            set y[] = H(x[])
            solve DH(x[]) Δx[] = -y[] for Δx[]
            if (DH(x[]) is ill-conditioned) then
               restart with new x[]
            endif
            set x[] = x[] + Δx[]
      until (||Δx[]|| < Er ||x[]|| + Ea and ||H(x[])|| < ε)
      output x[]
end newton_raphson
```

Figure A.2: Pseudo-code for a routine that implements the Newton-Raphson algorithm. Brackets indicate that a variable is an n-vector.

Remarks:

1. When E_r is set to 0, the relative/absolute test reduces to the absolute test (A.4).
2. Set $E_a = 0$ to obtain

$$\frac{\|x^{(i+1)} - x^{(i)}\|}{\|x^{(i)}\|} < E_r \tag{A.6}$$

   which shows that the relative tolerance is a limit on the fractional change in $x^{(i)}$ per iteration.
3. The absolute tolerance E_a is required whenever $x^{(i)}$ is converging to 0 since, in this case, a purely relative test might never be satisfied.

Since $H(x^{(i)})$ is converging to 0, a relative test is useless and a simple absolute test, $\|H(x^{(i)})\| < E_a$, is used. In the special case where H is a sum of $p > 1$ functions

$$H(x) = H_1(x) + \cdots + H_p(x), \tag{A.7}$$

one can use the test

$$\|H(x^{(i)})\| < E_r \left\{\|H_1(x^{(i)})\| + \cdots + \|H_p(x^{(i)})\|\right\} + E_a. \tag{A.8}$$

If more precise control of the error of x is required, a relative-absolute test can be performed on each component of $x^{(i)}$ or $H(x^{(i)})$, with different tolerance values for the different components.

Appendix B

The Variational Equation

Many of the algorithms in this book require $D_{x_0}\phi_t(x_0, t_0)$, the derivative of a trajectory with respect to the initial condition.

Consider the nth-order system

$$\dot{x} = f(x, t), \qquad x(t_0) = x_0 \tag{B.1}$$

with solution $\phi_t(x_0, t_0)$, that is,

$$\dot{\phi}_t(x_0, t_0) = f(\phi_t(x_0, t_0), t), \qquad \phi_{t_0}(x_0, t_0) = x_0. \tag{B.2}$$

Differentiate (B.2) with respect to x_0 to obtain

$$\begin{aligned} D_{x_0}\dot{\phi}_t(x_0, t_0) = D_x f(\phi_t(x_0, t_0), t)\, D_{x_0}\phi_t(x_0, t_0), \\ D_{x_0}\phi_{t_0}(x_0, t_0) = I. \end{aligned} \tag{B.3}$$

Define $\Phi_t(x_0, t_0) := D_{x_0}\phi_t(x_0, t_0)$. Then (B.3) becomes

$$\dot{\Phi}_t(x_0, t_0) = D_x f(\phi_t(x_0, t_0), t)\, \Phi_t(x_0, t_0), \qquad \Phi_{t_0}(x_0, t_0) = I \tag{B.4}$$

which is the *variational equation.*

Remarks:

1. The variational equation is a matrix-valued time-varying linear differential equation. It is the linearization of the vector field along the trajectory $\phi_t(x_0, t_0)$. Note that if the trajectory changes, so does the variational equation.

2. Since the initial condition is the identity matrix, $\Phi_t(x_0, t_0)$ is the state transition matrix of the linear system (B.4). It follows that a perturbation δx_0 of x_0 evolves as

$$\delta x(t) = \Phi_t(x_0, t_0)\, \delta x_0. \tag{B.5}$$

   The perturbation δx may be interpreted in two ways: as an infinitesimal perturbation of the original system (B.1) or as a (finite) vector-valued solution of the linearized system (B.4).

3. For autonomous systems, the variational equation simplifies to

$$\dot{\Phi}_t(x_0) = D_x f(\phi_t(x_0))\, \Phi_t(x_0), \qquad \Phi_0(x_0) = I. \tag{B.6}$$

4. For non-autonomous systems with additive forcing, $\dot{x} = f(x) + g(t)$, the variational equation is

$$\dot{\Phi}_t(x_0, t_0) = D_x f(\phi_t(x_0, t_0))\, \Phi_t(x_0, t_0), \qquad \Phi_{t_0}(x_0, t_0) = I. \tag{B.7}$$

   Note that g does not appear explicitly, though it is implicit in $\phi_t(x_0, t_0)$.

Calculating the solution to the variational equation Since the variational equation depends on both ϕ and Φ, they are usually calculated at the same time. To perform this simultaneous integration, the variational equation is appended to the original system to obtain the combined system

$$\left\{ \begin{array}{c} \dot{x} \\ \dot{\Phi} \end{array} \right\} = \left\{ \begin{array}{c} f(x,t) \\ D_x f(x,t)\, \Phi \end{array} \right\} \tag{B.8}$$

which is integrated from the initial condition

$$\left\{ \begin{array}{c} x(t_0) \\ \Phi(t_0) \end{array} \right\} = \left\{ \begin{array}{c} x_0 \\ I \end{array} \right\}. \tag{B.9}$$

Appendix C

Differential Topology and Structural Stability

C.1 Differential topology

The field of differential topology is an extremely useful tool for understanding the behavior of dynamical systems. Throughout the book, we avoid using terms from differential topology whenever possible, but there are situations in which they are the only terms available. Thus, in this section, we present a brief review of some of the basic definitions from differential topology. Guillemin and Pollack [1974] is an excellent book for those readers interested in studying this topic further.

Invertible maps A map $f: X \to Y$ is *one to one* if no two points in X map to the same point in Y.

The map $f: X \to Y$ is *onto* if for every point $y \in Y$ there exists at least one point in X that is mapped to y by f.

A map that is one to one and onto is called *invertible.* If f is invertible, then the inverse map, $f^{-1}: Y \to X$, is also one to one and onto.

Smooth maps Let $U \subset \mathbb{R}^m$ be an open set. A map $f: U \to \mathbb{R}^p$ is *smooth* if it possesses continuous partial derivatives of all orders.

Let $X \subset \mathbb{R}^m$ be an arbitrary set. A map $f: X \to \mathbb{R}^p$ is *smooth* if it can be locally extended to a smooth map, that is, if around each point $x \in X$, there exists an open neighborhood $U \subset \mathbb{R}^m$ and a smooth map $F: U \to \mathbb{R}^p$ such that $f = F$ on X.

Diffeomorphisms A smooth map $\phi: X \to Y$ is a *diffeomorphism* if it is invertible and if the inverse map $\phi^{-1}: Y \to X$ is also smooth.

Remarks:

1. A diffeomorphism is invertible and differentiable.
2. If ϕ is a diffeomorphism, so is ϕ^{-1}.

If there exists a diffeomorphism between two sets X and Y, then X and Y are called *diffeomorphic.*

Manifolds A *k-dimensional manifold M* is a set of points that locally resembles $\mathbb{R}^k$. More precisely, M is a k-dimensional manifold if for each point $x \in M$, there exists an open neighborhood U of x such that U is diffeomorphic to some open neighborhood in $\mathbb{R}^k$.

Fact: If X is a k_x-dimensional manifold and Y is a k_y-dimensional manifold, then the Cartesian product $X \times Y$ is a $(k_x + k_y)$-dimensional manifold.

Examples:

1. Any open subset of $\mathbb{R}^k$ is a k-dimensional manifold.
2. The circle S^1 is a one-dimensional manifold.
3. The sphere S^2 is a two-dimensional manifold.
4. A square is not a manifold. It is not smooth at the corners.
5. A figure eight is not a manifold. It does not resemble any Euclidean space at the point where it intersects itself.

Embeddings A set $U \subset \mathbb{R}^k$ is *compact* if it is closed and bounded. Examples of compact sets are the circle and the sphere.

For $f: X \to Y$, the *pre-image* of a set $Z \subset Y$, denoted by $f^{-1}(Z)$, is the set of all points in X that f maps into Z, that is,

$$f^{-1}(Z) = \{x \in X : f(x) \in Z\}. \tag{C.1}$$

A map f is *proper* if the pre-image of every compact set in Y is compact.

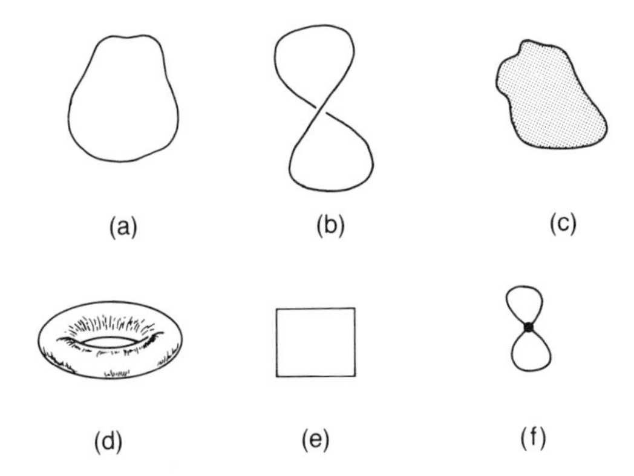

Figure C.1: (a) through (d) are manifolds; (e) and (f) are not.

Let X and Y be two manifolds with $\dim(X) < \dim(Y)$. A map $f: X \to Y$ is an *immersion* if its derivative $Df(x)$ is of full rank for every $x \in X$.

Let X be a compact manifold. If $f: X \to Y$ is an immersion, proper, and one to one, then it is called an *embedding*.

Fact: An embedding $f: X \to Y$ maps X diffeomorphically onto a manifold $X' \subset Y$.

To illustrate the importance of this fact, Fig. C.2 shows two examples of maps that are not embeddings. The function in Fig. C.2(a) maps the circle S^1 onto a figure eight. Though the map is an immersion and proper, it fails to be an embedding because it is not one to one.

The map in Fig. C.2(b) takes the real line $\mathbb{R}$ into a figure eight. In this case, the map is an immersion and is one to one, but it is not proper because the inverse image of the figure eight, a compact set, is the entire real line which is not compact.

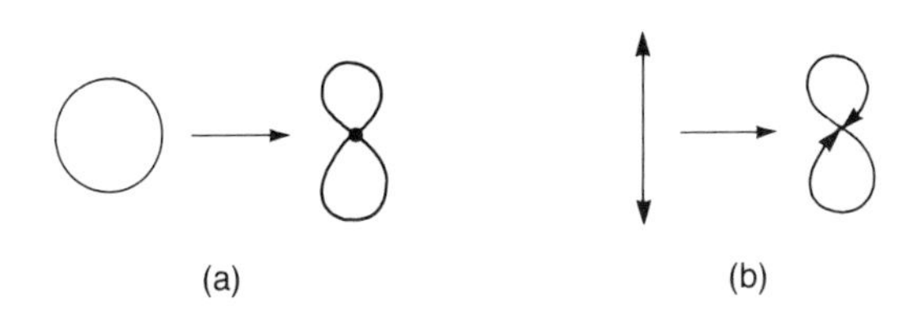

Figure C.2: Two maps that are not embeddings.

C.2 Structural stability

The field of structural stability is a difficult one and it is beyond the scope of this book to define it rigorously. We are satisfied with an intuitive, informal approach. Readers wishing a more thorough treatment are referred to Guckenheimer and Holmes [1983] and Chow and Hale [1982].

Informally, a dynamical system is *structurally stable* if there exists no infinitesimal perturbation of the system that alters the qualitative behavior of the system's solutions.

Typical examples of a qualitative change are a change in the stability of a limit set or the creation or disappearance of a limit set. More precisely, two systems are considered to have the same qualitative behavior if they are topologically equivalent.

A map $h: X \to Y$ is a *homeomorphism* if it is continuous, invertible, and its inverse is continuous.

Two discrete-time systems, $f, g: \mathbb{R}^n \to \mathbb{R}^n$ are *topologically equivalent* if there exists a homeomorphism $h: \mathbb{R}^n \to \mathbb{R}^n$ such that $h \circ f = g \circ h$.

Two vector fields, $f, g: \mathbb{R}^n \to \mathbb{R}^n$, are *topologically equivalent* if there exists a homeomorphism $h: \mathbb{R}^n \to \mathbb{R}^n$ that takes trajectories of f into trajectories of g and preserves the sense of direction of the trajectories with respect to time. This means that for any t and x, there exists a $\hat{t}$ such that

$$h \circ \phi_t^f(x) = \phi_{\hat{t}}^g \circ h(x) \tag{C.2}$$

and that if $t_2 > t_1$, then $\hat{t}_2 > \hat{t}_1$. In (C.2), the superscript on ϕ indicates to which vector field the flow belongs.

To illustrate the idea of structural stability, consider the function $f: \mathbb{R}^2 \to \mathbb{R}$ defined by

$$f(x, \alpha) := x^2 + \alpha \tag{C.3}$$

where $\alpha \in \mathbb{R}$ is a parameter that can be varied. Let the solution of this "system" be the zeros of $f(\cdot, \alpha)$. As Fig. C.3 shows, $f(\cdot, \alpha)$ has two zeros for $\alpha < 0$, one zero for $\alpha = 0$, and no zeros for $\alpha > 0$.

Though this is not a dynamical system, we can still apply the concept of structural stability by defining $f(\cdot, \alpha)$ and $f(\cdot, \hat{\alpha})$ to be equivalent if they have the same number of zeros. For $\alpha = 0$, f is not structurally stable because there exists an infinitesimal perturbation

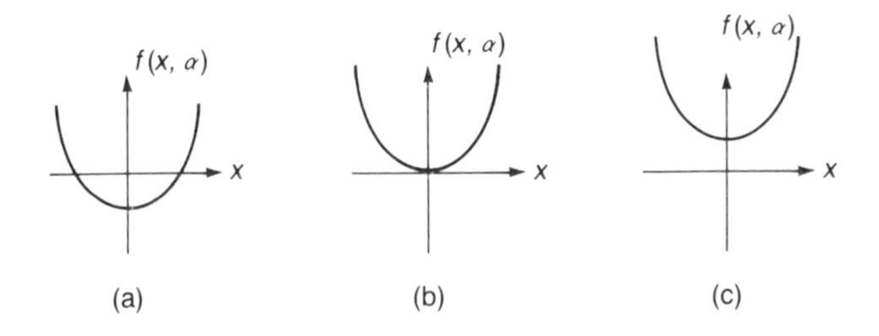

Figure C.3: A simple example of structural stability. $f(x, \alpha) = x^2 + \alpha$ for (a) $\alpha < 0$ (structurally stable); (b) $\alpha = 0$ (not structurally stable); and (c) $\alpha > 0$ (structurally stable).

of α that changes the number of zeros; in fact, every perturbation from $\alpha = 0$ changes the number of zeros. For every other value of α, f is structurally stable.

Structural stability is important in the simulation of dynamical systems for the following reason. Due to the finite precision of computers and the inevitable errors of floating-point arithmetic, dynamical systems are never represented exactly in a computer. The system that is actually being simulated is a perturbed version of the system described by the state equations, so the only systems that can be reliably simulated are structurally stable ones.

Additionally, as is shown in Chapter 8, structural stability plays a key role in the theory of bifurcations.

To build further intuition on structural stability, we resort to some ideas from differential topology.

Transversality Consider a point x lying on a k-dimensional manifold M. The *tangent space of M at x*, denoted by $T_x(M)$, is the k-dimensional linear space which, when its origin is shifted to x, is tangent to M at x. Two example tangent spaces are shown in Fig. C.4.

Remark: Let M_1 and M_2 be manifolds. The derivative of a map $f: M_1 \to M_2$ evaluated at a point $x \in M_1$ is a linear map from the tangent space at x to the tangent space at $f(x)$, that is, $Df(x): T_x(M_1) \to T_{f(x)}(M_2)$.

Two manifolds, $M_1 \subset \mathbb{R}^k$ and $M_2 \subset \mathbb{R}^k$, are *transversal* if for each $x \in M_1 \bigcap M_2$, $T_x(M_1) + T_x(M_2)$ spans $\mathbb{R}^k$.

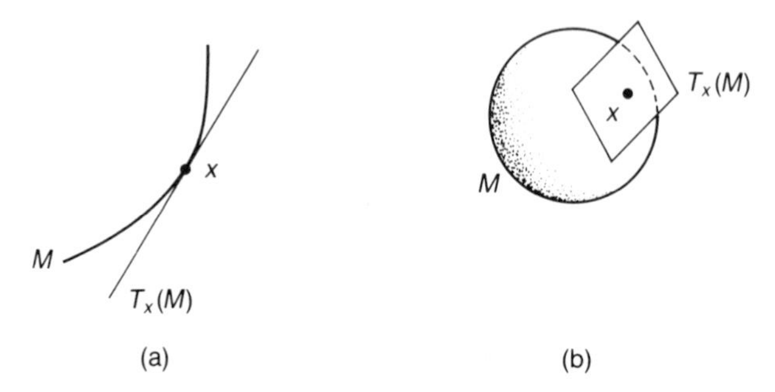

Figure C.4: Two examples of tangent spaces. In these drawings, the origin of $T_x(M)$ is shifted to x.

Remarks:

1. It is a consequence of the definition that two manifolds that do not intersect are transversal.
2. The intersection of transversal manifolds is a manifold.
3. Let M_1 and M_2 be manifolds in $\mathbb{R}^k$ with $\dim(M_1)+\dim(M_2) < k$. Then M_1 and M_2 are transversal if and only if their intersection is empty.

Several examples of transversal and non-transversal manifolds are shown in Fig. C.5. Roughly speaking, if two manifolds are tangent at a point of intersection, then they do not intersect transversally.

Transversal manifolds are structurally stable in the sense that transversality persists under sufficiently small perturbations of the manifolds.

To understand this assertion, reconsider the example of Fig. C.3. In this example, the manifolds of interest are the x-axis and the parabola defined by $y = x^2 + \alpha$. For $\alpha \neq 0$, the manifolds intersect transversally and the system is structurally stable. At $\alpha = 0$, the manifolds are not transversal and the system is not structurally stable.

An examination of Fig. C.5 reveals that a slight perturbation of transversal manifolds does not change the nature of the intersection. On the other hand, the structure of the intersection of non-transversal manifolds is extremely sensitive and can be destroyed by an infinitesimal perturbation.

It is tempting to go further and assert that non-transversal manifolds are not structurally stable. This assertion is not true, however,

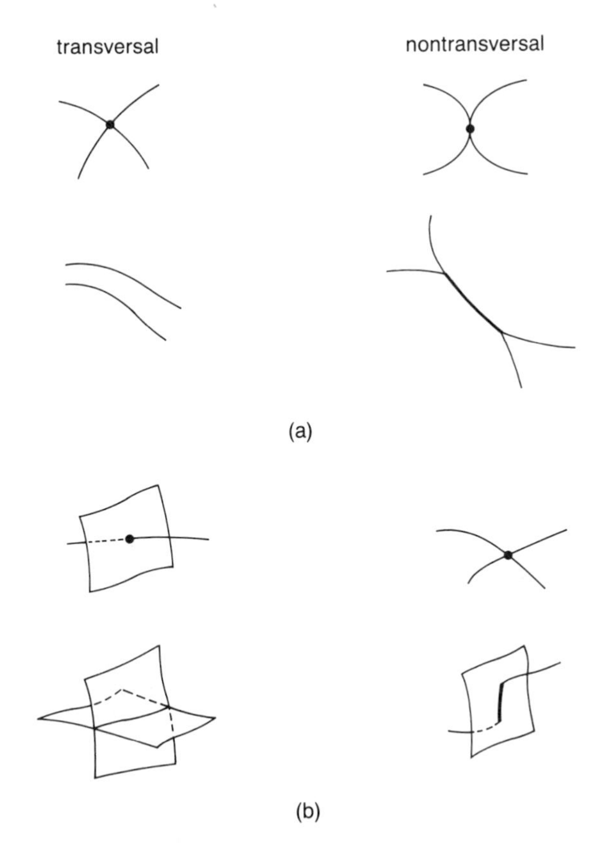

Figure C.5: Examples of transversal and non-transversal intersections. (a) in $\mathbb{R}^2$; (b) in $\mathbb{R}^3$.

because the set of allowable perturbations depends on the problem at hand. For instance, if the system (C.3) is changed to

$$\hat{f}(x, \alpha) := \alpha x^2, \tag{C.4}$$

then even though the parabola and the x-axis are not transversal, the number of zeros of $\hat{f}$ is structurally stable with respect to changes in α.

All is not lost, however, for restricting the allowed perturbations is rather meaningless in numerical simulations. Owing to the finite precision of a digital representation and to the errors of digital arithmetic, the actual perturbation of a system is quite complicated and is not restricted by the perturbation parameters that occur in the original system. One way to view the situation is that the actual equation being simulated is not $f(x)$, but $f(x) + \epsilon g(x)$ where g is a function that represents the errors inherent in the simulation and $\epsilon > 0$ is an additional perturbation parameter above and beyond any perturbation parameters present in the original function f. With this in mind, it is generally safe to assume that in numerical simulations, non-transversal manifolds are, indeed, structurally unstable.

Generic properties Informally, given a set Y, P is a generic property if almost all the members of Y exhibit property P. Before we present the precise definition of a generic property, we must first define dense sets.

A set $X \subset Y$ is *dense* if for each $y \in Y$, $U \bigcap X$ is non-empty for every open neighborhood U of y. A dense set $X \subset Y$ has the property that any element in Y can be approximated arbitrarily closely by elements in X. For example, the rational numbers are a dense subset of the real numbers.

A property P that refers to members of a set Y is a *generic property* if the subset of Y whose members exhibit property P contains a dense open subset of Y.

Remark: If P_1 and P_2 are generic properties of a set Y, then so is the property "P_1 and P_2."

One way to generate a dense open set in $\mathbb{R}^n$ is to delete a subset that is defined by a polynomial equality (e.g., the set of points in $\mathbb{R}^3$ that do not lie on the unit sphere, $x^2 + y^2 + z^2 = 1$, is a dense open set). It follows that properties that hold for all points except those in a subset defined by a polynomial equality are generic. An example of such a generic property is the invertibility of a matrix $A \in \mathbb{R}^{n \times n}$. Since A is invertible if and only if $\det(A) = 0$ and since the determinant of A is a polynomial, invertibility is a generic property.

Three other useful generic properties of matrices in $\mathbb{R}^{n \times n}$ are all eigenvalues are distinct; no eigenvalues lie on the imaginary axis (i.e., have real part equal to zero); and no eigenvalues lie on the unit circle (i.e., have magnitude equal to one).

The significance of a generic property is that unless there is reason to believe otherwise, it is reasonable to assume that the property holds. For instance, knowing nothing about a matrix A, it is reasonable to assume that none of the eigenvalues of A lie on the imaginary axis. If, on the other hand, it is known that A has special properties, that it is the Jacobian of a system undergoing a bifurcation, for instance, then the assumption may be unwarranted.

Appendix D

Results on the Poincaré Map

In this Appendix, we prove several useful theorems regarding the differentiability of Poincaré maps.

Consider a point $x_1 \in \mathbb{R}^n$ that is mapped by the flow to x_2 in T seconds, that is, $x_2 = \phi_T(x_1)$. Choose an $(n-1)$-dimensional hyperplane Σ_1 that contains x_1 and that is transversal to the flow at x_1 (see Fig. D.1). Similarly, choose Σ_2 transversal to the flow at x_2.

We will prove that there exists a diffeomorphism $P_{\Sigma_1\Sigma_2}$ that maps a neighborhood of x_1 on Σ_1 to a neighborhood of x_2 on Σ_2. We will also find an explicit expression for its derivative. Since all four of the Poincaré maps defined for autonomous systems are equivalent to $P_{\Sigma\Sigma}$—the choice of x_2 distinguishes one from another—these results prove that the Poincaré map is a local diffeomorphism, and they yield an expression for its derivative. The derivative is then used to

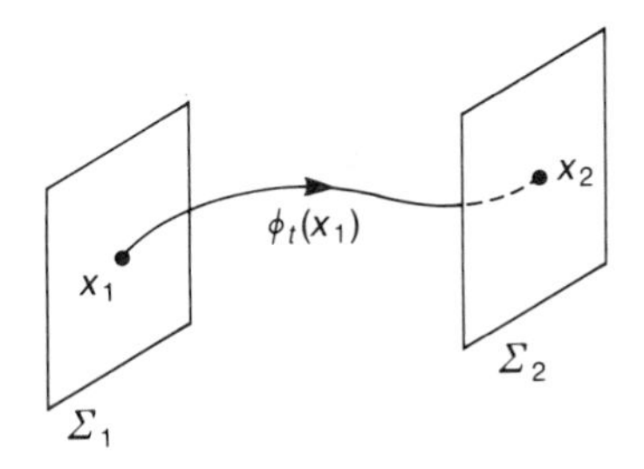

Figure D.1: The generalized Poincaré map $P_{\Sigma_1\Sigma_2}$ is defined by two transversal cross-sections, Σ_1 and Σ_2.

show that given a fixed point x^* of a Poincaré map, the characteristic multipliers are a subset of the eigenvalues of $\Phi_T(x^*)$.

Lemma D.1 (Hirsch and Smale [1974]) *There exist an open $U \in \mathbb{R}^n$ with $x_1 \in U$, and a unique C^1 map $\tau: U \to \mathbb{R}$, such that, for all $x \in U$, $\phi_{\tau(x)}(x) \in \Sigma_2$ and $\tau(x_1) = T$.*

Proof Without loss of generality, move the origin to x_2. Let $h \in \mathbb{R}^n$ be a vector orthogonal to Σ_2. Then $\langle h, y \rangle = 0$ iff $y \in \Sigma_2$.

To use the implicit function theorem, define $g: \mathbb{R}^n \times \mathbb{R} \to \mathbb{R}$ by $g(x, \tau) := \langle h, \phi_\tau(x) \rangle$. Then $D_\tau g(x_1, T) = \langle h, \dot{\phi}_T(x_1) \rangle$ is non-zero because $\dot{\phi}_T(x_1) = f(x_2) \notin \Sigma_2$. Therefore $\tau(x)$ exists such that $g(x, \tau(x)) = 0$, that is, $\phi_{\tau(x)}(x) \in \Sigma_2$. ■

Furthermore, differentiating g yields

$$D_x g(x_1, \tau(x_1)) + D_\tau g(x_1, \tau(x_1)) D\tau(x_1) = 0. \tag{D.1}$$

It follows that

$$D\tau(x_1) = \frac{-1}{\langle h, f(x_2) \rangle} h^T \Phi_T(x_1). \tag{D.2}$$

Lemma D.2 $f(x_2) = \Phi_T(x_1) f(x_1)$.

Proof Differentiate the state equation

$$\dot{\phi}_t(x_0) = f(\phi_t(x_0)) \tag{D.3}$$

to obtain

$$\ddot{\phi}_t(x_0) = Df(\phi_t(x_0))\, \dot{\phi}_t(x_0) \tag{D.4}$$

which is the variational equation. It follows that

$$\begin{aligned} \dot{\phi}_T(x_1) &= \Phi_T(x_1)\, \dot{\phi}_0(x_1) \\ f(x_2) &= \Phi_T(x_1)\, f(x_1). \end{aligned} \tag{D.5}$$

■

Definition D.1 $P_{\Sigma_1 \Sigma_2}: (U \bigcup \Sigma_1) \to \Sigma_2$ *is the generalized Poincaré map and is defined by*

$$P_{\Sigma_1 \Sigma_2}(x) = \phi_{\tau(x)}(x). \tag{D.6}$$

Theorem D.1 *$P_{\Sigma_1\Sigma_2}: T_{x_1}(\Sigma_1) \to T_{x_2}(\Sigma_2)$ is a local diffeomorphism at x_1 with*[1]

$$DP_{\Sigma_1\Sigma_2}(x_1) = \left[I - \frac{1}{\langle h, f(x_2)\rangle} f(x_2)h^T\right] \Phi_T(x_1). \tag{D.7}$$

Proof We need to show that $DP_{\Sigma_1\Sigma_2}(x_1)$ is non-singular.

$$\begin{aligned} DP_{\Sigma_1\Sigma_2}(x_1) &= D_{x_0}\phi_{\tau(x_1)}(x_1) + D_t\phi_{\tau(x_1)}(x_1)D\tau(x_1) \\ &= \Phi_T(x_1) + f(x_2)D\tau(x_1). \end{aligned} \tag{D.8}$$

By (D.2),

$$\begin{aligned} DP_{\Sigma_1\Sigma_2}(x_1) &= \Phi_T(x_1) - f(x_2)\frac{1}{\langle h, f(x_2)\rangle} h^T\Phi_T(x_1) \\ &= \left[I - \frac{1}{\langle h, f(x_2)\rangle} f(x_2)h^T\right] \Phi_T(x_1). \end{aligned} \tag{D.9}$$

$DP_{\Sigma_1\Sigma_2}(x_1)$ is singular iff there exists a $y \in T_{x_1}(\Sigma_1)$, $y \neq 0$, such that $DP_{\Sigma_1\Sigma_2}(x_1)\, y = 0$. Choose $y \in T_{x_1}(\Sigma_1)$, $y \neq 0$. Let $z := \Phi_T(x_1)\, y$. Then $z \neq 0$ since $\Phi_T(x_1)$ is non-singular (it is a state-transition matrix). Thus $DP_{\Sigma_1\Sigma_2}$ is non-singular iff

$$\left[I - \frac{1}{\langle h, f(x_2)\rangle} f(x_2)h^T\right] z \neq 0. \tag{D.10}$$

Suppose (D.10) doesn't hold. Then

$$z = \frac{\langle h, z\rangle}{\langle h, f(x_2)\rangle} f(x_2), \tag{D.11}$$

that is, z lies along $f(x_2)$. Since $z \in \Phi_T(x_1)T_{x_1}(\Sigma_1)$, it follows that $f(x_2) \in \Phi_T(x_1)T_{x_1}(\Sigma_1)$. Since $f(x_1)$ is transversal to $T_{x_1}(\Sigma_1)$, by Lemma D.2, $f(x_2) \notin \Phi_t(x_1)T_{x_1}(\Sigma_1)$. ■

[1] The domain of $DP_{\Sigma_1\Sigma_2}(x_1)$ is the $(n-1)$-dimensional linear space $T_{x_1}(\Sigma_1)$, and the domain of $\Phi_T(x_1)$ is the n-dimensional space $\mathbb{R}^n$. Thus the mathematically proper notation is

$$DP_{\Sigma_1\Sigma_2}(x_1) = \left[I - \frac{1}{\langle h, f(x_2)\rangle} f(x_2)h^T\right] \Phi_T(x_1)\,|\,T_{x_1}(\Sigma_1)$$

where $\Phi_T(x_1)\,|\,T_{x_1}(\Sigma_1)$ means the function $\Phi_T(x_1)$ with domain restricted to $T_{x_1}(\Sigma_1)$. Since this notation causes more clutter than clarity, we will not use it.

Corollary to Theorem D.1 *Let x^* be a fixed point of a Poincaré map*[2] *P defined by a cross-section Σ, and let T^* be the period of the underlying limit cycle. Then $P = P_{\Sigma\Sigma}$, and*

$$DP(x^*) = \left[I - \frac{1}{\langle h, f(x^*)\rangle} f(x^*)h^T\right] \Phi_{T^*}(x^*). \tag{D.12}$$

Theorem D.2 *Let x_1^* and x_2^* be any two points on a limit cycle. Let Σ_1 be an $(n-1)$-dimensional hyperplane passing through x_1^* transversal to $f(x_1^*)$. Likewise, define Σ_2 with respect to x_2^*. Then $DP_{\Sigma_1}(x_1^*)$ is similar to $DP_{\Sigma_2}(x_2^*)$.*

Corollary to Theorem D.2 *As long as Σ is transversal to the limit cycle, the eigenvalues of $DP(x^*)$ are independent of the choice of x^*, and of the position of Σ.*

Proof We need to find a non-singular Q such that

$$Q\,DP_{\Sigma_1}(x_1^*) = DP_{\Sigma_2}(x_2^*)\,Q. \tag{D.13}$$

Differentiate both sides of the identity

$$P_{\Sigma_1\Sigma_2}(P_{\Sigma_1}(x)) \equiv P_{\Sigma_2}(P_{\Sigma_1\Sigma_2}(x)) \tag{D.14}$$

to obtain

$$DP_{\Sigma_1\Sigma_2}(x_1^*)\,DP_{\Sigma_1}(x_1^*) = DP_{\Sigma_2}(x_2^*)\,DP_{\Sigma_1\Sigma_2}(x_1^*). \tag{D.15}$$

By Theorem D.1, $DP_{\Sigma_1\Sigma_2}(x_1^*)$ is non-singular. ■

Corollary to Lemma D.2 *$f(x^*)$ is an eigenvector of $\Phi_{T^*}(x^*)$ with eigenvalue 1.*

Proof Set $x_1 = x_2 = x^*$, and set $T = T^*$. ■

Theorem D.3 *Let x^* be a fixed point of P, and let the eigenvalues of $\Phi_{T^*}(x^*)$ be $\{m_1, \ldots, m_{n-1}, 1\}$. Then the eigenvalues of $DP(x^*)$ are $\{m_1, \ldots, m_{n-1}\}$.*

---

[2] P may be P_A, $P_\pm$, P_+, P_-, or a higher-order iterate of any of these (e.g., P_A^k).

Proof By Theorem D.2, we need to find the eigenvalues of $DP(x^*)$ for just one particular Σ.

Without loss of generality, move the origin to x^*. For simplicity, assume $m_i \neq 1$ for all i. Then $\Phi_{T^*}(x^*)$ has n eigenvectors, one of which, by the Corollary to Lemma D.2, is equal to $f(x^*)$. The remaining $n-1$ eigenvectors span an $(n-1)$-dimensional subspace Σ^* that is invariant under $\Phi_{T^*}(x^*)$, and which does not contain $f(x^*)$. The cross-section Σ^* defines a Poincaré map $P^*: \Sigma^* \to \Sigma^*$.

Let h^* be a vector orthogonal to Σ^*, that is, $\langle h^*, y\rangle = 0$ iff $y \in \Sigma^*$. For all $y \in \Sigma^*$,

$$\begin{aligned} D\tau(x^*)\, y &= \frac{-1}{\langle h^*, f(x^*)\rangle} h^{*T} \Phi_{T^*}(x^*)\, y \\ &= \frac{-1}{\langle h^*, f(x^*)\rangle} \langle h^*, z\rangle \\ &= 0 \end{aligned} \tag{D.16}$$

because $z \in \Sigma^*$ by invariance. Therefore, $DP^* = \Phi_{T^*}(x^*)$ (restricted to Σ^*), and, since Σ^* is spanned by the $n-1$ eigenvectors with non-unity eigenvalues, the proof is complete. ■

Appendix E

One Lyapunov Exponent Vanishes

In this Appendix, we present a proof, due to Haken [1983], that for any limit set other than an equilibrium point, one Lyapunov exponent is always zero.

Consider a system

$$\dot{x} = f(x), \qquad x(0) = x_0 \tag{E.1}$$

with $f: \mathbb{R}^n \to \mathbb{R}^n$ differentiable. Let the solution to (E.1) be $\phi_t(x_0)$.

Theorem E.1 (Haken [1983]) *If, given system (E.1),*

i) $\phi_t(x_0)$ *is bounded for* $t \geq 0$,

ii) $\phi_t(x_0)$ *does not tend toward an equilibrium point, and*

iii) f *has a finite number of zeros,*

then at least one of the Lyapunov exponents is zero.

Proof We will pick a particular perturbation δx_0, and show that

$$\limsup_{t \to \infty} \frac{1}{t} \ln \|\delta x(t; x_0, t_0)\| = 0 \tag{E.2}$$

from which it follows that at least one Lyapunov exponent is 0.

Choose δx_0 so that it lies in the direction of the trajectory,

$$\delta x_0 = \dot{\phi}_0(x_0) = f(x_0). \tag{E.3}$$

The perturbation δx_0 evolves according to the variational equation

$$\dot{\delta x} = Df(\phi_t(x_0))\,\delta x, \qquad \delta x(0) = \delta x_0. \tag{E.4}$$

Let the solution of (E.4) be $\delta x(t; x_0, \delta x_0)$.

To find $\delta x(t; x_0, \delta x_0)$, differentiate $\dot{\phi}_t(x_0) = f(\phi_t(x_0))$ to obtain

$$\ddot{\phi}_t(x_0) = Df(\phi_t(x_0))\,\dot{\phi}_t(x_0), \qquad \dot{\phi}_0(x_0) = f(x_0). \tag{E.5}$$

Comparison of (E.4) with (E.5) shows that

$$\delta x(t; x_0, \delta x_0) = \dot{\phi}_t(x_0) = f(\phi_t(x_0)). \tag{E.6}$$

Since f is continuous, and $\phi_t(x_0)$ is assumed to be bounded, there exists a $D > 0$ such that, for all $t > 0$, $\|f(\phi_t(x_0))\| < D$, that is, $\|\delta x(t; x_0, \delta x_0)\| < D$.

It follows that

$$\lambda = \limsup_{t \to \infty} \frac{1}{t} \ln \|\delta x(t; x_0, \delta x_0)\| \le \limsup_{t \to \infty} \frac{1}{t} \ln D. \tag{E.7}$$

Therefore $\lambda \le 0$.

We complete the proof by showing that $\lambda \not< 0$.

Suppose $\lambda < 0$. Then, by the definition of *lim sup*, for any $\epsilon > 0$, there exists a t_0 such that for all $t > t_0$,

$$\frac{1}{t} \ln \|\delta x(t; x_0, \delta x_0)\| < \lambda + \epsilon. \tag{E.8}$$

Choose ϵ such that $\tilde{\lambda} := \lambda + \epsilon < 0$. Then for all $t > t_0$,

$$\frac{1}{t} \ln \|\delta x(t; x_0, \delta x_0)\| < \tilde{\lambda}, \tag{E.9}$$

or, equivalently, for all $t > t_0$, $\|\delta x(t; x_0, \delta x_0)\| < e^{\tilde{\lambda} t}$.

Hence, $\|\delta x(t; x_0, \delta x_0)\| \to 0$ as $t \to \infty$.

From (E.6) it follows that $f(\phi_t(x_0)) \to 0$ as $t \to \infty$, which, together with assumption *iii)*, implies that $\phi_t(x_0)$ approaches an equilibrium point.

This contradicts assumption *ii)*, and therefore $\lambda \not< 0$. ∎

Appendix F

Cantor Sets

A Cantor set is a rather bizarre set that is quite useful in the study of chaotic systems.

The middle-third Cantor set The *middle-third Cantor set*, often referred to simply as the Cantor set and which we denote by $\mathcal{C}$, is generated recursively as follows (see Fig. F.1). Start with the unit interval $[0,1]$. Remove the middle third of this interval leaving two intervals, $[0,1/3]$ and $[2/3,1]$. Remove the middle third from each of these intervals leaving four intervals each with length $1/9$. The set resulting from applying this middle-third deletion process *ad infinitum* is the middle-third Cantor set $\mathcal{C}$.

Remarks:

1. The Cantor set is closed and has measure zero.
2. The complement (with respect to the unit interval) of the middle-third Cantor set is open and dense.

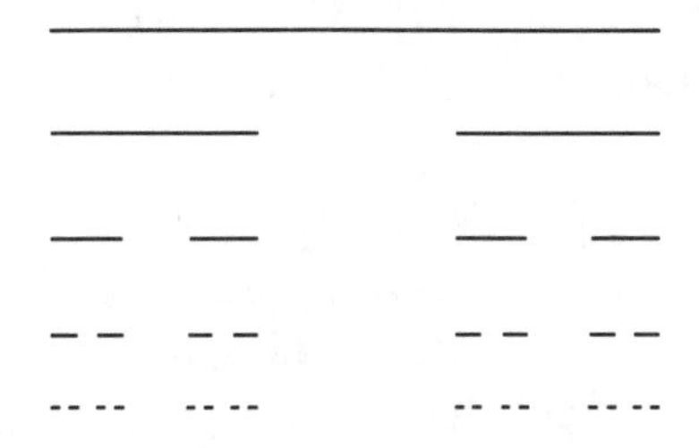

Figure F.1: Generating the middle-third Cantor set.

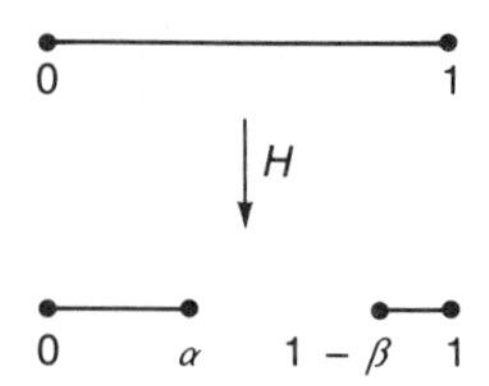

Figure F.2: The Cantor set generating function H applied to the unit interval. H replaces a set I with two smaller, but identical, sets, I_α and I_β.

3. The middle-third Cantor set is self-similar. More precisely, let I_k be one of the intervals remaining at the kth step in the iteration. Then $I_k \bigcap \mathcal{C}$ is identical to $\mathcal{C}$ except that it is scaled by the length of I_k.

Generalized Cantor sets Choose two positive numbers, α and β, such that $\alpha+\beta < 1$. Define a function H that replaces a set $I \subset [0,1]$ with two smaller sets $I_\alpha \subset [0,\alpha]$ and $I_\beta \subset [1-\beta,\beta]$ (see Fig. F.2). I_α is identical to I except it is smaller by a factor of α. I_β is identical to I except it is smaller by a factor of β. Given α and β, the associated *generalized Cantor set* $\mathcal{C}(\alpha,\beta)$ is generated by repeatedly applying H to the unit interval, that is, the Cantor set is

$$\lim_{k\to\infty} H^k([0,1]). \tag{F.1}$$

Remarks:

1. The middle-third Cantor set is $\mathcal{C} = \mathcal{C}(1/3, 1/3)$.
2. $C(\alpha,\beta)$ is invariant under H, that is, $H(C(\alpha,\beta)) = C(\alpha,\beta)$.

Dimension of Cantor sets

In this section, we highlight the differences between the types of dimension by calculating the dimension of a generalized Cantor set.

Family of dimensions Renyi [1970] defined an entire family of dimensions that includes capacity, information dimension, and correlation dimension as special cases. The definition depends on refining a covering of $N(\epsilon)$ volume elements where ϵ is the diameter of each volume element.

For $q = 0, 1, \ldots$, define the *generalized entropy* as

$$H_q(\epsilon) := \frac{1}{1-q} \ln \sum_{i=1}^{N(\epsilon)} P_i^q(\epsilon), \qquad \neq 1 \tag{F.2}$$

and

$$H_1(\epsilon) = -\sum_{i=1}^{N(\epsilon)} P_i(\epsilon) \ln P_i(\epsilon), \qquad q = 1 \tag{F.3}$$

where P_i is the relative frequency of visitation of a typical trajectory to the ith volume element.

The *Renyi dimensions* are defined as

$$D(q) := -\lim_{\epsilon \to 0} \frac{H_q(\epsilon)}{\ln \epsilon}, \qquad q = 0, 1, \ldots. \tag{F.4}$$

Remarks:

1. $D_{cap} = D(0)$.
2. $D_I = D(1)$.
3. $D_C = D(2)$.

To calculate the Renyi dimensions of the generalized Cantor set, we need to specify a probability density for the set. To keep with the self-similar nature of the set, choose a self-similar density as follows. With each application of H, the probability of being in I_α is p_α, and the probability of being in I_β is $p_\beta = 1 - p_\alpha$.

Renyi dimensions of the generalized Cantor set Apply H to $\mathcal{C}(\alpha, \beta)$ to get two rescaled versions of the Cantor set. Call the two rescaled versions $\mathcal{C}_\alpha$ and $\mathcal{C}_\beta$. Since $C(\alpha, \beta)$ is invariant under H, it is composed of the two parts, $\mathcal{C}_\alpha$ and $\mathcal{C}_\beta$. With this in mind, rewrite (F.2) as

$$\begin{aligned} e^{(1-q)H_q(\epsilon)} &= \sum_{i=1}^{N(\epsilon)} P_i^q(\epsilon) \\ &= \sum_{i_\alpha} P_i^q(\epsilon) + \sum_{i_\beta} P_i^q(\epsilon) \end{aligned} \tag{F.5}$$

where $\{i_\alpha\}$ are the indices of those volume elements covering $\mathcal{C}_\alpha$, and $\{i_\beta\}$ are the indices of those volume elements covering $\mathcal{C}_\beta$.

Owing to self-similarity, $\mathcal{C}_\alpha$ is a rescaled version of the entire Cantor set. Thus, if lengths in $\mathcal{C}_\alpha$ are scaled by $1/\alpha$, and if the probabilities $\{P_i(\epsilon)\}_{i_\alpha}$ are renormalized so that they sum to 1, then $\mathcal{C}_\alpha$ becomes $\mathcal{C}(\alpha, \beta)$. When the lengths are rescaled, ϵ becomes ϵ/α, and when the probabilities are renormalized, $P_i(\epsilon)$ becomes $P_i(\epsilon)/p_\alpha$. Thus, by (F.2)

$$\sum_{i_\alpha} P_i^q(\epsilon) = p_\alpha^q \, e^{(1-q)H_q(\epsilon/\alpha)}. \tag{F.6}$$

Likewise, for $\mathcal{C}_\beta$,

$$\sum_{i_\beta} P_i^q(\epsilon) = p_\beta^q \, e^{(1-q)H_q(\epsilon/\beta)}. \tag{F.7}$$

Substitute (F.6) and (F.7) into (F.5) to obtain

$$e^{(1-q)H_q(\epsilon)} = p_\alpha^q \, e^{(1-q)H_q(\epsilon/\alpha)} + p_\beta^q \, e^{(1-q)H_q(\epsilon/\beta)}. \tag{F.8}$$

From (F.4), for ϵ small,

$$k \, \epsilon^{-D(q)} = e^{H_q(\epsilon)} \tag{F.9}$$

where k is a constant of proportionality. Substitute this equation into (F.8) to obtain

$$\begin{aligned} k^{(1-q)} \epsilon^{-(1-q)D(q)} &= p_\alpha^q \, k^{(1-q)} \left(\tfrac{\epsilon}{\alpha}\right)^{-(1-q)D(q)} \\ &\quad + p_\beta^q \, k^{(1-q)} \left(\tfrac{\epsilon}{\beta}\right)^{-(1-q)D(q)} \end{aligned} \tag{F.10}$$

Divide both sides by $k^{(1-q)}\epsilon^{-(1-q)D(q)}$ to obtain

$$1 = p_\alpha^q \alpha^{(1-q)D(q)} + p_\beta^q \beta^{(1-q)D(q)} \tag{F.11}$$

which is an implicit equation for the qth-order Renyi dimension $D(q)$.

Remark: By carrying out a similar procedure for the case $q = 1$, it can be shown that

$$D(1) = \frac{p_\alpha \ln p_\alpha + p_\beta \ln p_\beta}{p_\alpha \ln \alpha + p_\beta \ln \beta} \tag{F.12}$$

Discussion To simplify the discussion, set $\alpha = \beta = 1/3$ and solve (F.11) for $D(q)$ to obtain

$$D(q) = \frac{\ln(p_\alpha^q + p_\beta^q)}{(1-q)\ln 3}, \qquad q \neq 1 \tag{F.13}$$

and simplify (F.12) to obtain

$$D(1) = -\frac{p_\alpha \ln p_\alpha + p_\beta \ln p_\beta}{\ln 3} \tag{F.14}$$

Remark: $D(q)$ is a decreasing function of q. It can be proved that for a given attractor, $D(q) \geq D(q')$ whenever $q < q'$.

In the numerator of (F.13), the probabilities, p_α and p_β, appear in the term $p_\alpha^q + p_\beta^q$. Without loss of generality, assume that $p_\alpha > p_\beta$. Then as $q \to \infty$, the p_α^q term dominates the p_β^q term. Thus, the higher-order Renyi dimensions tend to favor the more probable parts of the attractor and downplay the less probable.

Appendix G

List of Symbols

$x := y$	x is defined as y
$\overline{A}$	the complement of the set A
$\|x\|$	the magnitude of x
$\|\|x\|\|$	the norm of x
x^T	the transpose of x
$\langle x, y\rangle$	the inner product $(:= x^T y)$
$\{x_k\}_{k=k_1}^{k=k_2}$	the sequence of points $x_{k_1}, x_{k_1+1}, \ldots, x_{k_2}$
$\{x_k\}$	the sequence of all x_k
$\arg(x)$	the angle of the complex number x
B_L	the basin of attraction of the limit set L
$B_\epsilon(x)$	the ball of radius ϵ centered at x
$\mathrm{bd}(A)$	the boundary of the set A
$\mathbf{C}$	the complex numbers
$\mathcal{C}$	the middle-third Cantor set
$\mathcal{C}(\alpha, \beta)$	the generalized Cantor set
$\mathrm{cl}(A)$	the closure of the set A
δ_{jk}	the Kronecker delta
$\mathrm{diag}(a_1, \ldots, a_n)$	$n \times n$ diagonal matrix with diagonal entries $\{a_i\}$
D_C	correlation dimension
D_{cap}	capacity
D_I	information dimension
D_L	Lyapunov dimension
D_{nn}	kth nearest-neighbor dimension
Df	the derivative of f
$D_x f$	the derivative of f with respect to x
$\dim(M)$	the dimension of the manifold M
ϵ_r	local round-off error

$\bar{\epsilon}_r$	global round-off error
ϵ_t	local truncation error
$\bar{\epsilon}_t$	global truncation error
E_r	relative tolerance
E_a	absolute tolerance
f	a vector field
$\mathcal{F}$	the Fourier transform
$\mathcal{F}_K$	the Fourier transform truncated to K components
Γ	a limit cycle
$\mathrm{Im}[x]$	the imaginary part of x
λ_i	eigenvalue at an equilibrium point, Lyapunov exponent
m_i	characteristic multiplier, Lyapunov number
η_i	eigenvector
n	dimension of a continuous-time system
ϕ_t	a flow
$\phi_t(x_0)$	the point at time t on the trajectory emanating from x_0 at time $t = 0$—used with autonomous systems.
$\phi_t(x_0, t_0)$	the point at time t on the trajectory emanating from x_0 at time $t = t_0$—used with non-autonomous systems
$\phi_t^{(j)}(x_0)$	the jth component of $\phi_t(x_0)$
$\phi_t^{(j)}(x_0, t_0)$	the jth component of $\phi_t(x_0, t_0)$
$\Phi_t(x_0)$	the solution to the variational equation—autonomous case
$\Phi_t(x_0, t_0)$	the solution to the variational equation—non-autonomous case
P	a map
P_+	one-sided Poincaré map
P_-	one-sided Poincaré map
$P_\pm$	two-sided Poincaré map
P_A	Poincaré map of an autonomous system
P_N	Poincaré map of a non-autonomous system
$\mathbb{R}$	the real numbers
$\mathbb{R}^+$	the non-negative real numbers
$\mathrm{Re}[x]$	the real part of x
Σ	a hyperplane
S^1	the circle
$S^1 \times S^1$	the two-torus

T	the period
$T_x(M)$	the tangent space of the manifold M at the point x
x^*	a fixed point
x_{eq}	an equilibrium point
$W^s(L)$	the stable manifold of the limit set L
$W^u(L)$	the unstable manifold of the limit set L
$W^{u+}(L)$	an unstable half-manifold of the limit set L
$W^{u-}(L)$	an unstable half-manifold of the limit set L

Bibliography

[1] T.J. Aprille and T.N. Trick. A computer algorithm to determine the steady-state response of nonlinear oscillators. *IEEE Transactions on Circuit Theory*, CT-19(4):354–60, July 1972.

[2] D. Azzouz, R. Duhr, and M. Hasler. Bifurcation diagram in a piecewise-linear circuit. *IEEE Transactions on Circuits and Systems*, CAS-31:587–88, June 1984.

[3] D. Azzouz, R. Duhr, and M. Hasler. Transition to chaos in a simple nonlinear circuit driven by a sinusoidal voltage source. *IEEE Transactions fn Circuits and Systems*, CAS-30:913–14, December 1983.

[4] R. Badii and A. Politi. Statistical description of chaotic attractors: the dimension function. *Journal of Statistical Physics*, 40(5/6):725–50, 1985.

[5] A. Ben-Mizrachi and I. Procaccia. Characterization of experimental (noisy) strange attractors. *Physical Review A*, 29(2):975–77, 1984.

[6] S. Boyd and L.O. Chua. Dynamical systems need not have spectrum. *IEEE Transactions on Circuits and Systems*, CAS-32:968–69, September 1985.

[7] M. Broucke. One-parameter bifurcation diagram for Chua's circuit. *IEEE Transactions on Circuits and Systems*, CAS-34:208–9, February 1987.

[8] W.E. Caswell and J.A. Yorke. Invisible errors in dimension calculations: geometric and systematic effects. In G. Mayer-Kress, editor, *Dimensions and Entropies in Chaotic Systems*, pages 123–36, Springer-Verlag, New York, N.Y., 1986.

[9] H.D. Chiang, M.W. Hirsch, and F.F. Wu. Stability region of nonlinear autonomous dynamical systems. *IEEE Transacations on Automatic Control*, AC-33:16–27, January 1988.

[10] S.-N. Chow and Hale J.K. *Methods of Bifurcation Theory*. Springer-Verlag, New York, N.Y., 1982.

[11] L.O. Chua, M. Hasler, J. Neirynck, and P. Verbugh. Dynamics of a piecewise-linear resonant circuit. *IEEE Transactions on Circuits and Systems*, CAS-29:535–47, August 1982.

[12] L.O. Chua, M. Komuro, and T. Matsumoto. The double scroll family. *IEEE Transactions on Circuits and Systems*, CAS-33(11):1073–1118, November 1986.

[13] L.O. Chua, M. Komuro, and T. Matsumoto. The double scroll family—Part I: rigorous proof of chaos. *IEEE Transactions on Circuits and Systems*, CAS-33(11):1072–97, November 1986.

[14] L.O. Chua, M. Komuro, and T. Matsumoto. The double scroll family—part II: rigorous analysis of bifurcation phenomena. *IEEE Transactions on Circuits and Systems*, CAS-33(11):1097–1118, November 1986.

[15] L.O. Chua and P.-M. Lin. *Computer-Aided Analysis of Electronic Circuits: Algorithms & Computational Techniques.* Prentice-Hall, Englewood Cliffs, N.J., 1975.

[16] L.O. Chua and T. Lin. Chaos in digital filters. *IEEE Transactions on Circuits and Systems*, CAS-34:648–58, June 1988.

[17] L.O. Chua and T. Sugawara. Three-dimensional rotation instrument for displaying strange attractors. In F.M.A. Salam and M.L. Levi, editors, *Dynamical Systems Approaches in Systems and Circuits*, pages 75–116, SIAM, 1988.

[18] L.O. Chua and A. Ushida. Algorithms for computing almost-periodic steady-state response of nonlinear systems to multiple input frequencies. *IEEE Transactions on Circuits and Systems*, CAS-28:953–71, October 1981.

[19] L.O. Chua and A. Ushida. A switching-parameter algorithm for finding multiple solutions of nonlinear resistive circuits. *International Journal on Circuits and Systems*, 4:215–39, July 1976.

[20] J.-P. Collet, P. amd Eckmann. *Iterated Maps on the Interval as Dynamical Systems.* Birkhäuser, Boston, Mass., 1980.

[21] G. Dahlquist and Å. Björck. *Numerical Methods.* Prentice-Hall, Englewood Cliffs, N.J., 1974.

[22] C.A. Desoer. *Notes for a Second Course on Linear Systems.* van Nostrand Reinhold, New York, N.Y., 1970.

[23] E. Doedel. *AUTO: Software for continuation and bifurcation problems in ordinary differential equations.* Applied Mathematics, California Institute of Technology, Pasadena, Calif., 1986.

[24] J.J. Dongarra, J.R. Bunch, C.B. Moler, and G.W. Stewart. *LINPACK Users' Guide.* SIAM Publications, Philadelphia, Penn., 1979.

[25] J.-P. Eckmann and D. Ruelle. Ergodic theroy of chaos and strange attractors. *Reviews of Modern Physics*, 57(3):617–56, July 1985.

[26] T. Endo and L.O. Chua. Chaos from phase-locked loops. *IEEE Transactions on Circuits and Systems*, :987–1003, August 1988.

[27] J.D. Farmer, E. Ott, and J.A. Yorke. The dimensions of chaotic attractors. *Physica 7D*, 7D:153–80, 1983.

[28] M.J. Feigenbaum. Quantitative universality for a class of nonlinear transformations. *Journal of Statistical Physics*, 19:25–52, 1978.

[29] O. Frederickson, J.L. Kaplan, E.D. Yorke, and J.A. Yorke. The Liapunov dimension of strange attractors. *Journal of Differential Equations*, 49:185–207, 1983.

[30] E. Freire, L.G. Franquelo, and J. Aracil. Periodicity and chaos in an autonomous electronic system. *IEEE Transactions on Circuits and Systems*, CAS-31:237–47, 1984.

[31] J.H. Friedman, W. Stuetzle, and A. Schroeder. Projection pursuit density estimation. *Journal of the American Statistical Association*, 79(387):599–608, September 1984.

[32] B.S. Garbow, J.M. Boyle, J.J. Dongarra, and C.B. Moler. *Matrix Eigensystem Routines—EISPACK Guide Extension.* Volume 51 of *Lecture notes in Computer Science*, Springer-Verlag, New York, N.Y., 1977.

[33] C.W. Gear. *Numerical Initial Value Problems in Ordinary Differential Equations.* Prentice-Hall, Englewood Cliffs, N.J., 1971.

[34] J.A. Glazier and A. Libchaber. Quasi-periodicity and dynamical systems: an experimentalist's view. *IEEE Transactions on Circuits and Systems*, 35(7):790–809, July 1988.

[35] P. Grassberger and I. Procaccia. Measuring the strangeness of strange attractors. *Physica 9D*, 9D:189–208, 1983.

[36] J. Guckenheimer and P. Holmes. *Nonlinear Oscillations, Dynamical Systems, and Bifurcations of Vector Fields.* Springer-Verlag, New York, N.Y., 1983.

[37] V. Guillemin and A. Pollack. *Differential Topology.* Prentice-Hall, Englewood Cliffs, N.J., 1974.

[38] H. Haken. At least one Lyapunov exponent vanishes if the trajectory of an attractor does not contain a fixed point. *Phys. Lett.*, 94A(2):71–72, February 1983.

[39] M. Henon. On the numerical computation of Poincaré maps. *Physica 5D*, 5D:412–14, 1982.

[40] M.W. Hirsch and S. Smale. *Differential Equations, Dynamical Systems, and Linear Algebra.* Academic Press, New York, N.Y., 1974.

[41] J. Holzfuss and G. Mayer-Kress. An approach to error estimation in the application of dimension algorithms. In G. Mayer-Kress, editor, *Dimensions and Entropies in Chaotic Systems*, pages 114–22, Springer-Verlag, New York, N.Y., 1986.

[42] P.J. Huber. Projection pursuit. *The Annals of Statistics*, 13(2):435–75, 1985.

[43] F. Hunt and F. Sullivan. Efficient algorithms for computing fractal dimensions. In G. Mayer-Kress, editor, *Dimensions and Entropies in Chaotic Systems*, pages 74–81, Springer-Verlag, New York, N.Y., 1986.

[44] C. Kahlert. The chaos producing mechanism in Chua's circuit and related piecewise-linear dynamical systems. In *Proceedings of the 1987 European Conference on Circuit Theory and Design*, September 1987.

[45] C. Kahlert and L.O. Chua. Transfer maps and return maps for piecewise-linear three-region dynamical systems. *International Journal of Circuit Theory and Applications*, 15(1):23–49, January 1987.

[46] J.L. Kaplan and J.A. Yorke. *Chaotic behavior of multidimensional difference equations*, pages 228–37. *Lecture Notes in Mathematics*, Springer-Verlag, New York, N.Y., 1979.

[47] M.P. Kennedy and L.O. Chua. Van der Pol and chaos. *IEEE Transactions on Circuits and Systems*, CAS-33(10):974–80, October 1986.

[48] I.G. Kevrekidis, R. Aris, L.D. Schmidt, and S. Pelikan. Numerical computation of invariant circles of maps. *Physica 16D*, 16D:243–51, 1985.

[49] M. Koksal. On the state equations of nonlinear networks and the uniqueness of their solutions. *Proceedings of the IEEE*, :513–14, March 1986.

[50] M.A. Krasnosel'skiy, A.I. Perov, A.I. Povolotskiy, and P.P. Zabreiko. *Plane Vector Fields.* Academic Press, New York, N.Y., 1966.

[51] M. Kubíček and M. Marek. *Computational Methods in Bifurcation Theory and Dissipative Systems.* Springer-Verlag, New York, N.Y., 1983.

[52] K.S. Kundert. Sparse matrix techniques. In A.E. Ruehli, editor, *Circuit Analysis, Simulation and Design, Part 1*, chapter 6, North-Holland, Amsterdam, 1986.

[53] K.S. Kundert and A. Sangiovanni-Vincentelli. Simulation of nonlinear circuits in the frequency domain. *IEEE Transactions on Computer-Aided Design of Integrated Circuits and Systems*, CAD-5:521–35, October 1986.

[54] K.S. Kundert, A. Sangiovanni-Vincentelli, and T. Sugawara. *Techniques for Finding the Periodic Steady-State Response of Circuits*, chapter 6. Marcel Dekker, 1987.

[55] K.S. Kundert, G.B. Sorkin, and A. Sangiovanni-Vincentelli. Applying harmonic balance to almost-periodic circuits. *IEEE Transactions on Microwave Theory and Techniques*, MTT-36(2):366–78, February 1988.

[56] T. Matsumoto. A chaotic attractor from Chua's circuit. *IEEE Transactions on Circuits and Systems*, CAS-31:1055–58, December 1984.

[57] T. Matsumoto, L.O. Chua, and K. Kobayashi. Hyperchaos: laboratory experiment and numerical confirmation. *IEEE Transactions on Circuits and Systems*, CAS-33(11):1143–47, November 1986.

[58] T. Matsumoto, L.O. Chua, and M. Komuro. Birth and death of the double scroll. *Physica D*, 24:97–124, 1987.

[59] T. Matsumoto, L.O. Chua, and M. Komuro. The double scroll bifurcations. *International Journal of Circuit Theory and Applications*, 14(1):117–46, 1986.

[60] T. Matsumoto, L.O. Chua, and S. Tanaka. Simplest chaotic nonautonomous circuit. *Physical Review A*, 30:1155–57, August 1984.

[61] T. Matsumoto, L.O. Chua, and K. Tokumasu. Double scroll via a two-transistor circuit. *IEEE Transactions on Circuits and Systems*, CAS-33:828–35, August 1986.

[62] T. Matsumoto, L.O. Chua, and R. Tokunaga. Chaos via torus breakdown. *IEEE Transactions on Circuits and Systems*, CAS-34(3):240–53, March 1987.

[63] G. Mayer-Kress, editor. *Dimensions and Entropies in Chaotic Systems*, Springer-Verlag, New York, N.Y., 1986.

[64] A.I. Mees. *Dynamics of Feedback Systems*. John Wiley & Sons, New York, N.Y., 1981.

[65] A.I. Mees and C. Sparrow. Some tools for analyzing chaos. *Proceedings of the IEEE*, 75(8):1058–70, August 1987.

[66] J. Milnor. On the concept of attractor. *Commun. Math. Phys.*, 99:177–95, 1985.

[67] B. Noble and J.W. Daniel. *Applied Linear Algebra*. Prentice-Hall, Englewood Cliffs, N.J., 1979.

[68] E. Ott, E.D. Yorke, and J.A. Yorke. A scaling law: how an attractor's volume depends on noise level. *Physica 16D*, 16D:62–78, 1985.

[69] T. Parker and L.O. Chua. A computer-assisted study of forced relaxation oscillations. *IEEE Transactions on Circuits and Systems*, CAS-30(11):518–533, August 1983.

[70] T. Parker and L.O. Chua. The dual double scroll equation. *IEEE Transactions on Circuits and Systems*, CAS-34:1059–73, September 1987.

[71] T. Parker and L.O. Chua. Efficient solution of the variational equation for piecewise-linear differential equations. *International Journal of Circuit Theory and Applications*, 14:305–14, 1986.

[72] L.-Q. Pei, F. Guo, S.-X. Wu, and L.O. Chua. Experimental confirmation of the period-adding route to chaos. *IEEE Transactions on Circuits and Systems*, CAS-33:438–44, April 1986.

[73] A.W. Pettis, T.A. Bailey, A.K. Jain, and R.C. Dubes. An intrinsic dimensionality estimator from near-neighbor information. *IEEE Transactions on Pattern Analysis and Machine Intelligence*, PAM-1(1):25–37, 1979.

[74] A. Renyi. *Probability Theory.* North-Holland, Amsterdam, 1970.

[75] A. Rodriguez-Vasquez, J.L. Huertas, and L.O. Chua. Chaos in a switched-capacitor circuit. *IEEE Transactions on Circuits and Systems*, CAS-32:1083–85, October 1985.

[76] A. Rodriguez-Vasquez, J.L. Huertas A. Rueda, B. Perez-Verdu, and L.O. Chua. Chaos from switched capacitors: discrete maps. *Proceedings of the IEEE*, :1090–1106, August 1987.

[77] O.E. Rössler. An equation for hyperchaos. *Phys. Lett.*, 71A:155, 1979.

[78] D. Ruelle. Strange attractors. *The Mathematical Intelligencer*, 2:126–37, 1980.

[79] T. Saito. A chaos generator based on a quasi-harmonic oscillator. *IEEE Transactions on Circuits and Systems*, CAS-32:320–31, 1985.

[80] L.F. Shampine and H.A. Watts. The art of writing a Runge-Kutta code, part I. In *Mathematical Software III*, pages 257–75, Academic Press, New York, 1977.

[81] L.F. Shampine and H.A. Watts. The art of writing a Runge-Kutta code, part II. *Applied Mathematics and Computation*, 5:93–121, 1979.

[82] C.P. Silva and L.O. Chua. The overdamped double-scroll family. *International Journal of Circuit Theory and Applications*, :223–302, July 1988.

[83] B.T. Smith, J.M. Boyle, J.J. Dongarra, B.S. Garbow, Y. Ikebe, V.C. Kleme, and C.B. Moler. *Matrix Eigensystem Routines—EISPACK Guide.* Volume 6 of *Lecture notes in Computer Science*, Springer-Verlag, New York, N.Y., 1976.

[84] R.L. Somorjai. Methods for estimating the intrinsic dimensionality of high-dimensional point sets. In G. Mayer-Kress, editor, *Dimensions and Entropies in Chaotic Systems*, pages 137–49, Springer-Verlag, New York, N.Y., 1986.

[85] F. Takens. *Detecting strange attractors in turbulence*, pages 366–81. *Lecture Notes in Mathematics*, Springer-Verlag, New York, N.Y., 1980.

[86] Y.S. Tang, A.I. Mees, and L.O. Chua. Synchronization and chaos. *IEEE Transactions on Circuits and Systems*, CAS-30:620–26, 1983.

[87] A. Ushida and L.O. Chua. Frequency domain analysis of nonlinear circuits driven by multi-tone signals. *IEEE Transactions on Circuits and Systems*, CAS-31:766–79, September 1984.

[88] A. Ushida and L.O. Chua. Frequency-domain analysis of nonlinear circuits driven by multi-tone signals. *IEEE Transactions on Circuits and Systems*, CAS-31:766–78, September 1984.

[89] A. Ushida and L.O. Chua. Tracing solution curves of nonlinear equations with sharp turning points. *International Journal of Circuit Theory and Applications*, 12:1–21, January 1984.

[90] A. Ushida, L.O. Chua, and T. Sugawara. A substitution algorithm for solving nonlinear circuits with multi-frequency components. *International Journal of Circuit Theory and Applications*, 15:327–55, October 1987.

[91] A. Ushida, L.O. Chua, and T. Sugawara. A substitution algorithm for solving nonlinear circuits with multi-frequency components. *International Journal of Circuit Theory and Applications*, 15:327–55, October 1987.

[92] L.P. Šilnikov. A case of the existence of a denumerable set of periodic motions. *Dokl. Sov. Math.*, 6:163–66, 1965.

[93] A. Wolf, J.B. Swift, H.L. Swinney, and J.A. Vastano. Determining Lyapunov exponents from a time series. *Physica 16D*, 16D:285–317, 1985.

[94] S. Wu. Chua's circuit family. *Proc. of the IEEE*, :1022–32, August 1987.

[95] L. Yang and Y. Liao. Self-similar bifurcation structure from Chua's circuit. *International Journal of Circuit Theory and Applications*, :189–92, April 1987.

[96] Y.S. Young. Entropy, Lyapunov exponents, and Hausdorff dimension in differentiable dynamic systems. *IEEE Transactions on Circuits and Systems*, CAS-30(8):599–607, August 1983.

[97] G.Q. Zhong and F. Ayrom. Experimental confirmation of chaos from Chua's circuit. *International Journal of Circuit Theory and Applications*, 13:93–98, January 1985.

Index

D

E

I

K

P

U

V

W